Technical Analysis and Applications with MATLAB

by

William D. Stanley, Ph.D., P.E.

Eminent Professor Emeritus
Old Dominion University

THOMSON

DELMAR LEARNING

Australia Canada Mexico Singapore Spain United Kingdom United States

THOMSON

DELMAR LEARNING

Technical Analysis and Applications with MATLAB

William D. Stanley, Ph.D., P.E.

Vice President, Technology and Trades SBU:
Alar Elken

Editorial Director:
Sandy Clark

Senior Acquisitions Editor:
Steve Helba

Senior Development Editor:
Michelle Ruelos Cannistraci

Marketing Director:
Dave Garza

Senior Channel Manager:
Dennis Williams

Marketing Coordinator:
Casey Bruno

Production Director:
Mary Ellen Black

Senior Production Manager:
Larry Main

Senior Project Editor:
Christopher Chien

Art/Design Coordinator:
Francis Hogan

Technology Project Manager:
Kevin Smith

Technology Project Specialist:
Linda Verde

Senior Editorial Assistant:
Dawn Daugherty

ISBN: 14018-6481-3

NOTICE TO THE READER

Technical Analysis and Applications with MATLAB

Contents

Preface

Although there are a number of fine computational software packages available, the MATLAB® program appears to be one that is evolving in the engineering world as a major tool for mathematical analysis. Various institutions are now using MATLAB both for interactive mathematical analysis and as a form of programming language.

This book was created with one particular broad goal: to combine various technical and engineering mathematical concepts at a basic level with the use of MATLAB for support and analysis. There exist numerous books dealing with various aspects of MATLAB, but most can be roughly divided into two groups: MATLAB programming books and advanced engineering, science, or mathematics books that have MATLAB sections.

Most of the books in the first category assume that the reader is already familiar with the mathematical methods, and the primary concentration is on the programming techniques. Most books in the second category are generally devoted to special subject areas such as communication or control systems and are usually written at a somewhat advanced level.

This book is different in that it assumes only a modest mathematical background on the part of the reader. Moreover, it introduces the mathematical concepts with a somewhat traditional approach, and once the concepts are understood, MATLAB is then used as the primary tool for performing subsequent analysis.

The text has the following major objectives:

1. Introduce the MATLAB Command Window at the outset as the primary mode for computational analysis and functional manipulation in much the same manner as a scientific calculator.
2. Develop the capability for creating MATLAB M-files as a valid approach for software programming to support engineering and scientific applications.
3. Introduce matrix notation and terminology very early in order to maximize the use of MATLAB software in analyzing and manipulating data.
4. Develop competency in the use of MATLAB to prepare professional plots that can be exported to other programs.
5. Review applied differential and integral calculus concepts at a level suitable for readers having only a modest background in calculus and develop competency in using MATLAB for performing both symbolic and numerical differentiation and integration.

6. Provide an applications-oriented treatment of constant coefficient linear ordinary differential equations from the classical approach, from the Laplace transform approach, and with the use of MATLAB to obtain a solution.
7. Provide an introductory treatment of probability and statistical analysis.
8. Provide a treatment of curve fitting and correlation, with emphasis on the concepts' MATLAB application.
9. Provide a treatment of spatial vector analysis operations in the rectangular coordinate system.
10. Provide a treatment of complex numbers, including the associated MATLAB operations.
11. Provide a treatment of Fourier analysis, with the emphasis on using MATLAB, particularly the fast Fourier transform (FFT).
12. Demonstrate practical applications of the various mathematical concepts and the associated MATLAB operations to various engineering application areas.

The text has numerous worked-out examples and problems utilizing both analytical methods and MATLAB to the extent possible. Each chapter begins with a set of objectives that establishes chapter goals. This is followed with the major development utilizing simple explanations and with numbers that are easy to manipulate by hand. In most cases, this is followed by a treatment of MATLAB as related to the particular analysis methods. Many application problems relevant to real-life engineering or engineering technology are presented.

This book may be used as either the primary text for a course or as a supplement to other courses, depending on the situation. The history associated with the development of the book may provide some insight into one of the expected uses for it. At Old Dominion University, most of the book is used in a four semester-hours course at the junior level for engineering technology students. Students at this level have typically taken a single four-semester-hours calculus course at the second-year level. Even so, the chapters reviewing calculus have proven to be very helpful since most students do not master the practical aspects of calculus in one course. If more coverage of the review material is desired and if the entire book is used, it could serve as the basis for two semesters with three credits per semester.

One other expected use of the book is as a supplement to various courses in which MATLAB is used as a supporting tool. The writing style within the book should allow students to obtain reasonable proficiency with MATLAB without much assistance from the instructor. Practicing engineers and technologists who wish to learn how to use MATLAB, but need to review and strengthen their basic mathematical skills should also find the book very useful.

Aside from the MATLAB treatment, one of the guiding factors in determining the level of the mathematical treatment is that of the Fundamentals of Engineering (FE) examination. While not the major objective of the text, it is expected that a mastery of the topics should provide a strong preparation for the mathematical content of this examination.

While the first four chapters should be completed in sequence as a basis for the overall approach, instructors may be able to select topics as appropriate throughout the remainder of the book. All of the MATLAB examples and problems within the text can be completely solved with the Student Version, which contains the Symbolic Math Toolbox. The Professional Version can also be used, provided that the Symbolic Math Toolbox is installed. Go to www.mathworks.com for information on purchasing MATLAB.

ORGANIZATION

Chapter 1   MATLAB Primer

This short chapter provides an introduction to the MATLAB Command Window and the basic operations that can be performed in it. In a sense, it will show the student how to use MATLAB in much the same manner as a calculator. It will also provide an introduction to the notations and conventions utilized throughout the book. In addition, the M-file is introduced as the basis for developing complete programs.

Chapter 2   Matrices

Because all computations in MATLAB utilize matrix concepts, it is essential that students learn some basic matrix terminology very early. This may seem very radical, but as long as the development is limited to the notation, conventions, and basic arithmetic operations of matrices, they may be readily introduced at this point. Only those basic operations essential to utilizing MATLAB are provided.

Chapter 3   Matrix Algebra with MATLAB

The algebraic techniques of Chapter 2 are now implemented with MATLAB. This includes all of the basic algebraic operations on matrices and vectors. The primary focus is on using the Command Window to perform the operations. Among the topics covered is the solution of simultaneous linear equations using MATLAB.

Chapter 4   Curve Plotting with MATLAB

The powerful features of MATLAB for plotting and developing curves are introduced in this chapter. This includes linear plots, logarithmic plots, bar plots, and stem plots. After completing this chapter, the student should be able to prepare professional-looking plots and export them to other programs.

Chapter 5   Common Functions and their Properties

This chapter provides a treatment of the concepts of functions as they are applied in engineering. Common functions appearing in engineering and scientific applications are covered and, in some cases, reviewed from earlier courses. This includes power functions, linear equations, exponential functions, logarithmic functions, and trigonometric functions. In this chapter, the MATLAB commands for generating the various functions are covered immediately following the analytical formulas, within the same sections. Three-dimensional and polar plots are introduced in this chapter.

Chapter 6   Differential Calculus

The concept of the derivative is developed in this chapter, with a special emphasis on piecewise linear functions. Piecewise linear functions appear in electrical circuits and in moment and structural applications. Differentiation of such functions is quite simple and provides an intuitive visualization of the process. A table of common derivatives is provided and various applications of differentiation are covered.

Chapter 7   Integral Calculus

Integral calculus is first introduced through the antiderivative concept; that is, how to determine a function whose derivative is the given function. A major approach, however, reverts back to the concept of piecewise linear functions used in introducing the derivative. The area under the curve of a piecewise-linear function may be determined by simple arithmetic operations, which provides an intuitive understanding of the integration process. A table of common integrals is provided. Various applications of integration are covered.

Chapter 8   Calculus Operations with MATLAB

This chapter is devoted to a study of how differentiation and integration are performed with MATLAB. Symbolic operations of calculus, which in a sense are sophisticated "look-up" tables, are introduced and the means for simple plotting of the functions are provided. Numerical methods for both differentiation and integration using MATLAB are covered.

Chapter 9   Differential Equations: Classical Methods

The classical approach to solving constant coefficient linear ordinary differential equations is developed. Both the homogeneous solution and the particular solution are covered. Practical physical interpretations of the meanings of these functions are discussed. Applications of these differential equations to various technical areas such as mechanical and electrical systems are also covered.

Chapter 10   Differential Equations: Laplace Transform Methods

This chapter complements Chapter 9 and utilizes the Laplace transform to solve constant coefficient linear ordinary differential equations. Those forms that arise most frequently in practical engineering and science applications are stressed.

Chapter 11   Solution of Differential Equations with MATLAB

This chapter shows how MATLAB can be used to solve a differential equation and plot the results in a few short steps. The emphasis is on the symbolic techniques and on plotting the results.

Chapter 12   Introduction to Statistics

A brief applied approach to probability and statistics is provided. The concepts of probability density functions, probability distribution functions, and statistical parameters are developed. A few of the most common probability density functions, such as the gaussian function, are explored. Some of the basic MATLAB commands covering these operations are provided.

Chapter 13   Curve-Fitting and Correlation

The application of MATLAB to provide the best-fitting curves for experimental data is explored. Linear regression, higher-order regression, and multiple regression are covered. Correlation techniques and the associated MATLAB operations are introduced.

Chapter 14   Introduction to Spatial Vector Analysis

The concept of spatial vectors is introduced. The rectangular coordinate system is used as the basis for the work of this chapter. Various vector operations such as the dot product, the cross product, and the triple scalar product are introduced. The corresponding MATLAB operations are also covered.

Chapter 15   Complex Variables

While some limited applications of complex numbers were introduced earlier in the text, this chapter provides a more comprehensive treatment of the subject. Various arithmetic operations associated with complex numbers are developed and applicable MATLAB operations are covered.

Chapter 16   Fourier Analysis with MATLAB

A basic treatment of Fourier analysis is provided. The three basic forms of the Fourier series are discussed and the Fourier transform is introduced. The emphasis then shifts to the use of the fast Fourier transform (FFT) capabilities of MATLAB for computing various spectral forms.

END-OF-CHAPTER PROBLEMS

There are three types of problems contained within the text, although many chapters will have only one or two of the types. The categories are as follows:

1.   General Problems

General problems are those that are to be solved "by hand" without the use of MATLAB special commands. Depending on the discretion of the instructor (if the book is being used as a course), the use of a calculator or even the MATLAB Command Window for performing arithmetic manipulations may be acceptable. However, the reader should not be expected to use the special MATLAB commands to simplify the operations for these problems.

General problems may include drill problems using "simple numbers," or they may include application problems. In many cases it is difficult to make a distinction, but the major theme associated with these problems is that they only utilize MATLAB as an aid rather than a primary focus.

2.   MATLAB Problems

MATLAB problems are those in which the Command Window may be employed along with special commands that are covered within the text. However, some of these problems are best solved by developing an M-file that may be used for various data input. In those problems in which an M-file is the expected outcome, the problem statement will say so, but there will be many cases in which either the Command Window or an M-file can be used.

3.   Derivation Problems

While this text is not oriented toward the derivation of mathematical results, a few chapters contain some derivation problems. These problems emphasize the derivation of specific concepts utilizing some of the basic laws of mathematics or engineering.

CD

The CD enclosed with the text provides three important supplements to the text.

1.   Power Point Slides

There are over 600 slides covering most of the material contained within the text. These slides are arranged by chapter and are in the Microsoft ppt format.

2.   Voice plus Power Point Slides

Separate files providing a voice track plus the slides are included. The voice accompanying the slides is that of the author and this material has been used to supplement and even replace the lectures in some cases.

3.   M-files

Selected m-files are provided to assist both the reader and an istructor with programming assignments. The files include the programs within the text for the first three chapters and selected files for programming assignments at the ends of most chapters.

Acknowledgments

The author would like to express his deep appreciation to the following people for their important contributions in the development and production of this book:

From Thomson Delmar Learning:
    Steve Helba, Senior Acquisitions Editor
    Michelle Ruelos Cannistraci, Senior Development Editor
    Dennis Williams, Senior Channel Manager
    Larry Main, Senior Production Manager
    Christopher Chien, Senior Project Editor
    Francis Hogan, Art/Design Coordinator

From Publishing Synthesis, Ltd.:
    Otto Barz
    George Ernsberger

Finally, both the author and Thomson Delmar Learning would like to thank the following reviewers for their valuable suggestions:
    Don Abernathy, DeVry University, Irving, TX
    Jalaluddin Ahmad, University of Northern Iowa, Cedar Falls, IA
    Abul Azad, Northern Illinois University, DeKalb, IL
    Seyed Mohammad Jalali, DeVry University, Long Beach, CA
    David Loker, Penn State Erie, Erie, PA
    Wieslaw Marszalek, DeVry University, North Brunswick, NJ
    David Oveissi, DeVry University, Arlington, VA
    Cree Stout, York Technical College, Rock Hill, SC

Technical Analysis and Applications with MATLAB

MATLAB Primer

1

1–1 OVERVIEW AND OBJECTIVES

This introductory chapter is relatively short and has as its main objective the introduction of MATLAB® to the reader. This early introduction has the purpose of presenting the reader with the basic operations of MATLAB so that the program may be used throughout the book to support the mathematical analysis.

If you are already familiar with MATLAB, you may be able to skip over Chapter 1 and move on to Chapter 2. However, it is recommended that you skim through this chapter as a minimum since some of the notation and conventions utilized throughout the book will be explained here.

MATLAB is a general software package for technical computing, mathematical analysis, and system simulation. It is a product of The MathWorks, Inc., and is widely employed in industry, government, and education. In addition to the basic MATLAB program, there are numerous supplements called **toolboxes** that provide software applications for specific specialty areas.

The MathWorks, Inc., has an excellent record of maintaining relatively seamless compatibility as new versions are released. I have been working with MATLAB for more than a decade at the time of this writing and have numerous **M-files** (programs) written early in that period to support research work with the National Aeronautics and Space Administration (NASA). They continue to run flawlessly on the latest versions of MATLAB. Depending on the age of this book and its ultimate market endurance, there may be some changes that occur in the MATLAB environment, but it can be reasonably expected that the company will continue to maintain the fine record established thus far.

Objectives

After completing this chapter, the reader should be able to:

1. Describe the **Desktop Layout** and the various windows associated with it.

    Moreover, the reader should be able to perform the following operations with MATLAB:

® MATLAB is a registered trademark of The MathWorks, Inc., 24 Prime Way, Natick, MA 01760-1500; on-line at *http://www.mathworks.com*.

2. Enter scalar values in the **Command Window** and perform operations such as clearing the screen (**clc**), clearing values in memory (**clear**), and determining variables using the **who** and **whos** commands.
3. Perform addition (+) and subtraction (–).
4. Perform multiplication (*) and division (/).
5. Perform exponentiation (^).
6. Perform the square root operation (**sqrt**).
7. Explain the hierarchy of arithmetic operations.
8. Explain *nesting* and apply it to arithmetic operations.
9. Write an *M-file program* that can be saved and used again for different input data.

1-2   DESKTOP LAYOUT

Very little will be said about installing or activating the program since that process may vary somewhat from one computer to another. Anyone using this text is assumed to know how to turn on a computer and click on an icon or go to a program menu to activate a program. Upon activating the program, the **Desktop Layout** appears on the screen.

The discussion that follows is based on **Version 7, Release 14**, along with the **Symbolic Mathematics Toolbox**. The **Student Version** comes with that particular toolbox, which means that all examples in the text should work with the **Student Version**. Other toolboxes are available from the company at a modest cost for students. The **Student Version** also contains the program **Simulink**, which provides a block diagram approach for simulating systems. However, the latter program is not required to support this book.

Windows on Desktop Layout

The default screen based on the version employed is shown in Figure 1-1. Brief descriptions of the windows will be provided in the next few paragraphs.

The upper left area toggles between the **Workspace** and **Current Directory** windows. The **Workspace** provides a list of the variables used in the current work session. The **Current Directory** provides a list of the MATLAB programs available in the given directory. The **Command History** on the lower left shows a continuous record of the commands used in the analysis.

In the **Professional Version**, the window on the right is called the **Command Window**, a term we will use in all subsequent references. In the **Student Version**, it is called the **Student Version Command Window**. This is the area in which all commands are entered for using MATLAB on an interactive basis. Commands for executing **M-file** programs may also be entered in that window.

The different windows may be restructured if desired, and such actions are generated with the **Desktop** option on the upper toolbar. Once you become familiar with the program,

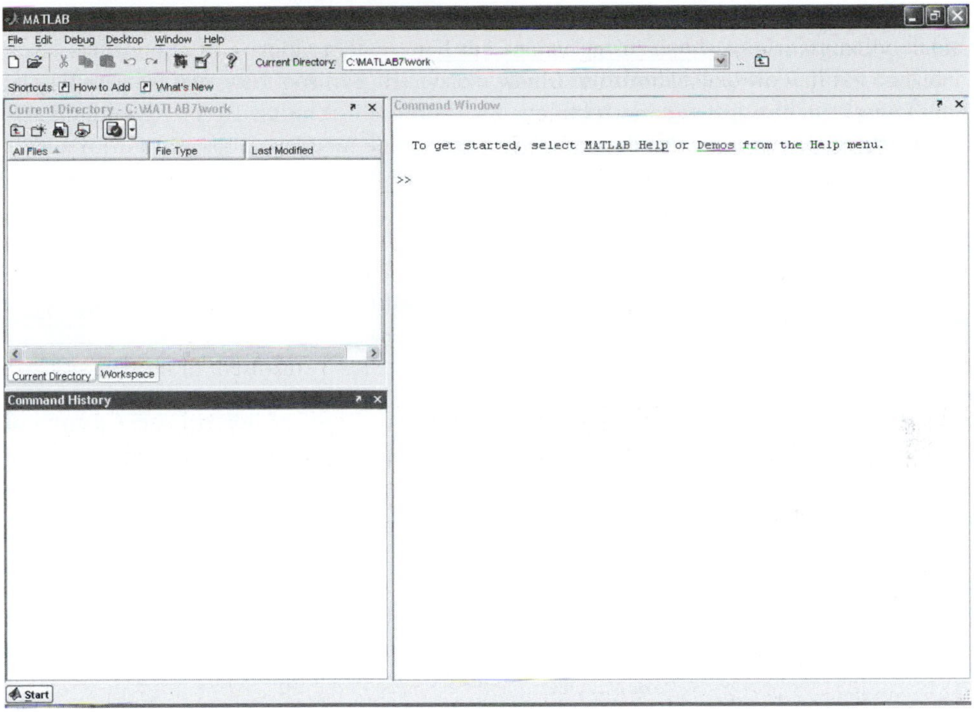

Figure 1-1 Default screen display obtained after activating MATLAB

you may wish to experiment with different options. To restore the original layout, first left-click on **Desktop**. Next, highlight **Desktop Layout** and then left-click on **Default**.

Prompt

MATLAB automatically inserts the **prompt** >> at the beginning of each line in the **Command Window** of the **Professional Version**. The corresponding **prompt** for the **Student Version** is **EDU>>**. We will assume the shorter prompt of the **Professional Version** throughout the text, and the presence of the prompt on a line will alert the reader to the fact that a MATLAB command and possible results will follow.

Command Window

The **Command Window** may be used on an interactive basis by simply typing in the commands after >> for the **Professional Version** or after **EDU>>** for the **Student Version** on a line-by-line basis. Depress **Enter** on the keyboard after each line is entered, and any results generated by the command will immediately appear on the screen unless the addition of a semicolon has suppressed the display (to be discussed later) or unless the command is part of a sequence.

If you desire to have the **Command Window** stand alone, move the mouse arrow to the small arrow on the upper right-hand side of the **Command Window**. You will see the statement **Undock Command Window**. Next, left-click on the arrow, and the window will now separate from the **Desktop Layout**. It can now be maximized to occupy the entire screen.

1-3   GETTING STARTED WITH THE COMMAND WINDOW

Let us begin by assuming that MATLAB has been activated and that the **Desktop Layout** appears on the screen. The focus at this point will be directed toward the **Command Window**. If desired, you could maximize the **Command Window** and let it take over the screen, but it is probably better for the moment to leave it as it is. There are numerous menu options for changing type formats, number of digits appearing on the screen, and other things, but again it is better to wait until you have experience with the simple things first.

The main identifier for the **Professional Version** is that each instruction line in the **Command Window** begins with the prompt >>, which is automatically inserted by MATLAB. To clearly illustrate that point, all entries in the **Command Window** will begin with that identifier. Following the typing of a command, the command is activated by depressing **Enter** on the computer. Results following a command will be shown in the text exactly as they appear on the screen.

Notational Conventions

In the MATLAB examples, we will usually display the results exactly as they appear on the screen. This means that the type style will be different from the standard mathematical style in the remainder of the book, but the symbols should be perfectly clear.

To illustrate this concept, the mathematical format of a common variable in the text is x_1, with the italicized form of x and a subscript for the 1. In a MATLAB command, the variable will probably be indicated as x1, with a nonitalicized x and a 1 that is not subscripted. This change in format should not cause any problems, and it was important to keep the MATLAB notation as close as possible to the way it appears on a computer screen in the default mode. Occasionally there may be a variable that could be stated in either form, in which case an arbitrary choice will be made. Be alert to these different styles of notation, but remember that it is not a result of "sloppy" editing.

Command Window versus M-files

The **Command Window** will allow you to work almost as if using a calculator. You can enter a command and see the numerical results immediately. The important thing is that it is interactive. This is in contrast to the writing of **M-files**, which is performed in a different window. An **M-file** is, in reality, a computer program that can be written, saved, and activated as many times as desired. We will introduce the process of developing an **M-file** in Section 1-5.

As indicated earlier, for all commands written on the screen following either the >> or the **EDU** >> prompt, the command is activated by **Enter**. It will be assumed in all cases whether stated or not.

Clear Commands

At any time that you wish to clear the screen, the following command can be used:

$$>> \text{clc} \tag{1-1}$$

The variables previously defined remain in memory, but you will have a clear screen.

If you wish to delete all the variables in memory, the following command can be used:

$$>> \text{clear} \tag{1-2}$$

Suppose you wish to delete a variable x but retain all others in memory. This can be done by the command

$$>> \text{clear x} \tag{1-3}$$

1-4   BASIC ARITHMETIC OPERATIONS IN THE COMMAND WINDOW

In this section, we will take you back to some of the most basic arithmetic operations learned in elementary and junior high school. Please do not be insulted by this step. To use MATLAB properly, you need to understand how the most elementary operations are performed, and what better way is there to learn such operations than with simple arithmetic?

Spacing

As a general rule, blank spaces between distinct sections of a command line may be used to avoid a "crowded" appearance provided they do not disrupt a particular command function or number. Thus, x=5, x = 5, and x   =   5 are all acceptable. However, if the number is 51, you will get an error message if you try to put a space between the 5 and the 1. Likewise a variable x1 cannot be expressed as x 1. The same holds for various functions that will be introduced throughout the text.

Before inserting a space, ask whether adding a space will either change the meaning or create an ambiguity in the interpretation. If in doubt, it is probably better to avoid the blank space and have the symbols run consecutively as long as you can read the expression. To make it easier for the reader, we will frequently use spaces in the examples, and that should provide some guidance as to what is acceptable.

Entering Values

Numbers can be entered directly in the **Command Window** using the keyboard. For example, suppose we wish to enter the value 5. We simply type 5 and then depress **Enter**. The screen appears as follows:

$$>> 5 \tag{1-4}$$

ans =

5

If you do not assign a specific name to a quantity, MATLAB will give it the name **ans** (for answer). Suppose we desire to assign the name x to the value of 5. In that case, we type x = 5 and depress **Enter**. The screen then appears as follows:

$$>> x = 5 \tag{1-5}$$

x =

5

As long as we do not delete or redefine x, any subsequent references to it will be treated as the value 5. At any point, if we wish to review the value of the quantity, we simply type x and depress **Enter**. The screen then appears as follows:

$$>> x \tag{1-6}$$

x =

5

In anticipation of subsequent operations, let us define y as 3. The operation follows.

$$>> y = 3 \tag{1-7}$$

y =

3

Addition

Now suppose we wish to add the preceding two values to form z1 as the sum. We could write

$$>> z1 = 5 + 3 \tag{1-8}$$

$$z1 =$$

$$8$$

In this case we typed out the actual values again since they were simple and easily remembered. However, a more systematic approach based on the utilization of variables defined earlier is

$$\text{>> } z1 = x + y \tag{1-9}$$

$$z1 =$$

$$8$$

Thus, variables in memory may be manipulated in arithmetic expressions by using their names.

Subtraction

Next, suppose we wish to form z2 as y − x. We have

$$\text{>> } z2 = y - x \tag{1-10}$$

$$z2 =$$

$$-2$$

The value of z2 is a negative number as expected. To enter a negative number, simply place a hyphen (-) in the space preceding the positive number. Thus negative 2 would be entered as -2.

Note that addition and subtraction utilize the same symbols encountered in ordinary arithmetic. As will be seen shortly, this is not the case for multiplication and division, however.

Checking on the Variables

Suppose now that we wish to pause to see what quantities have been defined or computed up to this point and are in memory. The command **whos** will provide a list as follows:

$$\text{>> whos} \tag{1-11}$$

Name	Size	Bytes	Class
x	1x1	8	double array
y	1x1	8	double array
z1	1x1	8	double array
z2	1x1	8	double array

Grand total is 4 elements using 32 bytes.

Because MATLAB is matrix oriented, ordinary single-valued variables are considered as arrays with one row and one column, that is, 1x1 arrays as shown. Such quantities are called *scalars*. Each scalar value is stored as a double array number with eight bytes. An alternate command to **whos** is **who**. The latter command will list the variables, but not their dimensions. The information provided by the last two commands may also be found in the **Workspace** window.

Multiplication

Next, we will consider multiplication. Multiplication of scalars is denoted by placing an asterisk (*) between the numbers. Thus, if z3 is the product of x and y, as defined earlier, we could write

$$\gg z3 = 5*3 \tag{1-12}$$

$$z3 =$$

$$15$$

However, making use of the names of the variables, a better approach is

$$\gg z3 = x*y \tag{1-13}$$

$$z3 =$$

$$15$$

Division

Division with scalars can be achieved by using the forward slash /. Assume that z4 is determined by dividing x by y. We then have

$$\gg z4 = x/y \tag{1-14}$$

$$z4 =$$

$$1.6667$$

The actual value in memory is more accurate than the value displayed on the screen. Options for displaying numbers in a variety of fashions are available but will not be pursued at this point.

Alternate Division Form

In the form of equation 1-14, the order is numerator, forward slash, and denominator. An alternate MATLAB form is denominator, reverse slash, and numerator. Thus, the preceding division could also be expressed as

$$\text{>> z4 = y\backslash x} \qquad (1\text{-}15)$$

$$z4 =$$

$$1.6667$$

This latter form may seem awkward, but it turns out to be useful in dealing with matrix forms and will be considered in Chapter 3.

Exponentiation

Exponentiation of scalars is achieved by utilizing the **caret** or **circumflex** symbol ^. Suppose z5 is defined as the square of x. We have

$$\text{>> z5 = x\^{}2} \qquad (1\text{-}16)$$

$$z5 =$$

$$25$$

Suppressing the Listing of Computational Results

Suppose we desire to perform a computation but prefer not to have the value or values displayed on the screen. This is often the case when the results are intermediate and not the final values being sought. This situation also arises with dimensioned variables, when long lists would move the data on the screen out of the immediate area of interest. Suppression of the screen value for any operation can be achieved by placing a semicolon (;) at the end of the command. For example, let z6 represent y raised to a power of x. If we prefer not to have the value immediately shown on the screen, the operation will be

$$\text{>> z6 = y\^{}x,} \qquad (1\text{-}17)$$

where the preceding result was followed with **Enter**. Nothing appeared on the screen, but the value of z6 was computed and is in memory. At some later time, we can determine the value simply by typing z6 and pressing **Enter**.

$$\text{>> z6} \qquad (1\text{-}18)$$

$$z6 =$$

$$243$$

Entering in Exponential Form

The numbers entered thus far have all been simple values, but when very small or very large values are to be entered, the use of exponential forms can be employed. For example, a microwave frequency $f = 15$ GHz (1 GHz $= 10^9$ Hz) can be entered as

$$\text{>> f = 15e9} \tag{1-19}$$

$$f =$$

$$1.5000e+010$$

Note that MATLAB expresses the value according to the rules of standard scientific notation.

A constant that appears in statistical thermodynamics is Boltzmann's constant k. This value is
1.38×10^{-23} joules/kelvin (J/K). It can be entered as

$$\text{>> k = 1.38e-23} \tag{1-20}$$

$$k =$$

$$1.3800e-023$$

Square Root

The square root of a scalar quantity x is denoted by **sqrt(x)**. Let z7 represent this value. We can type and enter

$$\text{>> z7 = sqrt(x)} \tag{1-21}$$

$$z7 =$$

$$2.2361$$

Hierarchy

There is a hierarchy or order associated with expressions that are to be evaluated with computer software. The default hierarchy is usually in the following order: *parentheses, exponentiation, multiplication, division, addition, and subtraction,* although the order for multiplication and division can sometimes be confusing in software equations. When there is ambiguity in a sequence of multiplication and division operations, the order is from left to right.

In most standard mathematical equations, this chain of operations is usually evident from the manner in which the equation is written. For example, consider the algebraic equation

$$y = 5x^3 + 4 \tag{1-22}$$

It should be evident that the first operation required is that of raising x to the third power and then multiplying that value by 5. The result is then added to 4 to form y.

If x is a single value and this equation is written in MATLAB, it will appear as follows:

$$\gg y = 5*x^3 + 4 \qquad (1\text{-}23)$$

where it is assumed that x is in memory. The hierarchy remains the same as in the algebraic equation.

Consider next the equation

$$y = (5x)^3 + 4 \qquad (1\text{-}24)$$

This equation is quite different from the first one. In the latter equation, we are to multiply x by 5 and then take the third power of the result. This value is then added to 4.

The corresponding MATLAB equation is

$$\gg y = (5*x)^3 + 4 \qquad (1\text{-}25)$$

Thus, parentheses may be used to alter the order of computation. When in doubt about the order, parentheses are recommended to ensure that the desired order is obtained.

Nesting

To create complex orders of operation, we can place parentheses within parentheses, provided that we ultimately have the same number of left as right parentheses. The order of computation is from the innermost to the outermost. The process of putting parentheses within parentheses is called *nesting*.

To illustrate this point, let us consider a MATLAB algorithm in which we start with the value x defined earlier and proceed to determine a desired result y as follows:

1. Add 3 to x.
2. Square the result of step 1.
3. Multiply the result of step 2 by 6.
4. Add 8 to the result of step 3.
5. Take the square root of the result of step 4: the value obtained is y.

One way to accomplish the chore is to introduce four intermediate values, which we will denote as u1, u2, u3, and u4. The number following u in each case corresponds to the step number. The MATLAB commands follow.

$$\gg u1 = 3 + x ; \qquad (1\text{-}26)$$

$$\gg u2 = u1^2 ; \qquad (1\text{-}27)$$

$$\gg u3 = 6*u2 ; \qquad (1\text{-}28)$$

$$\gg u4 = 8+u3 ; \qquad (1\text{-}29)$$

$$>> y = \text{sqrt(u4)} \qquad (1\text{-}30)$$

$$y =$$

$$19.7990$$

Note that we suppressed the printing of the intermediate variables since their values were not of primary interest.

Now let us see how this operation can be performed in one step.

$$>> y = \text{sqrt(8+(6*((3+x)\^2)))} \qquad (1\text{-}31)$$

$$y =$$

$$19.7990$$

Note that the operation starts with the innermost set of parentheses and proceeds toward the outside. Note also that the number of left parentheses is equal to the number of right parentheses (four in each case). We have put more parentheses in this example than actually necessary in order to make a point: when in doubt, adding parentheses to ensure that the computations are performed in the desired order is fine. Can the reader see how one or more sets of parentheses in this expression could be eliminated?

Either approach in the preceding example is acceptable. You can choose to break up a complex expression into a series of simpler operations by introducing intermediate variables, or you can choose to nest some or all of the variables together, as you see fit. The important thing is to use the form that is the easiest for you to understand and check in the event of an error.

Some Constants in MATLAB

The constant π appears so often in scientific analysis that MATLAB has built it into the library, where it is denoted simply as **pi**. The value can be expressed as

$$>> \text{pi} \qquad (1\text{-}32)$$

$$\text{ans} =$$

$$3.1416$$

Again, we emphasize that the value in memory is far more accurate than the result displayed on the screen in the default numerical format.

Another constant that is built into MATLAB is **eps** (for *epsilon*). It is a very small value that can sometimes serve to avoid indeterminate forms such as 0/0. The value can be seen as

$$>> \text{eps} \qquad\qquad\qquad (1\text{-}33)$$

ans =

2.2204e-016

This is not a "standard" definition but rather one employed by MATLAB based on its numerical precision.

Imaginary Numbers

Consider the square root of -1.

$$>> \text{sqrt(-1)} \qquad\qquad\qquad (1\text{-}34)$$

ans =

0 + 1.0000i

The quantity $\sqrt{-1}$ is denoted in complex number theory as i; and the result from MATLAB agrees with that although it carries the answer out to the default number of decimal places. The quantity 0 in front means that there is no "real part" to the particular value. Later, we will see that a *complex number* in general has both a *real part* and an *imaginary part*.

In electrical or electronics engineering and technology, the quantity j is normally used to represent $\sqrt{-1}$ since i is used for current flow. MATLAB will actually accept either i or j, but it always expresses the result on the screen as **i**.

If you desire to enter a purely imaginary number, there are four ways it can be entered. Consider, for example, the number $5i$ or $i5$ (both forms are acceptable in mathematical equations). The four commands that follow illustrate different ways of entering this value:

$$>> \text{5i} \qquad\qquad\qquad (1\text{-}35)$$

ans =

0 + 5.0000i

$$>> \text{5j} \qquad\qquad\qquad (1\text{-}36)$$

ans =

0 + 5.0000i

$$\gg i*5 \qquad\qquad (1\text{-}37)$$

$$\text{ans} =$$

$$0 + 5.0000i$$

$$\gg j*5 \qquad\qquad (1\text{-}38)$$

$$\text{ans} =$$

$$0 + 5.0000i$$

Note that when either the **i** or the **j** is placed after the number, the asterisk representing multiplication is not required. However, when the **i** or the **j** is placed in front, the asterisk is required. Moreover, the printing on the screen always appears as an **i**, which follows the number.

Division by Zero

Division by zero will either yield **Inf** ("infinity") or **NaN** ("not a number"), depending on the form. As examples, if the command **1/0** is entered, the result is **Inf**, and if the command **0/0** is entered, the result is **NaN**.

Help File

One of the most valuable resources available in the MATLAB program is the **Help** file. It is virtually impossible in any single book to provide documentation on all the possible situations that might be encountered in using the many operations. Therefore, you should get in the habit of going to the **Help** file for assistance when required.

The Help File may be opened by left-clicking on the **Help** button on the upper toolbar. A small window will open, providing access to a number of features. The primary help documentation is opened by left-clicking on the **MATLAB Help** option. The next window that opens will display a number of different ways to access information, with which you may wish to experiment. I tend to utilize the **index** option more frequently, but that is somewhat of a personal preference. In that option, one or more key words may be typed and entered, and if suitable documentation is available, it will be provided. Certain words may also lead to similar terms that might be applicable in a given case. More information on the Help file is available in the Appendix to this book.

Numerical Formats

All computations in MATLAB are performed in double precision, so most obtained values usually have a much greater precision than the accuracy of the values of physical parameters being studied. However, the results may be displayed on the screen in various formats. In fact, the **Help** file lists about a dozen different formats, some of which are of interest only in specialized applications; for example, the hexadecimal format, which utilizes a base of 16, consisting of the numbers 0 through 9 plus the letters a through f.

We believe it would be more confusing than helpful to the reader to introduce all of these formats at this point in the text. Rather, by typing **format** in the **index** slot previously referred to in the **Help** file, you can investigate the various formats whenever the need arises. However, a few of the most common ones will be discussed here. The default format is denoted as **short,** and it is the one we have employed so far in the text. Moreover, we use it more than any other format throughout the remainder of the text since it is the default format and is convenient to use. It utilizes a scaled, fixed-point format with five digits.

In general, any of the formats may be activated by the command

$$\gg \text{format } \textit{type} \tag{1-39}$$

where *type* represents a particular code name for the format, some of which will be illustrated in the steps that follow. At any time, the default short format may be restored by either of the following commands:

$$\gg \text{format short} \tag{1-40}$$

or simply

$$\gg \text{format} \tag{1-41}$$

Now we will illustrate a few of the formats by using the constant **pi**. First, we have the default short format.

$$\gg \text{pi} \tag{1-42}$$

$$\text{ans} =$$

$$3.1416$$

Next, consider the **long** format:

$$\gg \text{format long} \tag{1-43}$$

$$\gg \text{pi}$$

$$\text{ans} =$$

$$3.14159265358979$$

The result is certainly as long or longer than we could want. Now consider the **short e** format.

$$\gg \text{format short e} \tag{1-44}$$

$$>> pi$$

$$ans =$$

$$3.1416e+000$$

This latter format is useful for very large and very small values where scientific notation is desired. In fact, the short format may be temporarily changed to this format if required. For example, refer back to equations (1-19) and (1-20) to see how the short format automatically provides exponential forms when necessary.

The **bank** format assumes the format of dollars and cents and will be recommended for use in MATLAB Problems 1-23, 1-24, 2-31, and 3-32.

If you are using MATLAB as you follow along with the text, it is appropriate now to switch back to the default **short** format by giving the command

$$>> format \qquad\qquad\qquad\qquad (1\text{-}45)$$

1-5 PROGRAMMING WITH M-FILES

Thus far, all of the work has been performed within the **Command Window** on an interactive basis. This pattern will continue throughout the text since most commands will be introduced on an individual basis, and the results may be readily observed. In general, the **Command Window** is usually best for fast results when the procedure may not need to be repeated and where only a few lines of code are involved.

Consider next the situation where a computational procedure may need to be repeated for more than one set of input data and/or the code consists of a relatively long list of commands. In this case, it is usually more convenient to develop an **M-file**, which may be saved and used as often as desired. An **M-file** of this type (or **M-file script,** as it is often called) is a MATLAB-based computer program for performing a particular analysis. Thus, MATLAB may be considered as a valid form of programming language. In general, any of the commands that may be applied in the **Command Window** may be combined into an **M-file** program. Instructions for accepting input data for different cases may be added to the programs.

Creating an M-file

In general, any "text-only" or ASCII editor can be used to create an **M-file**. However, the simplest procedure is to use the built-in editor within MATLAB. To use the MATLAB editor, left-click on **File** and then left-click on **New**. Next, left-click on the **M-file** option. A window will open for the purpose of preparing the program. Comments may be freely added in the program by placing the percentage character **%** at the beginning of each comment line. Type all the commands required and enter each one individually. None of the

MATLAB commands will be executed during the entry process, and you may correct errors in typing that you notice. Basically, you are creating a program that will be executed later, and MATLAB will not perform any computations at this time. The prompt symbols will **not** appear at the beginning of each line of code, and they should **not** be added. Place only the appropriate command on each executable line, and place a **%** at the beginning of each comment line.

Saving a File

After a program has been completed, it should be saved. Left-click on **File** and then left-click on **Save As**. If the computer belongs to someone else or to an institution (e.g., a college computer), you may wish to save the file on a floppy disk. If it is your own computer, you may wish to save it in the MATLAB default **work** folder. In general, the name *must begin with a letter*, but it may contain numbers within the name. However, the name may *not* contain any symbols that could be confused with an arithmetic operation such as +, -, *, /, and so on. The **underscore** symbol _ *may* be used to separate portions of the name. Finally, the name must be continuous; that is, spaces are not permitted in the name. If the MATLAB editor is used, it will automatically assign an extension of **.m** to the name as it is saved. If a different editor is used to create the file, the extension **.m** should be typed to the end of the name.

Setting a Path

Assume next that you have saved the program and desire to run it. Return to the **Command Window** for this purpose. The next procedure may be unnecessary if you saved the file in the **work** folder. If you saved the file to a folder outside the MATLAB directory or to a floppy disc, you may need to set a path from the MATLAB directory to that particular location. If in doubt, try to run the program using the procedure in the next paragraph and see if it runs. If the program is not recognized, you will need to set a path. First, left-click on **File**. When the menu opens, left-click on **Set Path** and then left-click on **Add Folder**. Search through the file structure on the right to identify the path required to reach the appropriate folder. Then left-click on **OK**. If necessary, use the **Help** file to assist in this process.

Running a Program

Within the **Command Window**, type the name of the program after **>>** or **EDU >>** at the beginning of the first line and enter it. You should *not* add the extension **.m** since MATLAB assumes it is there by default. The program will either run as it should or (more likely on the first trial) an error statement will appear telling you that something is wrong with one or more lines of code. If there is an error, you will need to open the file and attempt to identify and correct the problem. MATLAB does not necessarily catch all possible errors on the first attempt. In long programs, it may take several attempts to catch all the possible errors.

Functions

A second type of **M-file** is that of a **function**. MATLAB contains a large number of built-in functions, many of which will be considered throughout the text. However, special functions may be created and added to the user's library for convenience.

Extending a Line

Most commands used in **M-files** within the text will easily fit on one line. However, if it is necessary to use more than one line for a command, place three periods in succession **...** at the end of the first line and any other lines that are to be continued. This combination of periods is referred to as an **ellipsis**.

Programming Example 1-1

To illustrate how an **M-file** is developed, a simple example will be considered in this example. A program will be written to determine the roots of a quadratic equation. We will also provide options for creating input to the program.

Solution

It should be noted that the roots of any polynomial of any reasonable degree can be determined within the **Command Window** with one line of code, a process that will be covered in Chapter 5. The purpose here is simply to learn the mechanics of creating a program, and that is best done with a simple example. Imagine, if you please, that you are writing this program for first-year algebra students to use.

Consider the determination of the roots of a quadratic equation of the form

$$ax^2 + bx + c = 0 \tag{1-46}$$

From basic algebra, the two roots of this equation are given by

$$x_1, x_2 = \frac{-b \pm \sqrt{b^2 - 4ac}}{2a} \tag{1-47}$$

In developing an **M-file**, you should decide if you want to provide an option for selecting input values when the program runs. One could always go back and modify the file each time the data are changed, but a more elegant approach is to let the program ask you for the input data, which is one of the objectives of this exercise.

In the present case, the three input variables are a, b, and c, and the desired two output variables are x_1 and x_2. In the default MATLAB nonitalicized format, these five quantities will be denoted simply as a, b, c, x1, and x2. These latter forms will be utilized in all steps that follow.

You may want to refer to the complete **M-file** in Figure 1-2 during the discussion that follows. The name of the **M-file** is **p_example_1_1.m**. This rather odd name stands for "Programming Example 1-1." Remember, the **M-file** name must be continuous and cannot

```
% M-file program p_example_1_1.m.
% It will be used to determine the two roots of a quadratic equation.
% The input variables will be the three coefficients of the equation.
% The output variables will be the two roots
a = input('Enter the coefficient of the second-degree term. ')
b = input('Enter the coefficient of the first-degree term. ')
c = input('Enter the constant term. ')
disc = sqrt(b^2-4*a*c);
x1 = (-b+disc)/(2*a)
x2 = (-b-disc)/(2*a)
```

Figure 1-2   M-file of Programming Example 1-1

contain a hyphen, since that would be confused for subtraction. Throughout the text, **M-file** names will frequently utilize the underscore _ in places that might otherwise use the hyphen if it were permitted.

The first few lines are comments to assist the programmer and any user of the program. Comments may be inserted anywhere in the program. Just remember that the symbol **%** is required at the beginning of each comment line. Incidentally, a user could be misled by the fact that the name of the program is given in the first comment line. This means nothing as far as execution is concerned. The program must be saved by the desired name before it can be run.

A common way to allow the insertion of data as a program is run is by means of an **input** statement. Using the input variable **a** as the basis, the general format is as follows:

$$a = input \,(\text{'anything you want to say for identification'}) \qquad (1\text{-}48)$$

When the program is run, the statement contained between the apostrophes (´ ´) will appear on the screen and the program will pause until you enter the value of **a**. All you should type on the screen is the number representing the value of **a**. (Do not enter **a =**) After you have entered the value of **a**, the program will move on to the next input statement, and after all inputs are provided, the analysis will follow.

For this particular case, the actual command is

$$a = input \,(\text{'Enter the coefficient of the second-degree term. '}) \qquad (1\text{-}49)$$

Although not necessary, a space was provided after the period on the screen statement. The reason is to have the cursor move one space before the number is entered, which provides a "cleaner" entry point. Two similar statements were provided for **b** and **c,** which follow the same format.

The quantity underneath the radical in Equation 1-47 is called the *discriminant*. It can be positive, zero, or negative. When it is negative, the square root is an imaginary number and will be followed by **i,** as discussed earlier. With many software packages, especially older versions, and with many calculators, the evaluation of the square root of a negative number may not be directly possible. MATLAB readily performs the square root of a negative number as well as that of a complex number.

Although the two roots could be determined with only two equations, it was decided to evaluate the discriminant separately. That step and the additional two steps to determine the two output variables follow.

$$\text{disc} = \text{sqrt(b\^2 - 4*a*c)}; \tag{1-50}$$

$$\text{x1} = \text{(-b + disc)/(2*a)} \tag{1-51}$$

$$\text{x2} = \text{(-b - disc)/(2*a)} \tag{1-52}$$

A choice was made to suppress the printing of the discriminant, but we obviously do not want to suppress the printing of the two output variables.

There are many additional features available in MATLAB to create more elaborate input and output formats, but the process shown here should be adequate for our purposes at this point.

MATLAB PROBLEMS

All of the exercises through 1-22 should be performed within the MATLAB **Command Window**. As was the case earlier in the chapter, some may seem trivial, but they are presented for the same purpose as much of this chapter; that is, practice and familiarization.

As further practice in converting from standard mathematical notation to MATLAB forms, all values and operations will be given in the standard form. You must then enter them in the MATLAB format. First, enter the values of x and y and then determine all of the operations in problems that follow. At any point that you need to clear the screen, use the **clc** command; however, if you clear the memory with **clear**, you will need to reenter the values of x and y.

$x = 5$
$y = 8$

1-1. $z_1 = x + y$

1-2. $z_2 = x - y$

1-3. $z_3 = xy$

1-4. $z_4 = \dfrac{x}{y}$

1-5. $z_5 = x^3$

1-6. $z_6 = 2^5$

1-7. $z_7 = (2^5)^2$

1-8. $z_8 = 2x + 3y$

1-9. $z_9 = 2x^2 + 3y^2$

1-10. $z_{10} = (2x + 3y)^2$

1-11. $z_{11} = 2\pi$

1-12. $z_{12} = \pi^2$

1-13. $z_{13} = 2\pi^2$

1-14. $z_{14} = (2\pi)^2$

1-15. $z_{15} = -1.6 \times 10^{-19} \times \pi^3$

1-16. $z_{16} = 3 \times 10^8 \times 5 \times 10^{-6}$

1-17. $z_{17} = \sqrt{x^2 - 4}$

1-18. $z_{18} = \sqrt{4 - x^2}$

1-19. The following algorithm is used to generate a variable z_{19}, starting with x.

    1. Square x.
    2. Subtract 12 from the result of 1.
    3. Take the square root of the result of 2.
    4. Add 9 to the result of 3.
    5. Square the result of 4, and the answer is z_{19}.

    Determine the solution utilizing four intermediate variables on a step-by-step basis.

1-20. The following algorithm is used to generate a variable z_{20}, starting with y.

    1. Take the square root of y.
    2. Add 4 to the result of 1.
    3. Form the square of the result of 2.
    4. Subtract 8 from the result of 3.
    5. Multiply the result of 4 by π, and the answer is z_{20}.

Determine the solution utilizing four intermediate variables on a step-by-step basis.

1-21. Solve Problem 1-19 with one equation using nesting.

1-22. Solve Problem 1-20 with one equation using nesting.

1-23. The future value resulting from compound interest on an initial investment is given by

$$F = I\left(1 + \frac{r}{n}\right)^{nN}$$

where

   I = initial value deposited in dollars

   r = annual interest rate expressed as a fraction (e.g., 5% is expressed as 0.05).

   n = number of periods within a year for compounding

   N = number of years

   F = final value in dollars at the end of N years

a. Write an M-file program to determine the final value.
   Inputs: The four values (I, r, n, and N) required for the evaluation.
   Output: The final value, F.
b. Based on a deposit of $1,000 and an annual interest rate of 5% compounded annually, determine the future value at the end of 10, 20, 30, and 40 years.
c. Repeat (b) based on an interest rate of 10% compounded annually.
d. Repeat (c) if the interest is compounded quarterly.
e. *Special Exercise:* Assume that you can invest $1 in a special account that has an annual interest rate of 100% for one year only. Determine the final value in one year if the interest is compounded (1) yearly, (2) monthly, (3) daily, (4) hourly, (5) by the minute, and (6) by the second. Does the value start to look like something familiar? What do you think it would be if it were continuously compounded? (*Hint:* When polled for a value, it may be entered as an arithmetic operation. For example, the number of hours in a year could be entered as 24*365.)
   Suggestion: For parts (a) through (d), set the *bank* format in the Command Window before executing the program. This format will express the answer in dollars and cents and is set by the following command:

   >> format bank

   For part (e), restore the default short format by the command

   >> format

1-24. Assume that one makes an equal deposit at the beginning of each year for N years and that compounding occurs annually at the end of each year. The future value of the net amount is given by

$$F = A(1+r)\left[\frac{(1+r)^N - 1}{r}\right]$$

where

A = amount deposited at the beginning of each year in dollars

r = annual interest rate expressed as a fraction

N = number of years

F = final value in dollars at the end of N years

a. Write an M-file program to determine the final value.
   Inputs: The three values (A, r, and N) required for the evaluation.
   Output: The final value, F.
b. Based on an annual investment of $1,000 and an annual interest rate of 5%, determine the final value at the end of 10, 20, 30, and 40 years.
c. Repeat (b) if the annual interest rate is 10%.
   Suggestion: Set the *bank* format in the Command Window before executing the program. This format will express the answer in dollars and cents and is set by the following command:

>> format bank

After completion of the problem, the default format can be reset by the command

>> format

Matrices

2

2-1 OVERVIEW AND OBJECTIVES

Matrices provide an orderly way of arranging values or functions to enhance the analysis of systems in a systematic manner. Their use is very important in simplifying large arrays of equations and in determining their solutions. In particular, computer algorithms for manipulating large arrays are simplified greatly by matrix notation and operations.

A particular reason for introducing matrices at this point is that MATLAB is heavily matrix oriented. You do not have to be an expert in matrix theory to utilize the powerful features of MATLAB, but it is very helpful to understand some of the terminology and manipulations. This chapter will deal with those basic elements of matrix theory that can be used for that purpose. The term *linear algebra* is often used to represent the general theory of matrices and the associated algebraic operations.

Objectives

After completing this chapter, the reader should be able to:

1. Define and recognize a *matrix* and determine its *size*.
2. Define a *row matrix* or *row vector* and a *column matrix* or *column vector*.
3. Define a *square matrix*.
4. Determine the *transpose* of a matrix.
5. State the criteria that must be satisfied for two matrices to be added or subtracted.
6. Form the sum or difference between two applicable matrices.
7. State the criteria that must be satisfied for two matrices to be multiplied together and determine the size of the product matrix.
8. Determine the product of two applicable matrices.
9. Determine the determinant of a 2×2 matrix or a 3×3 matrix, and discuss the algorithm for determining the determinant of higher-order square matrices.
10. Define a *minor* and a *cofactor*.
11. Define the *identity matrix*.
12. Define the *inverse matrix*.
13. Determine the *inverse matrix* of a 2×2 matrix or a 3×3 matrix, and discuss the algorithm for determining the inverse matrix of higher-order square matrices.

14. Formulate an array of simultaneous linear equations in matrix form and solve for the unknown vector for the case of two equations with two unknowns or three equations with three unknowns.

15. Apply matrix multiplication to perform a *linear transformation* of variables from one basis to another.

2-2   MATRIX FORMS AND TERMINOLOGY

A *matrix* is a rectangular array of constants or variables. For simplicity, we will use constants in describing the properties. The *size* of a matrix is described by two integers, which we will initially designate as m and n. The usual way this is stated is as an $m \times n$ matrix. The first integer, m, is the number of *rows,* and the second integer, n, is the number of *columns*. We will use **boldface** type to indicate the name of a matrix in standard mathematical equations (but not in MATLAB). The form of an $m \times n$ matrix **A** having all elements as constants can be expressed as

$$\mathbf{A} = \begin{bmatrix} a_{11} & a_{12} & a_{13}....a_{1n} \\ a_{21} & a_{22} & a_{23}....a_{2n} \\ \vdots & & \vdots \\ a_{m1} & a_{m2} & a_{m3}...a_{mn} \end{bmatrix} \tag{2-1}$$

When it is desired to emphasize the size of the matrix, we will employ the notation $\mathbf{A}_{m,n}$.

Don't confuse a matrix with a *determinant*. While they look similar, the determinant is often denoted by vertical bars | | while the matrix is usually denoted by brackets [ ]. Moreover, a determinant has a single value while a matrix does not have a single value (unless it is a trivial, 1×1 matrix).

Square Matrix

If the number of rows is equal to the number of columns; that is, if $m = n$, the matrix is said to be *square*. In this case, the notation $\mathbf{A}_{m,m}$ could be used to denote the matrix if the size is to be delineated.

Vectors

A matrix having only one row is called a *row matrix*. A matrix having only one column is called a *column matrix*. In many applications, a matrix of either form is called a *vector*. The *row matrix* is called a *row vector* and the *column matrix* is called a *column vector*. We will tend to use lower-case boldface names for vectors, although there may be some exceptions. Thus, a typical row or column vector might be indicated as **b**. A 1×1 vector is called a *scalar* and is the form of most variables considered in simple algebraic forms. Boldface notation will not be used for a scalar except when it appears as a limiting case of a more general matrix.

Transpose

The transpose of a matrix $\mathbf{A}$ will be denoted by $\mathbf{A}'$ and is obtained from $\mathbf{A}$ by interchanging the rows and columns. Thus, if $\mathbf{A}$ has a size of $m \times n$, $\mathbf{A}'$ will have a size of $n \times m$. Note that the transpose of a row vector is a column vector, and vice versa. (Many references use $\mathbf{A}^T$ to represent the transpose, but the choice of $\mathbf{A}'$ was made to be compatible with MATLAB.)

If the transpose is performed twice on a matrix, the original matrix is restored. Thus,

$$(\mathbf{A}')' = \mathbf{A} \tag{2-2}$$

Example 2-1

Calculate the size of a matrix given by

$$\mathbf{A} = \begin{bmatrix} 2 & -3 & 5 \\ -1 & 4 & 6 \end{bmatrix} \tag{2-3}$$

Solution

There are 2 rows and 3 columns, so it is a 2×3 matrix. It could be expressed as $\mathbf{A}_{2,3}$.

Example 2-2

Calculate the size of a matrix given by

$$\mathbf{B} = \begin{bmatrix} 2 & 1 \\ 7 & -4 \\ 3 & 1 \end{bmatrix} \tag{2-4}$$

Solution

There are 3 rows and 2 columns, so it is a 3×2 matrix. It could be expressed as $\mathbf{B}_{3,2}$.

Example 2-3

Calculate the size of a matrix given by

$$\mathbf{C} = \begin{bmatrix} 2 & -1 & 3 \\ 4 & 6 & 1 \\ -5 & 2 & 1 \end{bmatrix} \tag{2-5}$$

Solution

There are 3 rows and 3 columns, so it is a 3×3 **square** matrix. It could be expressed as $\mathbf{C}_{3,3}$.

Example 2-4

Express the integer values of time in seconds (s) from 0 to 5 s as a row vector.

Solution

Let **t** represent the row vector, which in this case will be a 1×6 matrix. We have

$$\mathbf{t} = [0\ 1\ 2\ 3\ 4\ 5] \tag{2-6}$$

It should be noted in passing that this is very similar to the format that will be used later with MATLAB to display an array of numbers in row form. The main difference in notation is that in MATLAB, the name would not be in boldface style and would be indicated simply as t.

Example 2-5

A certain physical system has three displacement variables: x_1, x_2, and x_3. Express the set of variables as a column vector.

Solution

Let **x** represent the column vector, which in this case will be a 3×1 matrix. We have

$$\mathbf{x} = \begin{bmatrix} x_1 \\ x_2 \\ x_3 \end{bmatrix} \tag{2-7}$$

If this column vector were expressed in MATLAB form, the name would not be in boldface type and we would use nonitalicized x forms; that is, x, x1, x2, and x3.

Example 2-6

Determine the transpose of the matrix of Example 2-1.

Solution

The matrix is repeated here for convenience.

$$\mathbf{A} = \begin{bmatrix} 2 & -3 & 5 \\ -1 & 4 & 6 \end{bmatrix} \tag{2-8}$$

For the transpose, the first row becomes the first column and the second row becomes the second column.

$$\mathbf{A'} = \begin{bmatrix} 2 & -1 \\ -3 & 4 \\ 5 & 6 \end{bmatrix} \tag{2-9}$$

2-3   ARITHMETIC OPERATIONS WITH MATRICES

Some of the basic arithmetic operations with matrices will be discussed in this section. In some ways, they follow the regular operations associated with scalars, but there are some peculiarities that must be carefully observed.

Addition and Subtraction

Matrices can be added or subtracted if and only if they are of the same size. Let $\mathbf{A_{m,n}}$ and $\mathbf{B_{m,n}}$ represent two matrices of the same size. Let $\mathbf{C_{m,n}}$ represent either the sum or the difference, which, for illustrative purposes, we will express as

$$\mathbf{C_{m,n}} = \mathbf{A_{m,n}} \pm \mathbf{B_{m,n}} \tag{2-10}$$

For addition, the corresponding elements are added, and for subtraction, the corresponding elements are subtracted. The resulting matrix has the same size as either of the matrices being combined. Addition and subtraction will be illustrated in Examples 2-7 and 2-8.

Multiplication

Multiplication of two matrices is a bit tricky and requires a little practice. In order for it to be possible, *the number of columns of the first matrix must equal the number of rows of the second matrix.* This means that matrix multiplication is not commutative except in very special cases. In general

$$\mathbf{AB} \neq \mathbf{BA} \tag{2-11}$$

Thus, the order of listing is very important, and if the first has the dimensions of $m \times n$, the second should have the dimensions of $n \times k$, where m and k are arbitrary but n is common between the two. The result of the product has the size $m \times k$. Let $\mathbf{C}$ represent the product. We have

$$\mathbf{A_{m,n} B_{n,k}} = \mathbf{C_{m,k}} \tag{2-12}$$

Any element c_{ij} in the matrix $\mathbf{C}$ can be determined from the following algorithm:

$$c_{ij} = \sum_{r=1}^{n} a_{ir} b_{rj} \tag{2-13}$$

There is a mechanical way to assist in this process. Take the first row of the first matrix and the first column of the second matrix, successively multiply terms from left to right in the first matrix and top to bottom in the second matrix, and then add the products. The result will be c_{11}. Repeat this process for all combinations of rows of the first matrix and columns of the second. The intersection point in each case is the position of the element in the product matrix. This process will be illustrated in Examples 2-9, 2-10, and 2-11.

Division (Matrix Inversion)

There is no such thing as division in matrix theory. The process of matrix inversion is the closest thing to division, and it will be considered in a later section. It turns out that MATLAB employs a symbolic form that resembles what might be considered loosely as "matrix division," but it is strictly a command for matrix inversion followed by multiplication.

Example 2-7

Two matrices are given by

$$A = \begin{bmatrix} 3 & 2 & 1 \\ -4 & 5 & 6 \end{bmatrix} \tag{2-14}$$

and

$$B = \begin{bmatrix} 4 & 9 & x \\ 6 & -3 & y \end{bmatrix} \tag{2-15}$$

Determine

$$C = A + B \tag{2-16}$$

Solution

The sum is possible because both matrices have the same size. We simply add the corresponding elements and obtain

$$C = \begin{bmatrix} 7 & 11 & 1+x \\ 2 & 2 & 6+y \end{bmatrix} \tag{2-17}$$

Don't be disturbed by the + signs in the third column. The second matrix had a combination of variables and constants, so the result will have the same property.

Example 2-8

For the two matrices of Example 2-7, determine a matrix **D** given by

$$D = A - B \tag{2-18}$$

Solution

In this case, the elements of **B** are subtracted from the corresponding elements of **A**. We have

$$\mathbf{D} = \begin{bmatrix} -1 & -7 & 1-x \\ -10 & 8 & 6-y \end{bmatrix} \tag{2-19}$$

Example 2-9

Consider the matrix **A** of Example 2-1 and the matrix **B** of Example 2-2. Determine what possible orders, if any, the two matrices can be multiplied together and indicate the size(s) of the resulting product(s).

Solution

For convenience, the two matrices will be repeated here.

$$\mathbf{A} = \begin{bmatrix} 2 & -3 & 5 \\ -1 & 4 & 6 \end{bmatrix} \tag{2-20}$$

$$\mathbf{B} = \begin{bmatrix} 2 & 1 \\ 7 & -4 \\ 3 & 1 \end{bmatrix} \tag{2-21}$$

The product $\mathbf{A}_{2,3}\mathbf{B}_{3,2}$ is possible because the number of columns of the first (3) is equal to the number of rows of the second (also 3). The result will be a 2×2 matrix. The product $\mathbf{B}_{3,2}\mathbf{A}_{2,3}$ is also possible because the number of columns in the first (2) is equal to the number of rows of the second (also 2). The result will be a 3×3 matrix. In this case, multiplication in either order is possible, but the results will be drastically different as will be seen in the next two examples.

Example 2-10

Using the matrices of Example 2-9, determine $\mathbf{C}_{2,2} = \mathbf{A}_{2,3}\mathbf{B}_{3,2}$.

Solution

We write

$$\mathbf{C} = \begin{bmatrix} 2 & -3 & 5 \\ -1 & 4 & 6 \end{bmatrix} \begin{bmatrix} 2 & 1 \\ 7 & -4 \\ 3 & 1 \end{bmatrix} \tag{2-22}$$

To determine c_{11}, we multiply the first row of the first matrix by the first column of the second matrix on a term-by-term basis, and then add the results. This yields

$$c_{11} = (2)(2) + (-3)(7) + (5)(3) = 4 - 21 + 15 = -2 \tag{2-23}$$

To determine c_{12}, we multiply the first row of the first matrix by the second column of the second matrix in a similar manner. We have

$$c_{12} = (2)(1) + (-3)(-4) + (5)(1) = 2 + 12 + 5 = 19 \tag{2-24}$$

To determine c_{21}, we multiply the second row of the first matrix by the first column of the second matrix.

$$c_{21} = (-1)(2) + (4)(7) + (6)(3) = -2 + 28 + 18 = 44 \tag{2-25}$$

Finally, to determine c_{22}, we multiply the second row of the first matrix by the second column of the second matrix.

$$c_{22} = (-1)(1) + (4)(-4) + (6)(1) = -1 - 16 + 6 = -11 \tag{2-26}$$

The matrix **C** is then given by

$$\mathbf{C} = \begin{bmatrix} -2 & 19 \\ 44 & -11 \end{bmatrix} \tag{2-27}$$

Note that in determining the various coefficients, there was always an equal number of terms in the portion of the first matrix being multiplied by the portion of the second matrix. This is a result of the fact that the number of columns of the first matrix must equal the number of rows of the second matrix.

Example 2-11

Using the matrices of Example 2-9, determine $\mathbf{D}_{3,3} = \mathbf{B}_{3,2}\mathbf{A}_{2,3}$.

Solution

The required multiplication is

$$\mathbf{D} = \begin{bmatrix} 2 & 1 \\ 7 & -4 \\ 3 & 1 \end{bmatrix} \begin{bmatrix} 2 & -3 & 5 \\ -1 & 4 & 6 \end{bmatrix} \tag{2-28}$$

In this case, we need to determine six different constants, each of which is determined by multiplying the two elements of a row of **B** by the two elements of a column of **A** and adding the products. We will not show the details here, but you are invited to verify that the result is

$$\mathbf{D} = \begin{bmatrix} 3 & -2 & 16 \\ 18 & -37 & 11 \\ 5 & -5 & 21 \end{bmatrix} \tag{2-29}$$

2-4 DETERMINANTS

The *determinant* of a matrix can be determined *only for a square matrix*. Unlike a matrix, a determinant has a single value. There are several different notations for a determinant. For a matrix **A**, the determinant can be represented as det(**A**) , by |**A**|, or by Δ. We will use the first notation more often.

The process of evaluating a determinant can be a messy process for anything larger than a 2×2 or 3×3 matrix. There are many "tricks" that appear in mathematics texts to simplify the process. The problem with some of these tricks is that they work fine when the values in the matrix are "clean" numbers, but the processes become more burdensome when the numbers represent "real" data.

With the ready access to such mathematical packages as MATLAB in the scientific community, there is very little reason or justification for trying to accomplish the chore by hand for anything but the simplest case. For that reason, we will limit our development here to the simple 2×2 and 3×3 determinants. In showing the development of the latter, we will introduce the concept of *expansion of minors*, which in theory could be used for any determinant.

The simplest matrix is a trivial 1×1 matrix. This is really a scalar, and its determinant is simply its value. Therefore, we need not consider it any further.

2×2 Case

Consider a 2×2 determinant. Let

$$\mathbf{A} = \begin{bmatrix} a_{11} & a_{12} \\ a_{21} & a_{22} \end{bmatrix} \tag{2-30}$$

The determinant of **A** is

$$\det(\mathbf{A}) = a_{11}a_{22} - a_{12}a_{21} \tag{2-31}$$

In words, the determinant of a 2×2 matrix is the product of the first element of the first row and the second element of the second row minus the product of the other two terms. A common error in evaluating determinants is that of keeping the signs straight, since some of the elements may have negative signs.

Minors

Prior to considering 3×3 determinants, it is helpful to define the term *minor*. Let M_{ij} represent the minor corresponding to the ith row and the jth column of a matrix **A**. If **A** is a square matrix of size $m \times m$, M_{ij} is a determinant of size $m - 1$ by $m - 1$ formed from the matrix by crossing out the ith row and the jth column and evaluating the resulting determinant.

Cofactors

A cofactor will be denoted as A_{ij}. It is directly proportional to the corresponding minor M_{ij} by the relationship

$$A_{ij} = (-1)^{i+j} M_{ij} \tag{2-32}$$

Thus, cofactors are alternately equal to the minors or the negatives of the minors.

The significance of minors and cofactors is that a higher-order determinant can be expanded into the sum of lower-order determinants. For example, a 3×3 determinant can be expanded into the sum of three 2×2 determinants, and a 4×4 determinant can be expanded into the sum of four 3×3 determinants. One can readily see how the number of smaller determinants can quickly grow as successive expansion is performed.

3×3 Case

Now, consider a 3×3 determinant. As indicated in the previous paragraph, it can be expanded into three 2×2 determinants. One can expand along a row or a column. We will choose the former and express the determinant as

$$\det(\mathbf{A}) = \begin{vmatrix} a_{11} & a_{12} & a_{13} \\ a_{21} & a_{22} & a_{23} \\ a_{31} & a_{32} & a_{33} \end{vmatrix} = a_{11}(A_{11}) + a_{12}(A_{12}) + a_{13}(A_{13}) \tag{2-33}$$

The individual cofactors can be expressed as

$$A_{11} = \begin{vmatrix} a_{22} & a_{23} \\ a_{32} & a_{33} \end{vmatrix} = a_{22}a_{33} - a_{23}a_{32} \tag{2-34}$$

$$A_{12} = -\begin{vmatrix} a_{21} & a_{23} \\ a_{31} & a_{33} \end{vmatrix} = -a_{21}a_{33} + a_{23}a_{31} \tag{2-35}$$

$$A_{13} = \begin{vmatrix} a_{21} & a_{22} \\ a_{31} & a_{32} \end{vmatrix} = a_{21}a_{32} - a_{22}a_{31} \tag{2-36}$$

The determinant can then be expressed as

$$\det(\mathbf{A}) = a_{11}(a_{22}a_{33} - a_{23}a_{32}) + a_{12}(-a_{21}a_{33} + a_{23}a_{31}) + a_{13}(a_{21}a_{32} - a_{22}a_{31}) \tag{2-37}$$

Clearly, this is as far as we need to go.

Singular Matrix

If $\det(\mathbf{A}) = 0$, the matrix is said to be *singular*. If the matrix represents the coefficients corresponding to simultaneous linear equations, this means that the equations are not independent and cannot be solved uniquely.

Example 2-12

Determine the determinant of the matrix

$$\mathbf{A} = \begin{bmatrix} 3 & 2 \\ -4 & 5 \end{bmatrix} \tag{2-38}$$

Solution

The determinant is

$$\det(\mathbf{A}) = \begin{vmatrix} 3 & 2 \\ -4 & 5 \end{vmatrix} = (3)(5) - (2)(-4) = 15 + 8 = 23 \tag{2-39}$$

Note the cancellation of the minus signs in the second term in this example.

Example 2-13

Determine the nine minors for the following matrix

$$\mathbf{A} = \begin{bmatrix} 1 & 2 & -1 \\ -1 & 1 & 3 \\ 3 & 2 & 1 \end{bmatrix} \tag{2-40}$$

Solution

Although tedious, the procedure is straightforward. Starting with a_{11}, cross out the first row and the first column and form a 2×2 determinant from the remaining four elements. We will move across the first row and then repeat the procedure on the second and third rows. The nine minors are calculated as follows:

$$M_{11} = \begin{vmatrix} 1 & 3 \\ 2 & 1 \end{vmatrix} = -5, \; M_{12} = \begin{vmatrix} -1 & 3 \\ 3 & 1 \end{vmatrix} = -10, \; M_{13} = \begin{vmatrix} -1 & 1 \\ 3 & 2 \end{vmatrix} = -5$$

$$M_{21} = \begin{vmatrix} 2 & -1 \\ 2 & 1 \end{vmatrix} = 4, \; M_{22} = \begin{vmatrix} 1 & -1 \\ 3 & 1 \end{vmatrix} = 4, \; M_{23} = \begin{vmatrix} 1 & 2 \\ 3 & 2 \end{vmatrix} = -4 \tag{2-41}$$

$$M_{31} = \begin{vmatrix} 2 & -1 \\ 1 & 3 \end{vmatrix} = 7, \; M_{32} = \begin{vmatrix} 1 & -1 \\ -1 & 3 \end{vmatrix} = 2, \; M_{33} = \begin{vmatrix} 1 & 2 \\ -1 & 1 \end{vmatrix} = 3$$

Example 2-14

Determine the cofactors of the nine minors determined in Example 2-13.

Solution

The cofactors are directly related to the minors with only a sign change for alternate values. The relationship is

$$A_{ij} = (-1)^{i+j} M_{ij} \tag{2-42}$$

All this means is that if $i + j$ is an even integer, the cofactor is equal to the minor, and if $i + j$ is an odd integer, the cofactor is equal to the negative of the minor. Thus, from (2-41), we can write down the cofactors as

$$
\begin{aligned}
A_{11} &= -5 & A_{12} &= 10 & A_{13} &= -5, \\
A_{21} &= -4 & A_{22} &= 4 & A_{23} &= 4, \\
A_{31} &= 7 & A_{32} &= -2 & A_{33} &= 3
\end{aligned}
\tag{2-43}
$$

These various cofactors will be used in a matrix inversion example at the end of the next section.

Example 2-15

Determine the determinant of the matrix of Example 2-13 by expansion in minors along the first row.

Solution

The pertinent expression was developed in Equation 2-33 and is repeated here for convenience.

$$\det(\mathbf{A}) = \begin{vmatrix} a_{11} & a_{12} & a_{13} \\ a_{21} & a_{22} & a_{23} \\ a_{31} & a_{32} & a_{33} \end{vmatrix} = a_{11}(A_{11}) + a_{12}(A_{12}) + a_{13}(A_{13}) \tag{2-44}$$

We need only three of the nine minors for determining the determinant. However, the other six will be used later. Substitution of the values from Examples 2-13 and 2-14 leads to

$$\det(\mathbf{A}) = (1)(-5) + (2)(10) + (-1)(-5) = 20 \tag{2-45}$$

2-5   INVERSE MATRIX

We will now turn our attention to the concept of the *inverse matrix*, which will be one of the last major matrix manipulations of this chapter. Determination of the inverse matrix is one of the most difficult processes to perform manually but is quite simple to perform with MATLAB or other mathematical software. As in the case of the determinant formulation, we will limit how far we go with the manual process. First, we need to introduce some definitions.

Identity Matrix

The identity matrix is to a matrix what the number 1 is to a scalar. It is a square matrix with the value 1 along the so-called *main diagonal* and 0 at all other positions. It will be denoted by **I** and will have the form

$$\mathbf{I} = \begin{bmatrix} 1\ 0\ 0....0 \\ 0\ 1\ 0....0 \\ 0\ 0\ 1....0 \\ \vdots\ \ \ \ \ \ \vdots \\ 0\ 0\ 0\ ....1 \end{bmatrix} \tag{2-46}$$

The size of the identity matrix can be adjusted for any particular case as long as the number of rows is equal to the number of columns.

Matrix of Cofactors

A matrix of cofactors is a square matrix in which each element of the original matrix is replaced by its cofactor. It will be denoted by $\mathbf{A}_{co}$, and it can be represented by

$$\mathbf{A}_{co} = \begin{bmatrix} A_{11}\ A_{12}....A_{1m} \\ A_{21}\ A_{22}....A_{2m} \\ \vdots\ \ \ \ \ \ \ \ \ \vdots \\ A_{m1}\ A_{m2}....A_{mm} \end{bmatrix} \tag{2-47}$$

Adjoint Matrix

The *adjoint matrix is defined as the transpose of the matrix of cofactors.* It will be denoted adj(**A**) and can be expressed as

$$\mathrm{adj}(\mathbf{A}) = \mathbf{A}'_{co} = \begin{bmatrix} A_{11}\ A_{21}....A_{m1} \\ A_{12}\ A_{22}....A_{m2} \\ \vdots\ \ \ \ \ \ \ \ \ \vdots \\ A_{1m}\ A_{2m}....A_{mm} \end{bmatrix} \tag{2-48}$$

Definition of Inverse Matrix

The inverse of a matrix will be considered here only for the case of a square matrix, which is the usual application. (There is a concept called the *pseudo-inverse* of a nonsquare matrix, but it is beyond the scope of the present treatment.) For a square matrix $\mathbf{A}$, the inverse is denoted as $\mathbf{A}^{-1}$, or sometimes as inv($\mathbf{A}$). It is defined by the relationship

$$\mathbf{AA}^{-1} = \mathbf{A}^{-1}\mathbf{A} = \mathbf{I} \tag{2-49}$$

Stated in words, the inverse matrix is a matrix of the same size as the given matrix such that multiplication by the given matrix results in the identity matrix. Note that multiplication in either order will produce the same result. The inverse matrix should not be confused with the simple reciprocal of a number (except for the trivial case of a 1×1 matrix).

Determining the Inverse Matrix

Using definitions that have been introduced, a formula for determining the inverse matrix follows.

$$\mathbf{A}^{-1} = \frac{\text{adj}(\mathbf{A})}{\det(\mathbf{A})} \tag{2-50}$$

It is assumed, of course, that $\det(\mathbf{A}) \neq 0$; that is, the matrix is nonsingular.

The formula looks deceptively simple, but remember that there are quite a few operations required to reach the adjoint and the determinant. Note that the determinant in the denominator is a scalar and all elements of the matrix in the numerator will be divided by this value.

Inverse of a 2 × 2 Matrix

The inverse of a 2×2 matrix arises often in many simple problems and is worthy of separate treatment. Without showing the details, the relationship follows.

$$\begin{bmatrix} a_{11} & a_{12} \\ a_{21} & a_{22} \end{bmatrix}^{-1} = \frac{\begin{bmatrix} a_{22} & -a_{12} \\ -a_{21} & a_{11} \end{bmatrix}}{\det(\mathbf{A})} = \begin{bmatrix} \dfrac{a_{22}}{\det(\mathbf{A})} & \dfrac{-a_{12}}{\det(\mathbf{A})} \\ \dfrac{-a_{21}}{\det(\mathbf{A})} & \dfrac{a_{11}}{\det(\mathbf{A})} \end{bmatrix} \tag{2-51}$$

Thus, to invert a 2×2 matrix, switch the elements of the main diagonal, change the signs of the other two elements, and divide all elements by the determinant of the matrix.

Example 2-16

Determine the inverse matrix of

$$\mathbf{A} = \begin{bmatrix} 2 & 3 \\ 4 & 5 \end{bmatrix} \tag{2-52}$$

Solution

Using the form developed in Equation 2-51, we first determine det(A)as

$$\det(A) = (2)(5) - (3)(4) = -2 \qquad (2\text{-}53)$$

The inverse matrix is then

$$A^{-1} = \begin{bmatrix} 2 & 3 \\ 4 & 5 \end{bmatrix}^{-1} = \frac{\begin{bmatrix} 5 & -3 \\ -4 & 2 \end{bmatrix}}{-2} = \begin{bmatrix} -2.5 & 1.5 \\ 2 & -1 \end{bmatrix} \qquad (2\text{-}54)$$

We will leave as an exercise for the reader to show that either the product AA^{-1} or $A^{-1}A$ results in the identity matrix.

Example 2-17

Determine the inverse matrix of A of Example 2-13.

Solution

The matrix is repeated here for convenience.

$$A = \begin{bmatrix} 1 & 2 & -1 \\ -1 & 1 & 3 \\ 3 & 2 & 1 \end{bmatrix} \qquad (2\text{-}55)$$

Most of the computations required have already been made in preceding examples. The determinant was determined in Example 2-15 to be

$$\det(A) = 20 \qquad (2\text{-}56)$$

The cofactors were computed in Example 2-14 and listed in equation (2-43). The matrix of cofactors can then be expressed as

$$A_{co} = \begin{bmatrix} -5 & 10 & -5 \\ -4 & 4 & 4 \\ 7 & -2 & 3 \end{bmatrix} \qquad (2\text{-}57)$$

The adjoint matrix is determined by taking the transpose of the matrix of cofactors.

$$\mathrm{adj}(A) = \begin{bmatrix} -5 & -4 & 7 \\ 10 & 4 & -2 \\ -5 & 4 & 3 \end{bmatrix} \qquad (2\text{-}58)$$

Finally, the inverse matrix $\mathbf{A}^{-1}$ is determined from Equation 2-50 as

$$\mathbf{A}^{-1} = \frac{\text{adj}(\mathbf{A})}{\det(\mathbf{A})} = \frac{\begin{bmatrix} -5 & -4 & 7 \\ 10 & 4 & -2 \\ -5 & 4 & 3 \end{bmatrix}}{20} = \begin{bmatrix} -0.25 & -0.2 & 0.35 \\ 0.5 & 0.2 & -0.1 \\ -0.25 & 0.2 & 0.15 \end{bmatrix} \tag{2-59}$$

This matrix will appear again in an example in the next section.

2-6 APPLICATIONS OF MATRICES

There are numerous practical applications of matrices, some of which will be encountered later in the text in conjunction with more advanced concepts. Also, remember that a primary reason for introducing matrices so early is their association with MATLAB. However, let us explore some of the basic applications in this section.

Simultaneous Linear Equations

One of the most basic applications of matrices is in solving simultaneous linear equations. Assume an array of m linear equations with m unknowns. Let $x_1, x_2. \ldots x_m$ represent the m variables and assume the following form for the equations:

$$
\begin{aligned}
a_{11}x_1 + a_{12}x_2 + \ldots a_{1m}x_m &= b_1 \\
a_{21}x_1 + a_{22}x_2 + \ldots a_{2m}x_m &= b_2 \\
\vdots \qquad\qquad \vdots \quad\ \vdots \\
a_{m1}x_1 + a_{m2}x_2 + \ldots a_{mm}x_m &= b_m
\end{aligned}
\tag{2-60}
$$

It is assumed that all the constants are known but it is desired to determine the variables.

Through the properties of matrix multiplication, the array of Equation 2-60 may be expressed as

$$
\begin{bmatrix}
a_{11} & a_{12} \ldots a_{1m} \\
a_{21} & a_{22} \ldots a_{2m} \\
\vdots & \quad \vdots \\
a_{m1} & a_{m2} \ldots a_{mm}
\end{bmatrix}
\begin{bmatrix}
x_1 \\ x_2 \\ \vdots \\ x_m
\end{bmatrix}
=
\begin{bmatrix}
b_1 \\ b_2 \\ \vdots \\ b_m
\end{bmatrix}
\tag{2-61}
$$

In case there is some confusion about this operation, note that the left-hand side involves an m by m matrix multiplied by an $m \times 1$ matrix (i.e., a column vector), and the result should be an $m \times 1$ matrix (also a column vector).

Now let

$$\mathbf{A} = \begin{bmatrix} a_{11} & a_{12} \dots a_{1m} \\ a_{21} & a_{22} \dots a_{2m} \\ \vdots & \vdots \\ a_{m1} & a_{m2} \dots a_{mm} \end{bmatrix} \tag{2-62}$$

$$\mathbf{x} = \begin{bmatrix} x_1 \\ x_2 \\ \vdots \\ x_m \end{bmatrix} \tag{2-63}$$

$$\mathbf{b} = \begin{bmatrix} b_1 \\ b_2 \\ \vdots \\ b_m \end{bmatrix} \tag{2-64}$$

The entire array of equations can now be written in matrix form as

$$\mathbf{Ax} = \mathbf{b} \tag{2-65}$$

The result is deceptively simple looking, but it obscures the fact that each of the quantities involved is multidimensional in nature. Nevertheless, this type of representation lends itself to the handling of large quantities of data.

Now how do we solve the equation? If these were ordinary one-dimensional algebraic equations, we would simply divide both sides by a constant to determine the unknown. With a matrix equation such as this, we *premultiply* both sides by the inverse of the matrix $\mathbf{A}$, that is, by $\mathbf{A}^{-1}$. Performing this operation we have

$$\mathbf{A}^{-1}\mathbf{Ax} = \mathbf{A}^{-1}\mathbf{b} \tag{2-66}$$

Note that if we premultiply on the left we must also premultiply on the right. Now the product of a square matrix and its inverse is the identity matrix, that is,

$$\mathbf{A}^{-1}\mathbf{A} = \mathbf{I} \tag{2-67}$$

Finally, it can be readily shown that the product of the identity matrix times a vector is the vector; that is,

$$\mathbf{Ix} = \mathbf{x} \tag{2-68}$$

Putting together the preceding several steps, we obtain

$$\mathbf{x} = \mathbf{A}^{-1}\mathbf{b} \tag{2-69}$$

Thus the solution to a matrix equation of the form of Equation 2-65 is determined by taking the inverse of the square matrix and multiplying it by the vector on the right. It is easy to remember Equation 2-69 since it looks similar to what we would do in a simple one-dimensional algebraic formula, but we must remember that we are dealing with an inverse matrix.

It is likely that the reader is familiar with other methods for solving simultaneous linear equations. The beauty of the matrix approach is that all the unknowns are obtained in one "big sweep," and the process may be readily implemented in algorithmic form on a computer. We will see the process applied with MATLAB in the next chapter.

Linear Transformations of Linear Variables

Another application of matrices is to perform a transformation on one set of dependent linear variables to obtain a different set as a basis for representation. The notation for this operation can become a bit unwieldy, so we will restrict the development to two variables at each stage, but the pattern that emerges should be sufficient to convey the general approach.

Let us begin with two independent variables x_1 and x_2. Assume that there are two dependent variables, y_1 and y_2, that are related to x_1 and x_2 by the linear equations

$$y_1 = b_{11}x_1 + b_{12}x_2 \tag{2-70}$$

$$y_2 = b_{21}x_1 + b_{22}x_2 \tag{2-71}$$

We have seen that an array of this type can be written in matrix-vector form as

$$\begin{bmatrix} y_1 \\ y_2 \end{bmatrix} = \begin{bmatrix} b_{11} & b_{12} \\ b_{21} & b_{22} \end{bmatrix} \begin{bmatrix} x_1 \\ x_2 \end{bmatrix} \tag{2-72}$$

or

$$\mathbf{y} = \mathbf{Bx} \tag{2-73}$$

where

$$\mathbf{y} = \begin{bmatrix} y_1 \\ y_2 \end{bmatrix} \tag{2-74}$$

$$\mathbf{B} = \begin{bmatrix} b_{11} & b_{12} \\ b_{21} & b_{22} \end{bmatrix} \tag{2-75}$$

and

$$\mathbf{x} = \begin{bmatrix} x_1 \\ x_2 \end{bmatrix} \tag{2-76}$$

Now suppose that there are two dependent variables, z_1 and z_2, that are linear functions of y_1 and y_2 according to the equations

$$z_1 = a_{11} y_1 + a_{12} y_2 \qquad (2\text{-}77)$$

$$z_2 = a_{21} y_1 + a_{22} y_2 \qquad (2\text{-}78)$$

This array can be written in matrix form as

$$\begin{bmatrix} z_1 \\ z_2 \end{bmatrix} = \begin{bmatrix} a_{11} & a_{12} \\ a_{21} & a_{22} \end{bmatrix} \begin{bmatrix} y_1 \\ y_2 \end{bmatrix} \qquad (2\text{-}79)$$

or

$$\mathbf{z} = \mathbf{Ay} \qquad (2\text{-}80)$$

where

$$\mathbf{z} = \begin{bmatrix} z_1 \\ z_2 \end{bmatrix} \qquad (2\text{-}81)$$

and

$$\mathbf{A} = \begin{bmatrix} a_{11} & a_{12} \\ a_{21} & a_{22} \end{bmatrix} \qquad (2\text{-}82)$$

Suppose that we now wish to obtain expressions for z_1 and z_2 directly in terms of x_1 and x_2 and eliminate the variables y_1 and y_2 in the process. Without knowing anything about matrices, this could be done algebraically by taking the expressions for y_1 and y_2 from Equations 2-70 and 2-71 and substituting them in Equations 2-77 and 2-78. However, we can also do this with matrices.

Take the expression for $\mathbf{y}$ in terms of $\mathbf{x}$ from Equation 2-73 and substitute in Equation 2-80. We then have

$$\mathbf{z} = \mathbf{ABx} \qquad (2\text{-}83)$$

This can be expressed as

$$\mathbf{z} = \mathbf{Cx} \qquad (2\text{-}84)$$

where

$$\mathbf{C} = \mathbf{AB} \qquad (2\text{-}85)$$

Stated in words, the transformation from the vector $\mathbf{x}$ to the vector $\mathbf{z}$ is simply the matrix product of the two individual matrices. Note the order that is required. We have changed the basis of representation of z in terms of y to z in terms of x.

It is questionable whether much is gained when both forms have only two variables and there are only two transformations. However, this process can be generalized to many dimensions and many transformations. The main requirement is that the successive products must be compatible in terms of the number of rows and the number of columns.

Example 2-18

A certain physical system has three unknown variables, x_1, x_2, and x_3, that can be described by the following three linear equations:

$$x_1 + 2x_2 - x_3 = -8 \tag{2-86}$$

$$-x_1 + x_2 + 3x_3 = 7 \tag{2-87}$$

$$3x_1 + 2x_2 + x_3 = 4 \tag{2-88}$$

Solve for the three unknowns using matrix algebra.

Solution

The array can be written as

$$\begin{bmatrix} 1 & 2 & -1 \\ -1 & 1 & 3 \\ 3 & 2 & 1 \end{bmatrix} \begin{bmatrix} x_1 \\ x_2 \\ x_3 \end{bmatrix} = \begin{bmatrix} -8 \\ 7 \\ 4 \end{bmatrix} \tag{2-89}$$

This equation can be expressed in matrix form as

$$\mathbf{Ax = b} \tag{2-90}$$

By now, the values of the elements of Equation 2-90 should be clear, so they will not be written out. The solution of Equation 2-90 is

$$\mathbf{x = A^{-1}b} \tag{2-91}$$

Now we will explain the scheme that has been evolving. The matrix **A** is the same as the matrix introduced in Example 2-13, and its inverse was determined in Example 2-17. Therefore, referring back to Equation 2-59, we can immediately write down the form of Equation 2-91 as

$$\mathbf{x} = \begin{bmatrix} x_1 \\ x_2 \\ x_3 \end{bmatrix} = \begin{bmatrix} -0.25 & -0.2 & 0.35 \\ 0.5 & 0.2 & -0.1 \\ -0.25 & 0.2 & 0.15 \end{bmatrix} \begin{bmatrix} -8 \\ 7 \\ 4 \end{bmatrix} \tag{2-92}$$

The multiplication yields

$$\mathbf{x} = \begin{bmatrix} 2 \\ -3 \\ 4 \end{bmatrix} \tag{2-93}$$

Thus, $x_1 = 2$, $x_2 = -3$, and $x_3 = 4$.

Example 2-19

Assume that a given engineering system has two inputs, x_1, and x_2 and two outputs, y_1 and y_2. The two outputs are related to the two inputs by the equations

$$y_1 = 2x_1 - 3x_2 \tag{2-94}$$

$$y_2 = 4x_1 - 2x_2 \tag{2-95}$$

Assume that the variables y_1, and y_2 control a system with two outputs, given by the equations z_1, and z_2

$$z_1 = 5y_1 - 2y_2 \tag{2-96}$$

$$z_2 = 4y_1 + 3y_2 \tag{2-97}$$

Determine the two values of z in terms of the two values of x.

Solution

The values of y in vector form can be expressed in terms of the values of x as

$$\begin{bmatrix} y_1 \\ y_2 \end{bmatrix} = \begin{bmatrix} 2 & -3 \\ 4 & -2 \end{bmatrix} \begin{bmatrix} x_1 \\ x_2 \end{bmatrix} \tag{2-98}$$

The values of z in vector form can be expressed in terms of the values of y as

$$\begin{bmatrix} z_1 \\ z_2 \end{bmatrix} = \begin{bmatrix} 5 & -2 \\ 4 & 3 \end{bmatrix} \begin{bmatrix} y_1 \\ y_2 \end{bmatrix} \tag{2-99}$$

Substitution of Equation 2-98 in Equation 2-99 yields

$$\begin{bmatrix} z_1 \\ z_2 \end{bmatrix} = \begin{bmatrix} 5 & -2 \\ 4 & 3 \end{bmatrix} \begin{bmatrix} 2 & -3 \\ 4 & -2 \end{bmatrix} \begin{bmatrix} x_1 \\ x_2 \end{bmatrix} \tag{2-100}$$

After the multiplication of the two square matrices, we obtain

$$\begin{bmatrix} z_1 \\ z_2 \end{bmatrix} = \begin{bmatrix} 2 & -11 \\ 20 & -18 \end{bmatrix} \begin{bmatrix} x_1 \\ x_2 \end{bmatrix} \tag{2-101}$$

The resulting equations are

$$z_1 = 2x_1 - 11x_2 \tag{2-102}$$

$$z_2 = 20x_1 - 18x_2 \tag{2-103}$$

2-7 PROGRAMMING: CONDITIONAL IF STATEMENTS

The entire next chapter will be devoted to matrix operations with MATLAB, but the particular coverage of this section has no direct relationship to that topic. However, it is a good point at which to introduce some useful MATLAB operations for programming. This section introduces **conditional statements**.

Conditional Statements

A series of conditional statements would not usually be performed in the **Command Window**, but they are often either required or at least very desirable in programs that can be run as **M-files**. They are used when a program has more than one path or option to follow based on certain specified input conditions or possible results of part of the program. We will refer to a set of conditional statements as an **if routine.**

If there are only two sets of possibilities, the syntax is as follows:

if *expression*
        statements1
else   (2-104)
        statements2
end

The term *expression* performs some type of equality or inequality analysis. If equality is the focus, two equal signs (= =) are required. One can also utilize statements such as "greater than" (>), "greater than or equal to" (>=), and so on. The MATLAB **Help** file can be used to determine all the possible formats.

The part called *statements1* contains one or more operations to be performed if *expression* is true. If *expression* is not true, the flow will move to **else** and *statements2* will become operational. Note the **end** statement at the end of the operation.

Suppose you desire to have three or more possibilities that continue in a linear fashion. We will illustrate with four, which should provide sufficient insight to generalize.

if *expression1*
        statements1
elseif *expression2*
        statements2
elseif *expression3*   (2-105)
        statements3
else
        statements4
end

Note that between the **if** statement and the **else** statement, we require as many **elseif** statements as necessary. Again an **end** statement is required at the end.

If **nesting** is required, meaning that further testing is required within a particular path, a separate **end** statement is required for each path. We will not consider that situation at this time.

Incidentally, other operations may precede or follow the **if routine**. It is only within this part of a program that some type of decision must be made. One must be careful to ensure that all of the expressions are compatible with each other and that there is no ambiguity in the flow.

Programming Example 2-1

Considering a program that serves no useful purpose other than to illustrate the concept of the **if routine**, assume that there is an event that charges an admission price according to the following age ranges:

Age	Price
65	free admission
under 12	$2
12–64	$5
66 and over	$3

Write an M-file program that provides the admission price for a given age.

Input: age
Output: price

Solution

Refer to Figure 2-1 for the M-file, which has the name **p_example_2_1.m**. After a series of comment statements, the input statement asks for the age and uses **age** as the variable. The output is to be designated as **price,** and the **if routine** begins the process with the one equality statement and the outcome.

$$\text{if age} == 65 \tag{2-106}$$
$$\text{price} = \text{´free admission´}$$

Note the two equal signs in the **if** statement. If this statement is true, the next command follows.

While a numerical value such as **price = 0** could have been used, we have chosen here to provide an output in the form of a **string variable**, which is denoted by inclusion between the apostrophes (´ ´). The outcome on the screen will then read **free admission**.

If the first statement is not true, the program moves to the next test.

elseif age < 12　　　　　　　　　　　　　　　　　　　　　　　　　(2-107)
　　　price = ´2 dollars´

```
% M-file program p_example_2_1.m
% It was written to illustrate the use of conditional statements.
% The input variable is the age of a person.
% The output variable is the price of admission in dollars for a particular event.
% People who are exactly 65 years of age get free admission.
age=input('Enter the age of a person as an integer number of years. ')
if age==65
    price='free admission'
elseif age<12
    price='2 dollars'
elseif age<=64
    price='5 dollars'
else
    price='3 dollars'
end
```

Figure 2-1　M-File of Programming Example 2-1

If neither of the first two statements is true, the program moves on to try the next one.

elseif age <= 64　　　　　　　　　　　　　　　　　　　　　　　　(2-108)
　　　price = ´5 dollars´

In this case, we have used a "less than or equal to" statement. It should be noted that either of the preceding tests could have used "less than" or "less than or equal to" depending on which number was used for the upper value. In the first case, a "less than or equal to 11" would have worked, and in the second case, a "less than 65" would have worked. However, both types were used to illustrate different formats.

Finally, if none of the preceding statements is true, the person must necessarily be 66 or older, and the final outcome is

else　　　　　　　　　　　　　　　　　　　　　　　　　　　　　(2-109)
　　　price = ´3 dollars´

General Problems

2-1. Calculate the size of a matrix given by

$$\mathbf{A} = \begin{bmatrix} 1 & 2 & -3 & 5 \\ -2 & 1 & 4 & -3 \end{bmatrix}$$

2-2. Calculate the size of a matrix given by

$$B = \begin{bmatrix} 2 & -1 \\ -3 & 5 \\ -2 & 1 \\ 4 & 2 \end{bmatrix}$$

2-3. Determine the transpose of the matrix of Problem 2-1.

2-4. Determine the transpose of the matrix of Problem 2-2.

In Problems 2-5 through 2-18, assume the following matrices:

$$A = \begin{bmatrix} -2 & 3 \\ 2 & -4 \end{bmatrix} \quad B = \begin{bmatrix} 2 & -2 \\ 4 & -5 \end{bmatrix} \quad C = \begin{bmatrix} 1 & 4 \\ 2 & 3 \end{bmatrix}$$

Perform the operation indicated in each problem.

2-5. $A + B$

2-6. $A + C$

2-7. $A - B$

2-8. $A - C$

2-9. AB

2-10. AC

2-11. BA

2-12. CA

2-13. ABC

2-14. CBA

2-15. $\det(A)$

2-16. $\det(B)$

2-17. A^{-1}

2-18. B^{-1}

2-19. For the matrices of Problems 2-1 and 2-2, show that the product **AB** is valid and determine it.

2-20. For the matrices of Problems 2-1 and 2-2, show that the product **BA** is valid and determine it.

In Problems 2-21 and 2-22, two vectors are defined as

$$\mathbf{A} = [1\ 3\ 5\ 7] \text{ and } \mathbf{B} = \begin{bmatrix} 2 \\ 4 \\ 6 \\ 8 \end{bmatrix}$$

2-21. Determine **AB**.

2-22. Determine **BA**.

2-23. Determine the determinant of the following matrix:

$$\mathbf{A} = \begin{bmatrix} 11 & -5 & 0 \\ -5 & 10 & -2 \\ 0 & -2 & 13 \end{bmatrix}$$

2-24. Determine the determinant of the following matrix:

$$\mathbf{B} = \begin{bmatrix} 5 & -2 & 0 \\ -2 & 11 & -5 \\ 0 & -5 & 11 \end{bmatrix}$$

2-25. Determine the inverse of the matrix **A** of Problem 2-23.

2-26. Determine the inverse of the matrix **B** of Problem 2-24.

2-27. The analysis of the forces in a basic static structural system involve some equilibrium equations:
  a. The algebraic sum of all the forces in any direction is zero.
  b. The algebraic sum of the moments about any pivot point is zero.
       The equations obtained for a certain two-dimensional system with the forces in newtons (N) are

$$0.8F_1 + 0.7071F_2 = F_3 + 100$$
$$0.6F_1 = 0.7071F_2$$
$$32F_1 + 3000 = 42.42F_2$$

Arrange the equations in matrix form and solve for the three variables using matrix algebra.

2-28. The analysis of the mesh currents in a dc circuit utilizes Kirchhoff's voltage law (KVL), which states that the algebraic sum of the voltages around each loop is zero. With voltages expressed in terms of three unknown currents I_1, I_2, and I_3, measured in amperes (A), the KVL equations for a certain three-mesh circuit are

$$-40 + 6I_1 + 5(I_1 - I_2) - 50 = 0$$
$$50 + 5(I_2 - I_1) - 64 + 3I_2 + 2(I_2 - I_3) + 12 = 0$$
$$-12 + 2(I_3 - I_2) + 11I_3 + 98 = 0$$

Arrange the equations in matrix form and solve for the three currents using matrix algebra.

2-29. The analysis of the node voltages in a dc circuit utilizes Kirchhoff's current law (KCL), which states that the algebraic sum of the currents leaving each node is zero. With currents expressed in terms of three unknown voltages, V_1, V_2, and V_3, measured in volts (V), the KCL equations for a certain circuit are

$$-10 + \frac{V_1}{6} + \frac{V_1}{5} + 15 + \frac{(V_1 - V_2)}{3} + 32 = 0$$
$$-32 + \frac{(V_2 - V_1)}{3} - 9 + \frac{V_2}{2} + \frac{(V_2 - V_3)}{4} = 0$$
$$\frac{(V_3 - V_2)}{4} + \frac{V_3}{7} - 21 = 0$$

Arrange the equations in matrix form and solve for the three voltages using matrix algebra.

2-30. The equations involving a certain steel girder involve three forces, F_1, F_2, and F_3, measured in newtons (N), and they are

$$20 + F_3 = 0.65F_2 + 0.7F_1$$
$$0.8F_1 = 0.9F_2$$
$$4F_1 = 110 + 0.9F_2$$

Arrange the equations in matrix form and solve for the three variables using matrix algebra.

MATLAB PROBLEM

2-31. Assume that an industrial distributor of a certain machine part charges a price based on the number of units purchased according to the following scale:

Number of Units	Price per Unit
First 100	$10
Next 900	$8
Next 9000	$6
All additional	$5

Write a program that will determine the cost for any arbitrary number of units.

Input: number of units purchased
Output: total cost in dollars
          average cost per unit

Suggestion: Set the *bank* format in the Command Window before executing the program. This format will express the answer in dollars and cents and is set by the following command:

$$>> \text{format bank}$$

The default short format is restored by the command

$$>> \text{format}$$

Matrix Algebra with MATLAB

3

Having provided a coverage of matrix concepts and algebraic manipulations in Chapter 2, we will now consider how the corresponding operations are performed in MATLAB. In a sense, all variables in MATLAB are considered as matrices. A simple scalar is considered as a 1×1 matrix. However, scalar quantities are easily manipulated with MATLAB, as demonstrated in Chapter 1. Once a MATLAB variable becomes multidimensional, however, it is necessary to consider some of the effects that will arise in matrix operations.

Objectives

After completing this chapter, the reader should be able to perform the following operations in the MATLAB Command Window:

1. Enter a matrix or a vector.
2. Perform addition and subtraction of matrices.
3. Determine the transpose of a matrix.
4. Perform multiplication of matrices based on the normal matrix definition.
5. Perform array multiplication of matrices.
6. Determine the determinant of a square matrix.
7. Determine the inverse of a square matrix.
8. Solve an array of linear simultaneous equations.

3-2 MATRIX AND VECTOR FORMS IN MATLAB

We will begin by discussing how one can enter a matrix in MATLAB. Consider Equation 2-3 of Example 2-1, which is repeated here for convenience.

$$\mathbf{A} = \begin{bmatrix} 2 & -3 & 5 \\ -1 & 4 & 6 \end{bmatrix} \tag{3-1}$$

Matrix Entry

In order for matrix values, some scalar values, and some commands to stand out within text discussions, they will sometimes be shown in boldface type. However, the type style for the corresponding MATLAB variables and operations will *not* be boldface style. Thus, it will be necessary to frequently switch between the different type styles.

MATLAB variables are *case-sensitive*, so **A** and **a** will represent different variables. When using A to represent the MATLAB entry for Equation 3-1, it is typed and entered as

$$>> A = [2 -3 5; -1 4 5] \tag{3-2}$$

$$A =$$

$$2 -3 5$$

$$-1 4 5$$

Note that each row can be entered with spaces between the values, although commas can be used if desired. We will use spaces for most cases. A semicolon is required at the end of each row. MATLAB lists the matrix in rectangular form when possible. If the matrix is too large to fit on the screen in rectangular form, the rows and columns will be identified on the screen.

Vector Entry

Consider now the case of a row vector such as

$$\mathbf{x} = [1 \ 4 \ 7] \tag{3-3}$$

The MATLAB variable will be denoted as x. It is typed and entered as

$$>> x = [1 \ 4 \ 7] \tag{3-4}$$

$$x =$$

$$1 \ 4 \ 7$$

Suppose we desired to enter **x** as a column vector. The format is

$$>> x = [1; 4; 7] \tag{3-5}$$

$$x =$$

$$1$$

$$4$$

$$7$$

Transpose

A transpose of a vector x can be obtained by following the variable with a prime (´), which means that an alternate way to create the column matrix is to simply take the transpose of the row vector. Thus, if we first return to the row form of x as given by Equation 3-4, we could use the command

$$>> x = x'$$ (3-6)

$$x =$$

$$1$$

$$3$$

$$5$$

This latter command says that we begin with the vector x, determine its transpose, and put it back in the memory location of x. Alternately, we could write

$$>> x = [1\ 3\ 5]'$$ (3-7)

$$x =$$

$$1$$

$$3$$

$$5$$

In this latter form, the vector is entered as a row vector and the transpose operation is applied immediately to the whole array.

Matrix Addition and Subtraction

Matrix addition and subtraction with MATLAB are achieved through the same commands as with scalars. Of course, it is necessary that the separate matrices have the same sizes. Thus if A and B are two matrices of the same size and we desire to add them to form a new matrix, C, the command is

$$>> C = A + B$$ (3-8)

A command for a matrix D that would be created by subtracting B from A is

$$>> D = A - B$$ (3-9)

If the matrices do not have the same sizes, the following error signal will appear:

$$??? \text{ Error using} ==> \pm$$

Matrix dimensions must agree.

A variety of different types of error signals appear whenever operations or commands are improper. Unless you are an excellent typist and pay extremely close attention to every detail, you can expect to see lots of error signals! Fortunately, they can be quickly corrected.

Matrix Multiplication

Matrix multiplication with MATLAB is achieved with the same command as with scalars, provided that the number of columns of the first matrix is equal to the number of rows of the second. Thus, if A is an m by n matrix and B is an n by k matrix, then MATLAB multiplication will yield an m by k matrix. When denoting the MATLAB product as E, the command is

$$>> E = A*B \tag{3-10}$$

Array Product of Two Matrices

There is another form of multiplication of matrices in which it is desired to multiply corresponding elements in a fashion similar to that of addition and subtraction. This operation arises frequently with MATLAB, and we will hereafter refer to the process as the **array product** to distinguish it from the standard matrix multiplication form.

For the array product to be possible, the two matrices must have the same size, as was the case for addition and subtraction. The resulting array product will have the same size. If F represents the resulting matrix, a given element of F, denoted by f_{ij}, is determined by the corresponding products from the two matrices as

$$f_{ij} = a_{ij}b_{ij} \tag{3-11}$$

The MATLAB command for performing this operation is

$$>> F = A.*B \tag{3-12}$$

This operation is commutative and could also be expressed as

$$>> F = B.*A \tag{3-13}$$

The requirement is that the first matrix must be followed with a period.

If there are more than two matrices for which array multiplication is desired, the periods should follow all but the last one in a successive multiplication chain, for example, A.*B.* C in the case of three matrices. Alternately, one can break up the operation into separate steps or employ nesting with parentheses. The command (A.*B).*C is an example of the latter form.

Determinant of a Matrix

The MATLAB command for the determinant of a matrix A is **det(A)**. If the resulting scalar value is denoted as a, the command is

$$>> a = \det(A) \tag{3-14}$$

Note that a is a scalar (1 x 1 "matrix").

Inverse Matrix

The inverse of a square matrix A in MATLAB is determined by the simple command **inv(A)**. Thus, if B is to represent the inverse of A, we would enter the command

$$>> B = \text{inv}(A) \tag{3-15}$$

Simultaneous Equation Solution

Assume that an array of simultaneous linear equations has been placed in matrix form as follows:

$$\mathbf{Ax = b} \tag{3-16}$$

The mathematical solution, as developed in Chapter 2, can be expressed as

$$\mathbf{x = A^{-1}b} \tag{3-17}$$

Assume now that the square matrix and the vector have been entered in MATLAB as A and b, respectively. The corresponding operation to determine the vector x is

$$>> x = \text{inv}(A)*b \tag{3-18}$$

MATLAB also provides an alternate command, which is recommended for large arrays. The alternate command is

$$>> x = A\backslash b \tag{3-19}$$

Let us now put these operations into practice using some of the arrays considered in Chapter 2. For most examples, the manipulations are very simple, but they should provide good practice for more complex operations later.

Note: Examples 3-1 through 3-6 follow in sequence based on the matrices given in Example 3-1, which follows.

Example 3-1

Consider the matrices of Examples 2-1, 2-2, 2-9, and 2-10, which are repeated here for convenience.

$$\mathbf{A} = \begin{bmatrix} 2 & -3 & 5 \\ -1 & 4 & 6 \end{bmatrix} \tag{3-20}$$

$$\mathbf{B} = \begin{bmatrix} 2 & 1 \\ 7 & -4 \\ 3 & 1 \end{bmatrix} \qquad (3\text{-}21)$$

Enter these variables in MATLAB.

Solution

Actually, we have already entered **A** earlier in the development, but the process will be repeated for practice. The MATLAB commands follow.

$$>> A = [2\ \text{-}3\ 5;\ \text{-}1\ 4\ 6]; \qquad (3\text{-}22)$$

$$>> B = [2\ 1;\ 7\ \text{-}4;\ 3\ 1]; \qquad (3\text{-}23)$$

Printing was suppressed, but the matrices are now in memory.

Example 3-2

Let C represent the transpose of B. Determine C.

Solution

The command and the screen results follow.

$$>> C = B' \qquad (3\text{-}24)$$

C =

2 7 3

1 -4 1

The original 3×2 matrix has now been converted to a 2×3 matrix.

Example 3-3

Let **D** represent the sum of **A** and **C**. Determine **D**.

Solution

The original matrices **A** and **B** were not compatible for addition, but **A** and **C** are each 2×3 in size and so can be added or subtracted. The result for **D** follows.

$$>> D = A + C \tag{3-25}$$

$$D =$$

$$4\ 4\ 8$$

$$0\ 0\ 7$$

Example 3-4

The matrix product **AB** was calculated manually in Example 2-10. Perform the operation with MATLAB.

Solution

The product was called **C** back in Example 2-10, but we have used that name for something else here. We do not anticipate using these results again, so the operation will be stated in its simplest form.

$$>> A*B \tag{3-26}$$

$$ans =$$

$$-2\ 19$$

$$44\ -11$$

The "answer" is the same as the result of Example 2-10, as given in Equation 2-27.

Example 3-5

The matrix product **BA** was calculated manually in Example 2-11. Perform the operation with MATLAB.

Solution

For the same reason as in the preceding example, the result will be written simply as

$$>> B*A \tag{3-27}$$

$$ans =$$

$$3 \quad -2 \quad 16$$

$$18 \quad -37 \quad 11$$

$$5 \quad -5 \quad 21$$

This result agrees with that of Example 2-11, as given in Equation 2-29.

Example 3-6

Let **E** represent the **array product** of **A** and **C**. Determine it with MATLAB.

Solution

Since the two matrices have the same size, the array product can be performed. We have

$$>> E = A.*C \tag{3-28}$$

$$E =$$

$$4 \quad -21 \quad 15$$

$$-1 \quad -16 \quad 6$$

The reader can readily verify that the products of the elements of A times the corresponding elements of C yield the result given. It can also be readily shown that the product C.*A produces the same result.

Note again that this is not matrix multiplication in the usual sense. Rather, we are considering a matrix formed from the corresponding products of the elements of two matrices having the same size .

Note: Examples 3-7 through 3-9 use the matrix given in Example 3-7, which follows.

Example 3-7

A 3×3 matrix was introduced in Example 2-13 and considered in several examples. It is repeated here for convenience.

$$\mathbf{A} = \begin{bmatrix} 1 & 2 & -1 \\ -1 & 1 & 3 \\ 3 & 2 & 1 \end{bmatrix} \tag{3-29}$$

Enter this matrix in MATLAB and determine its determinant.

Solution

The matrix is entered in MATLAB as follows:

$$>> A = [1\ 2\ -1;\ -1\ 1\ 3;\ 3\ 2\ 1] \tag{3-30}$$

A =

1 2 -1

-1 1 3

3 2 1

We will denote the determinant as **a**. We have

$$\gg a = \det(A) \tag{3-31}$$

a =

20

The result, of course, is a scalar and is the same value as determined in Equation 2-45.

Example 3-8

Determine the inverse matrix of **A** from Example 3-7.

Solution

Let Ainv represent the inverse. We have

$$\gg Ainv = inv(A) \tag{3-32}$$

Ainv =

-0.2500	-0.2000	0.3500
0.5000	0.2000	-0.1000
-0.2500	0.2000	0.1500

This result agrees with Equation 2-59.

Example 3-9

In Example 2-18, three simultaneous linear equations with three unknowns were formulated in matrix form and the result was determined manually. Determine the unknowns using MATLAB and matrix algebra.

Solution

The matrix **A** is in memory from preceding examples, and we need to define the vector **b**. In MATLAB terminology, we have

$$\text{>> b = [-8; 7; 4]} \tag{3-33}$$

$$b =$$

$$-8$$

$$7$$

$$4$$

While we have already computed inv(A) and could use it if desired, we will write out the entire expression here for completeness. The vector x is determined from the command

$$\text{>> x = inv(A)*b} \tag{3-34}$$

$$x =$$

$$2.0000$$

$$-3.0000$$

$$4.0000$$

The results are in obvious agreement with Equation 2-93.

3-3   PROGRAMMING: FOR LOOPS

In this section, the concept of the **for loop** will be introduced. Simple **for loops** can often be executed within the interactive environment of the Command Window, but complex loops, especially where there are two or more levels of nesting, are best performed through an M-file, where careful editing can be performed before execution.

Loop Format

A **for loop** is an operation that performs a repetitive type of operation based on a variable that assumes a sequence of values. For example, if it is desired to perform the same operation many times for different sets of variables, the operation can be placed in a **for loop**.

The syntax is as follows:

$$\text{for } \textit{variable} = \textit{expression} \qquad\qquad (3\text{-}35)$$
$$\textit{statements}$$
$$\text{end}$$

It is possible to place one or more inner loops within an outer loop by **nesting**. Each loop must have an **end** statement, and the sequence must be logical. It is best to illustrate the process with a simple example. The use of nesting will become clearer as you develop more sophistication with the software.

Programming Example 3-1

Write a simple program to generate $N!$ for an integer input, where

$$0! = 1 \text{ (by definition)}$$
$$1! = 1$$
$$2! = 2 \times 1 = 2 \qquad\qquad (3\text{-}36)$$
$$3! = 3 \times 2 \times 1 = 6$$
$$4! = 4 \times 3 \times 2 \times 1 = 24 \text{ (etc.)}$$

Solution

We could put an **if** statement up front to test to see if the input is a nonzero positive integer, but we will assume for simplicity that only positive integers will be entered.

The program is shown in Figure 3-1 and has the title **p_example_3_1.m**. A few comments are provided on the first few lines. We are then polled for the input by the statement

$$N = \text{input}(\text{'Enter the positive integer value of N. '}) \qquad\qquad (3\text{-}37)$$

Some loops may have all the operations contained within them. In this particular case, however, we need to provide the first value before entering the loop. We will choose a variable **n** to represent the loop variable, and we begin ahead of the loop by assigning it the value

$$n = 1; \qquad\qquad (3\text{-}38)$$

We now begin the loop using a loop index k. The first statement reads

$$\text{for } k = 2:N \qquad\qquad (3\text{-}39)$$

This tells MATLAB to repeat all statements between this point and the **end** statement a number of times, based on k beginning at 2 and going all the way to N. For each "cycle," the operation is

$$n = k*n; \qquad\qquad (3\text{-}40)$$

```
% The title of this program is p_example_3_1.m
% It will determine the value of N! for a positive integer value of N.
% The input variable is any positive integer N.
% The output variable is N_factorial.
N = input('Enter the positive integer value of N. ')
n = 1;
for k = 2:N
    n = k*n;
end
N_factorial = n
```

Figure 3-1   M-file of Programming Example 3-1

This means that that the previous value of n will be multiplied by the current value of k; the result is the new value of n.

After the **end** statement, we define N_factorial by the final value of n.

$$N\_factorial = n \tag{3-41}$$

It would have been perfectly fine to use N_factorial as the variable in the loop, but we would then have been required to omit the semicolon for screen suppression within the loop, and all possible values would have been repeated on the screen.

It should be noted that MATLAB already contains a function in its library that can be addressed as **factorial(N)**. However, the example developed here has a different name and was chosen to illustrate the use of the **for loop**.

MATLAB PROBLEMS

Many of the problems that follow are the same as or similar to problems at the end of Chapter 2. Whereas you were expected to solve the problems in Chapter 2 by hand, you should now use MATLAB. Unless otherwise directed, use MATLAB exclusively to solve the problems that follow.

In Problems 3-1 through 3-18, assume the following matrices:

$$\mathbf{A} = \begin{bmatrix} -2 & 3 \\ 2 & -4 \end{bmatrix} \quad \mathbf{B} = \begin{bmatrix} 2 & -2 \\ 4 & -5 \end{bmatrix} \quad \mathbf{C} = \begin{bmatrix} 1 & 4 \\ 2 & 3 \end{bmatrix}$$

Use MATLAB to perform the operation indicated in each case.

3-1.   **A + B**

3-2.   **A + C**

3-3. **A – B**

3-4. **A – C**

3-5. **AB**

3-6. **AC**

3-7. **BA**

3-8. **CA**

3-9. **ABC**

3-10. **CBA**

3-11. **ACB**

3-12. **BCA**

3-13. det(**A**)

3-14. det(**B**)

3-15. $(\mathbf{A})^{-1}$

3-16. $\mathbf{B}^{-1}$

3-17. $(\mathbf{AB})^{-1}$

3-18. $\mathbf{B}^{-1}\mathbf{A}^{-1}$

In Problems 3-19 and 3-20, let

$$\mathbf{A} = \begin{bmatrix} 1 & 2 & -3 & 5 \\ -2 & 1 & 4 & -3 \end{bmatrix} \text{ and } \mathbf{B} = \begin{bmatrix} 2 & -1 \\ -3 & 5 \\ -2 & 1 \\ 4 & 2 \end{bmatrix}$$

3-19. Determine **AB**

3-20. Determine **BA**

In Problems 3-21 and 3-22 let

$$\mathbf{A} = [1\ 3\ 5\ 7] \text{ and } \mathbf{B} = \begin{bmatrix} 2 \\ 4 \\ 6 \\ 8 \end{bmatrix}$$

3-21.   Determine **AB**

3-22.   Determine **BA**

3-23.   Determine the determinant of the following matrix:

$$\mathbf{A} = \begin{bmatrix} 11 & -5 & 0 \\ -5 & 10 & -2 \\ 0 & -2 & 13 \end{bmatrix}$$

3-24.   Determine the determinant of the following matrix:

$$\mathbf{B} = \begin{bmatrix} 5 & -2 & 0 \\ -2 & 11 & -5 \\ 0 & -5 & 11 \end{bmatrix}$$

For Problems 3-25 through 3-30, it is suggested that the format be changed by the following command:

>> format short e

The default short format can then be restored by the command

>> format

3-25.   Determine the inverse of matrix **A** of Problem 3-23.

3-26.   Determine the inverse of matrix **B** of Problem 3-24.

3-27.   Solve Problem 2-27 using MATLAB.

3-28.   Solve Problem 2-28 using MATLAB.

3-29.   Solve Problem 2-29 using MATLAB.

3-30.   Solve Problem 2-30 using MATLAB.

3-31.   Although you have seen how easy it is to solve simultaneous equations directly in the Command Window, assume that you wish to write a program that someone else can use to solve three simultaneous linear equations with three unknowns simply by entering the coefficients and the terms on the right-hand side of the equations. Thus, any matrix operations will be "hidden" within the program and not visible to the user, who is assumed to know nothing about matrix theory.

Consider the following format for the three equations:

$$a_{11}x_1 + a_{12}x_2 + a_{13}x_3 = b_1$$
$$a_{21}x_1 + a_{22}x_2 + a_{23}x_3 = b_2$$
$$a_{31}x_1 + a_{32}x_2 + a_{33}x_3 = b_3$$

Inputs: The 9 **a** coefficients and the 3 **b** coefficients

Hint: It would be rather boring to have all 12 coefficients entered separately. The following is a reasonable procedure for entry. A possible first statement is something like

a1=input('Enter the three coefficients of the first row
between brackets [ ] with spaces between values. ')

Similar statements could be used for **a2**, **a3**, and **b**. A square matrix can then be formed by the command

A=[a1; a2; a3]

The column vector **b** can be formed from the entered row vector by the command

b=b′

Outputs: x1, x2, and x3

Hint: x1=x(1), and so on.

3-32.   Assume that two men A and B are hired to work for 30 days. A is offered a very high salary of $1,000 per day, which of course would mean $30,000 for the 30 days. B is offered a changing pay scale that is defined as follows: On the first day, he is paid 1 cent; the second day, 2 cents; the third day, 4 cents; the fourth day, 8 cents; and so on. He is very disappointed because he perceives that his earnings will be substantially less than that of his friend A. To help him through this dilemma, you decide to write an M-file program to determine what his total earnings will be at the end of 30 days. While it is possible to determine the amount through a closed-form solution, you want to gain experience using a **for loop**. Using this concept, write an M-file program that will determine the total amount in dollars that B will earn in 30 days of work.

Suggestion: Set the *bank* format in the Command Window before executing the program. This format will express the answer in dollars and cents and is set by the following command:

>> format bank

The default short format is restored by the command

>> format

Curve Plotting with MATLAB

4

4-1 OVERVIEW AND OBJECTIVES

MATLAB provides some very powerful features for plotting and labeling curves. These operations can be performed as part of an overall mathematical analysis, or experimental data may be provided to the program for the primary purpose of plotting. Curves obtained from MATLAB plots can be exported to other programs for presentation purposes.

Many of the plotting operations are peculiar to certain mathematical concepts and will be introduced as the appropriate mathematical techniques are covered. However, the purpose of this chapter is to introduce the fundamental processes so that further developments can build on these concepts.

All plots in this chapter are two-dimensional. Three-dimensional plots will be considered near the end of Chapter 5.

Objectives

After completing this chapter, the reader should be able to perform the following operations as applied to the MATLAB Command Window:

1. State the basic requirements for curves to be plotted using MATLAB.
2. Apply both the step command and the linear spacing command for generating a linear vector.
3. Perform the **plot** command for plotting one vector as a function of another vector.
4. Label the horizontal and vertical axes and provide a title for a graph.
5. Apply the **grid** command, the **gtext** command, and the **axis** command for altering a figure.
6. Generate a vector with logarithmic spacing between elements.
7. Discuss the reasoning behind log-log plots.
8. Perform the **logspace** and the **loglog** commands in association with logarithmic plots.
9. Prepare a **stem** plot.
10. Prepare a **bar** plot.

4-2 BASIC CURVE PLOTTING PROCEDURES

MATLAB has the capability to generate plots of many types. This includes linear plots, line plots, logarithmic plots on both scales, logarithmic plots on one scale, stem plots, bar graphs, and three-dimensional plots. We will be using these capabilities throughout the text, so the present development is intended as an introduction, with many operations to follow in later chapters.

The reason for introducing plotting following the treatment of matrices is that plots are developed through the utilization of vectors. For example, a two-dimensional plot is achieved by plotting one vector versus a second vector.

Vector Lengths

A very important fact that should be emphasized at the outset is that to plot one vector against another, *the vectors must have the same number of elements.* Normally, if one is a row vector the other will also be a row vector, and likewise for column vectors. However, they are not required to be the same type. One can plot a column vector versus a row vector or vice versa, provided they have the same number of values. If the vectors have different lengths, it is possible to plot a partial vector versus the other vector by defining a partial vector to have the same number of elements as the other vector.

The Variables x and y

In the two-dimensional plotting commands, the horizontal axis will be referred to as the **x-axis** and the vertical axis will be referred to as the **y-axis**. However, the actual variables can be labeled with any quantities. It is only in the plot commands that **x** and **y** are used.

Creating a Linear Array

Whenever a plot is to be created from an equation and linear plots for both the dependent and independent variables are desired, the most convenient way to achieve the result is to create a linear array or vector for the values of the independent variable. MATLAB offers a number of different commands, along with several variations, that can be used for this purpose. For this explanation, assume that the independent variable is x. To obtain equal spacing, the following format can be used:

$$\gg x = x1{:}xstep{:}x2 \qquad\qquad (4\text{-}1)$$

where x1 is the first value, xstep is the step or "distance" between points, and x2 is the final value. We will assume that the final value is the initial value plus an integer multiple of the step size. Based on this assumption, the step process generates a number of points equal to

$$N = \frac{x_2 - x_1}{x_{step}} + 1 \qquad\qquad (4\text{-}2)$$

Note the seemingly odd requirement to add 1 to the number. To see why this is necessary, consider stepping from 0 to 3 in steps of 1. The first term on the right of Equation 4-2 yields $3/1 = 3$, but there are actually 4 steps (0, 1, 2, and 3).

A second command that achieves the same process is **linspace** (linear spacing), whose format reads as follows:

$$>> x = linspace(x1,x2,N) \qquad (4-3)$$

where x1 and x2 have the same meaning as in the first case and N is the total number of points in the vector. The number of points is related to the other parameters in the same manner as given by Equation 4-2.

Example 4-1

Whenever air resistance can be ignored, and at altitudes reasonably close to the Earth, the velocity in meters/second (m/s) of an object falling from rest can be expressed as a function of time in seconds by the equation

$$v = 9.8t \qquad (4-4)$$

Use MATLAB to plot the velocity over a time interval from 0 to 10 s.

Solution

First, it should be emphasized that this is a simple linear equation with a vertical intercept of 0, so we actually *need only two points to plot the curve*. However, our purpose is to learn how to use MATLAB for plotting, so we will utilize far more points than necessary as a learning process.

The best way to perform the task is to first define a row vector representing all values of the independent variable time at which points are to be determined. This could be done manually by actually typing the time values and then enclosing them by brackets to form the vector. This would be satisfactory for a few points, and it might be the only practical way when experimental data are used and the time values are random in nature. However, we are at liberty to choose the time points in a linear sequential fashion, and we will utilize a fairly large number of points to illustrate the process.

We will arbitrarily select a time step of 0.1 s. Hence, a command to generate the time vector is

$$>> t = 0:0.1:10; \qquad (4-5)$$

It is easy to show that this command results in 101 values of time. An alternate command to generate the same vector is

$$>> t = linspace(0,10,101); \qquad (4-6)$$

Without the semicolon at the ends of Equations 4-5 and 4-6, 101 values of time would "flash" down the screen.

Before performing any evaluation with the time vector, let us explore some of its properties. Since it is a row vector, there is an index number corresponding to each element. The index number can be any integer between 1 and 101. For example, suppose we wish to check the first value. We type and enter

$$>> t(1) \qquad\qquad (4\text{-}7)$$

$$ans =$$

$$0$$

As expected, the first value is zero. It should be stressed here that the first value in the vector is indexed by the value 1, and not 0. Thus, when the values of a vector start with 0, be careful about the index range. This can sometimes be a source of confusion.

Let us go all the way to the end and request the last value. We have

$$>> t(101) \qquad\qquad (4\text{-}8)$$

$$ans =$$

$$10$$

As expected, the last value is 10.

Suppose we desire to inspect the first four values. We can type the following command and enter it:

$$>> t(1:4) \qquad\qquad (4\text{-}9)$$

$$ans =$$

$$0 \ 0.1000 \ 0.2000 \ 0.3000$$

What we are getting on the screen are t(1), t(2), t(3), and t(4). If we typed t and entered it, we would get all 101 values. The variable t is an *indexed variable* or *dimensioned variable*, and an integer argument provides access to the appropriate value or values corresponding to the integer or integers.

Now suppose we wish to create a new time vector that consists of all values of time from t = 1 to t = 1.4 s. We can readily see that this will involve 5 values and the argument will begin at 11 and end at 15. Calling this new vector t1, we have

$$>> t1=t(11:15) \qquad\qquad (4\text{-}10)$$

$$t1 =$$

$$1.0000 \ 1.1000 \ 1.2000 \ 1.3000 \ 1.4000$$

The important point to note here is that colons (:) can be used to define a portion of a vector as a new, smaller vector. The integers used in the argument of the vector represent the beginning and ending values for the argument integers of the original vector. However, the new vector t1 has only five values and the corresponding integers for its argument are from 1 to 5; that is, t1(1)=1.0000, t1(2)=1.1000, and so on.

Now let us get back to the problem at hand, namely plotting the function defined in Equation 4-4. Since we have chosen to use 101 points, this means that Equation 4-4 must be evaluated at 101 points based on the complete independent vector t.

This brings us to one of the most powerful features of MATLAB. As a general rule, *MATLAB functions operating on a vector will generate a vector of output dependent variables at all values of the independent variable contained in the input vector.* This means that 1 command is capable of generating the 101 values of the dependent variable.

There is one precaution that must be observed. For each command, you must check to see if the mathematical operation would violate any rules of matrix algebra. If not, the command is entered in its simplest form. If there is any danger of MATLAB interpreting an operation as an invalid matrix command, one must work around it. Fortunately, there is a simple way around this in most cases by using the array product process introduced in the last chapter, and the process will be illustrated in the next example. For the present example, there is no problem, so we enter the command

$$\gg v = 9.8*t; \tag{4-11}$$

The value 9.8 is a scalar, and no rules of matrix algebra are violated. Therefore, 101 values of the velocity are generated in one step! As in the case of the t-vector, we have suppressed the printing on the screen but the 101 values of v are now in memory.

Now let us plot the function. The general command for a linear plot is **plot(x,y)** where **x** is the horizontal variable and **y** is the vertical variable. In this case **x** is **t** and **y** is **v**. Thus, the basic plotting command is

$$\gg \text{plot(t, v)} \tag{4-12}$$

No semicolons are required at the end of a plot command. Instead, the command causes a figure to appear as a window on the screen; this is shown in Figure 4-1. The figure appears to be correct, but it is pretty "raw" at this point. The only label provided thus far is the figure number and title, and the procedure to display that information will be discussed shortly. Let us see how to enhance the presentation.

Three commands that are usually desired for most figures are **xlabel, ylabel,** and **title.** Their names are self-explanatory, so let us see how to use them. First, we type and enter the **xlabel** command.

$$\gg \text{xlabel('Time, seconds')} \tag{4-13}$$

The desired statement for the screen is the quantity enclosed by (' '). Next, we type and enter the **ylabel** command.

$$\gg \text{ylabel('Velocity, meters/second')} \tag{4-14}$$

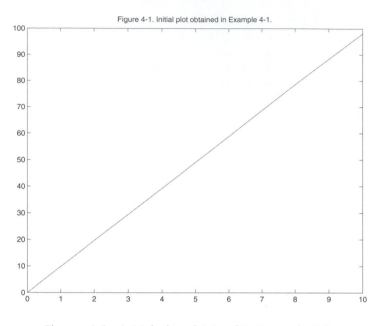

Figure 4-1. Initial plot obtained in Example 4-1.

Figure 4-1   Initial plot obtained in Example 4-1

This figure will follow the preceding one and the figure number will be entered as part of the title. The **title** command is then entered as

$$\gg \text{title('Figure 4-2. Velocity of falling object of Example 4-1.')} \qquad (4\text{-}15)$$

The plot with the addition of various labels is shown in Figure 4-2.

Suppose we desire to put a grid on the figure. This can be done with the command

$$\gg\text{grid} \qquad (4\text{-}16)$$

The resulting plot is shown in Figure 4-3. Along with the addition of the grid, a new title command was added reflecting the new figure number. It has the same form as Equation 4-15, but with the necessary changes.

The grid command is a command that toggles. Thus, entering the same command again would remove the grid lines. Alternately, one can use the commands **grid on** and **grid off**.

Example 4-2

The force $f_1(t)$ in newtons (N) in a certain industrial process increases with the square of time over a particular time interval and can be expressed as

$$f_1(t) = 0.25t^2 \qquad (4\text{-}17)$$

Use MATLAB to plot the function from $t = 0$ to $t = 10$ s.

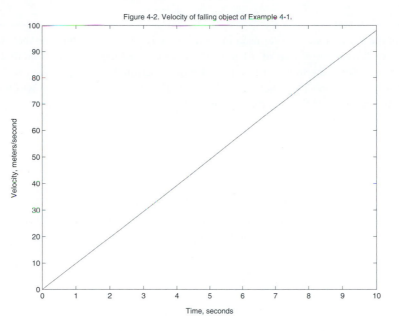

Figure 4-2. Velocity of falling object of Example 4-1.

Figure 4-2   Velocity of falling object of Example 4-1

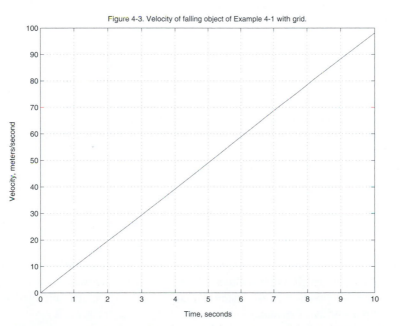

Figure 4-3. Velocity of falling object of Example 4-1 with grid.

Figure 4-3   Velocity of falling object of Example 4-1 with grid

Solution

This is the same time interval used in Example 4-1, so we will consider the same 101-point vector for t using a step of 0.1 s. If you have cleared the MATLAB memory or are starting here, refer back to either Equation 4-5 or Equation 4-6 for the code involved. The temptation is to express the function as f1 = 0.25*t*t or f1 = 0.25*t^2, but let us see what happens when either of those commands is performed.

$$>> f1 = 0.25*t*t \qquad (4\text{-}18)$$

??? Error using ==> *

Inner matrix dimensions must agree.

The last sentence is the MATLAB message on the screen. Let us try the other version.

$$>> f1 = 0.25*t\text{\textasciicircum}2 \qquad (4\text{-}19)$$

??? Error using ==> ^

Matrix must be square.

Again, MATLAB refuses the request. Now there are two strikes against us, but there is an obvious reason involving matrix concepts. In the first case, we were asking MATLAB to multiply two row matrices together, which is an illegal process in matrix theory. The second case was a different way of asking MATLAB to do the same thing, which was just as invalid.

The way around this is what we called the **array product** in the preceding chapter. We desire to perform the operation on a point-by-point basis rather than on the entire vector. This is achieved by following the first vector with a period (.) . The result can be applied to any matrices of the same size. Thus, to multiply **corresponding elements** of an *m* by *n* matrix **A** by the **corresponding elements** of an an *m* by *n* matrix **B** , we use the operation **A.*B** for the process. Understand that this is drastically different from standard matrix multiplication.

In the case at hand, we have two choices. We can enter

$$>> f1 = 0.25*t.*t; \qquad (4\text{-}20)$$

Alternately, we can enter

$$>> f1 = 0.25*t.\text{\textasciicircum}2; \qquad (4\text{-}21)$$

There are some other processes besides multiplication and exponentiation in which array forms must be used, which will be delineated when we encounter them. The main point for any mathematical operation is to remember that MATLAB is matrix oriented and that if there is any possibility of misinterpretation in a command, the array approach should be used to ensure proper interpretation.

Now let us plot this second-degree function. The command is

$$\gg \text{plot(t, f1)} \tag{4-22}$$

Some labeling is in order. The associated commands are

$$\gg \text{xlabel('Time, seconds')} \tag{4-23}$$

$$\gg \text{ylabel('Force, newtons')} \tag{4-24}$$

$$\gg \text{title('Figure 4-4. Force as a function of time in Example 4-2.')} \tag{4-25}$$

$$\gg \text{grid} \tag{4-26}$$

The resulting plot is shown in Figure 4-4.

Changing the Scaling

The figures of both the preceding example and this example had satisfactory scaling from the beginning, as provided by MATLAB. The next example will illustrate how the scaling can be changed if desired.

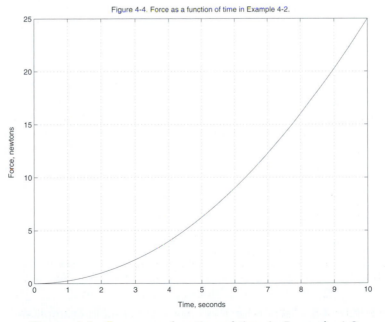

Figure 4-4 Force as a function of time in Example 4-2

Example 4-3

Assume that a force in newtons (N) is given by

$$f_2(t) = 0.25t^2 + 25 \qquad (4\text{-}27)$$

Plot the force over the time interval from 0 to 10 s.

Solution

The same 101-point time vector used in Examples 4-1 and 4-2 will be used here. Now that we have learned how to deal with matrix operations on a point-by-point basis, we can type and enter

$$\text{>> f2 = 25 + 0.25*t.^2;} \qquad (4\text{-}28)$$

Next we plot the function

$$\text{>> plot(t, f2)} \qquad (4\text{-}29)$$

MATLAB provides us with the initial plot shown in Figure 4-5 after some routine labeling. Now there is nothing inherently wrong with the plot given, and it may be the form that is best suited to our needs. However, the fact that the vertical scale starts with a value of 25 may not be desirable for a given presentation since it may convey erroneous impressions

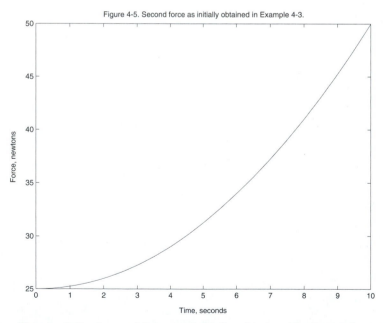

Figure 4-5. Second force as initially obtained in Example 4-3.

Figure 4-5   Second force as initially obtained in Example 4-3

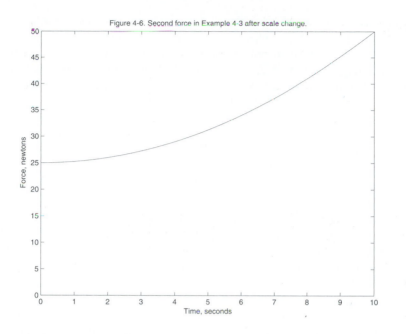

Figure 4-6. Second force in Example 4-3 after scale change.

Figure 4-6   Second force in Example 4-3 after scale change

about the relative change in the function over the given time interval. Suppose instead that we desire to have the vertical scale start at 0. Scales may be changed with the **axis** command. Assuming that the horizontal scale is fine from 0 to 10 and that we desire the vertical scale to range from 0 to 50, the command to change scales is

$$\gg \text{axis([0 10 0 50])} \qquad (4\text{-}30)$$

The resulting curve is shown in Figure 4-6. Note that even though the horizontal scale remains the same as before, it was necessary to enter it into the command.

While some of the resolution has been degraded, this curve displays a more realistic relative variation of the dependent variable.

4-3   MULTIPLE PLOTS ON THE SAME GRAPH

So far we have dealt with only one curve on a given plot. However, MATLAB can generate multiple curves on the same scale whenever it is convenient and reasonable. The most common situation is where there is one independent variable and two or more dependent variables, each of which is to be plotted as a function of the one independent variable. For multiple plots of this nature to make the most sense, the dependent variables should be of the same order of magnitude. For example, a plot of Bill Gates' income versus time on the same graph as the author's income versus time would not make good sense, since one curve would completely obscure the other one (You can guess which one).

We will illustrate the command with a hypothetical situation of three dependent variables, y_1, y_2, and y_3, each of which is a function of a single independent variable, x. We will assume that the vectors for the four variables have the same length. The plot command associated with this situation is as follows:

$$\gg \text{plot}(x, y1, x, y2, x, y3) \tag{4-31}$$

Note that it is necessary to show x preceding each of the dependent variables.

The preceding situation is based on the same independent variable for all of the curves. Actually, however, it is possible to have more than one independent variable, and the vector sizes in that case can be different. The only requirement is that the vector sizes of a given independent and dependent variable pair should be the same.

To illustrate different independent variables, assume that x1 and x2 are two independent vectors and y1 and y2 are two dependent vectors. Assume that x1 and y1 have the same size and assume that x2 and y2 have the same size. The following command will allow both curves on the same plot:

$$\gg \text{plot}(x1, y1, x2, y2) \tag{4-32}$$

An alternate way to put more than one curve on a given set of axes is through the **hold on** command. Assuming that one or more curves have already been placed on the axes but you anticipate adding one or more later, the following command can be used:

$$\gg \text{hold on} \tag{4-33}$$

You may then add one or more additional curves. After the curves are complete, the figure option may be opened for new curves by the command

$$\gg \text{hold off} \tag{4-34}$$

Labeling Different Curves

An obvious question that arises when there is more one curve is the process of labeling for identification. Actually, MATLAB has numerous options for creating different types of labels, including the use of different symbols on different curves and the use of different types of curves and shading. Once you become proficient in the basic operations of MATLAB, you will probably want to investigate some of the different forms. However, we do not want to overwhelm you with too many fine details at this point, so we will limit the development here to one particular form, the **gtext** command. Others will be considered at various points in the text.

The **gtext** command allows the user to specify wording that can be placed on the figure with crosshairs. The format is as follows:

$$\gg \text{gtext('wording to be placed on figure')} \tag{4-35}$$

Assuming that a figure has already been created, upon entering this command in the Command Window, a pair of crosshairs appears on the screen. They can be moved both horizontally and vertically with the mouse. The intersection point establishes the beginning location of the label. A left-click of the mouse will place the wording on the screen.

Example 4-4

Consider the functions f_1 and f_2 of Examples 4-2 and 4-3, respectively. Plot the two functions on the same linear graph and use **gtext** to provide some labels.

Solution

We will assume at the outset that the MATLAB variables t, f1, and f2, as defined in the preceding examples, are in memory. In that case, the plotting function to place both functions on the same curve is

$$>>\text{plot(t, f1, t, f2)} \qquad (4\text{-}36)$$

The two curves should appear on the screen. Although it should be obvious which one is which in this case, the default order for MATLAB is that the first one is in blue on the screen and the second one is in green. If a third curve were present, it would be in orange.

Some labels similar to the preceding examples are provided

$$>> \text{xlabel('Time, seconds')} \qquad (4\text{-}37)$$

$$>> \text{ylabel('Force, newtons')} \qquad (4\text{-}38)$$

$$>> \text{title('Force as a function of time')} \qquad (4\text{-}39)$$

Now we are ready to add some labels to the curves. Let us arbitrarily label the first one as "Force 1" and the second one as "Force 2." The first command to enter is

$$>> \text{gtext('Force 1')} \qquad (4\text{-}40)$$

A set of crosshairs will immediately appear on the screen and the location of this label will be placed close to the function f1 (the blue curve). We then type and enter

$$>> \text{gtext('Force 2')} \qquad (4\text{-}41)$$

The procedure for the first force is repeated, but the label is now put near f2 (the green curve).

Although we could add a grid, the figure does not look quite right with text over the grid lines, so we will omit it in this case. The result is shown in Figure 4-7.

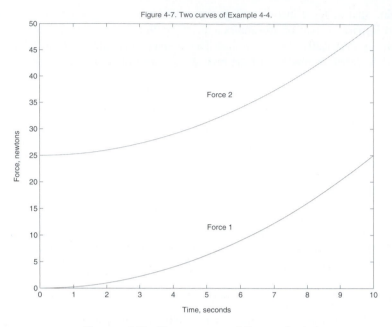

Figure 4-7. Two curves of Example 4-4.

Figure 4-7   Two curves of Example 4-4

4-4   LOG-LOG PLOTS

While the linear plot of both variables considered in the last two sections can probably be considered as the most basic of all two-dimensional continuous plots, it is by no means the only way in which data can be plotted. Two other forms of two-dimensional continuous plots are the *log-log plot* and the *semi-log plot*. For the latter case there are two possibilities, based on which of the scales is logarithmic and which is linear. We will consider log-log plots in this section and semi-log plots later in the text.

Logarithmic Forms in Both Variables

A *log-log plot* is one in which the physical dimensions for both the horizontal and vertical scales are proportional to the logarithms of the variables on the scales. A log-log plot is most useful when the dependent variable is related to the independent variable by an exponential power function. Consider the following relationship between an independent variable x and a dependent variable y:

$$y = Cx^k \qquad (4\text{-}42)$$

where C and k are constants. The quantity k may or may not be an integer. Except for the simple case of $k = 1$, a plot of y versus x on a linear plot would be displayed as a curved function. In some cases, it may be desired to more clearly delineate the behavior with respect to k, which is one of the features that can be best exhibited with a log-log plot.

To investigate this phenomenon, suppose we take the logarithms of both sides of Equation 4-42 to the base 10. Denoting this operation by $\log_{10}$, we have

$$\log_{10} y = \log_{10}(Cx^k) = \log_{10} C + k \log_{10} x \qquad (4\text{-}43)$$

The simplification shown in the right-hand side of the preceding equation was made utilizing the two identities:

$$\log_{10} ab = \log_{10} a + \log_{10} b \qquad (4\text{-}44)$$

$$\log_{10} x^k = k \log_{10} x \qquad (4\text{-}45)$$

An observation of Equation 4-43 reveals that it is of the form $y' = mx' + b'$, where the prime ($'$) indicates a quantity proportional to the logarithm of the respective variable. Graph paper that is denoted as **log-log** has the property that the dimensions of both the horizontal and vertical scales are proportional to the logarithms of the respective variables.

The result of the preceding development is that a function of the form of Equation 4-42 will appear as a straight line on a log-log plot. The slope of the curve will be the constant k, and the vertical intercept will be related to the constant C.

One of the useful properties of log-log plots is to assist in the evaluation of experimental data that might follow a pattern similar to Equation 4-42. From the slope of the curve it may be possible to infer the power law coefficient k.

Example 4-5

Plot the following function on a log-log scale over the domain of $0.1 \le x \le 10$.

$$y = x^2 \qquad (4\text{-}46)$$

Solution

It should be emphasized that you can never reach 0 on a logarithmic scale! The best way to make a log-log plot is to assume integer powers of 10 for the beginning and ending points of the independent variable. This will result in an integer number of *decades* for the independent variable. A *decade* is a ratio of 10 to 1.

For the case at hand, the left-hand value (0.1) is 10^{-1} and the right-hand value is 10^1. This will result in two decades for the independent variable (0.1 to 1) and (1 to 10).

The next step is to create a logarithmic scale for the independent variable x. The command has the form

$$\text{>> x = logspace(r1, r2, N)} \qquad (4\text{-}47)$$

The quantities r1 and r2 represent, respectively, the powers of 10 defining the beginning and ending values of the independent variable x. The integer value N is the number of points. For this particular example, r1 = -1 and r2 = 1. We will arbitrarily select N = 100. Thus, the command with suppression of the screen listing is

$$>> x = logspace(-1, 1, 100);$$ (4-48)

We now evaluate the dependent value y at all values of x by the command

$$>> y = x.^2;$$ (4-49)

The curve is plotted on a log-log scale by the command

$$>> loglog(x, y)$$ (4-50)

We will now add a grid and some labels by the commands that follow. Since this is a "pure" mathematics function, some of the labels are fairly basic and generic.

$$>> grid$$ (4-51)

$$>> xlabel('Independent Variable x')$$ (4-52)

$$>> ylabel('Dependent Variable y')$$ (4-53)

$$>> title('Figure 4-8. Plot of squared function on log-log scale in Example 4-5.')$$ (4-54)

Note that the grid and labeling commands are the same as for a linear plot.

The resulting graph is shown in Figure 4-8. The grid lines are spaced logarithmically. As expected, the curve is a straight line on the log-log scale.

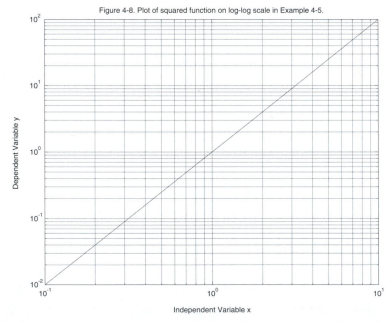

Figure 4-8   Plot of squared function on log-log scale in Example 4-5

At first glance, the straight line appears to have a slope of +1. Upon further investigation, however, the slope is actually +2. This fact is deduced from the observation that there are 4 *decades* on the vertical scale and 2 *decades* on the horizontal scale.

4-5 BAR AND STEM PLOTS

All plots are either *continuous* or *discrete*. All plots considered thus far have been continuous, but we now wish to introduce two types of discrete plots.

Actually, as we have seen, all data provided to MATLAB are necessarily discrete since a computer can only deal with discrete values. However, in all the cases considered, the data points have been so close together that the resulting plots were continuous for all practical purposes. Indeed, for the commands considered thus far, MATLAB automatically extrapolates between points and provides a very good representation on the assumption that the function is "reasonably well-behaved" and that there are a sufficient number of points.

In contrast, however, there are some types of data presentations where it is not desired to connect the points. In this section, we will investigate two different ways of presenting discrete data where the resulting plots do not extrapolate between points. The forms considered are the **bar** plot and the **stem** plot.

Bar Plot

Making the same assumption on vector sizes as previously stated, the bar plot is activated by the command

$$>> \text{bar(x, y)} \hspace{4cm} (4\text{-}55)$$

Stem Plot

The stem plot is activated by the command

$$>> \text{stem(x, y)} \hspace{4cm} (4\text{-}56)$$

Labeling is accomplished in the same manner as for previous plots. These forms will be illustrated in Examples 4-6 and 4-7, which follow.

Example 4-6

Assume that a small business opened at the beginning of 1993 and it is desired to show the sales for each year up through 2002, which constitutes a 10-year period. Assume that the following table gives the sales in thousands of dollars for each year:

Year	1993	1994	1995	1996	1997	1998	1999	2000	2001	2002
Sales	70	85	120	160	200	210	235	290	310	300

Construct a **bar** graph that displays the data.

Solution

To be as descriptive as possible, we will define the variables **year** and **sales** to represent the independent and dependent variables, respectively. First, let us consider the variable **year**. We could certainly type in the 10 values of the years in a row matrix, but there is a much easier way to do that. Since there is a difference of 1 between each of the successive values, the following command will suffice:

$$>> \text{year} = 1993{:}1{:}2002; \qquad (4\text{-}57)$$

This command will generate the 10 values for the year row vector in one step.

While the preceding command was pretty simple, we can go one step further. When a linear spaced vector has a step size of 1, we need only enter the first and last values. Thus, the following operation will work:

$$>> \text{year} = 1993{:}2002; \qquad (4\text{-}58)$$

Since the sales data are somewhat random, the values will have to be entered separately. We have

$$>> \text{sales} = [70\ 85\ 120\ 160\ 200\ 210\ 235\ 290\ 310\ 300]; \qquad (4\text{-}59)$$

The plot is now generated by the command

$$>> \text{bar(year, sales)} \qquad (4\text{-}60)$$

The result is shown in Figure 4-9. Additional labeling has been provided based on the concepts studied earlier in the chapter.

Example 4-7

Plot the data of Example 4-6 using a **stem** plot.

Solution

Assume that the year and sales vectors have been created as in Example 4-6. The command is simply

$$>> \text{stem(year, sales)}$$

The plot is shown in Figure 4-10 with the same additional labeling as in the preceding example.

For this "business-type" presentation, the bar plot is probably preferred. However, in dealing with sampled-data engineering systems and spectral forms, the stem plots can be very useful. We will encounter them later in the text.

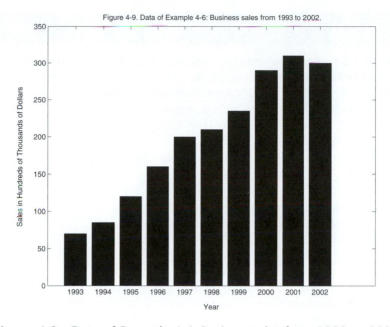

Figure 4-9 Data of Example 4-6: Business sales from 1993 to 2002

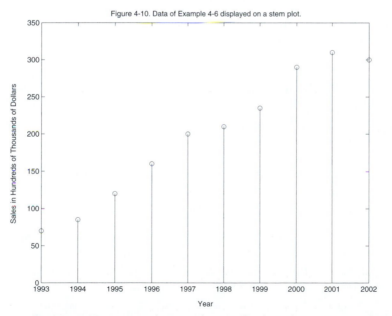

Figure 4-10 Data of Example 4-6 displayed on a stem plot

4-6   EDITING WITHIN THE FIGURE WINDOW

Thus far, all instructions for creating and modifying figures have been performed from the **Command Window** by means of codes. However, once a figure has been created, there are numerous modifications that may be performed directly within the **Figure Window**. The possibilities are so numerous that we believe that experimentation, along with assistance from the Help file, is probably the best way to learn how to use the options available there.

Among the options available in the Figure Window are the following:

> Edit Plot
> Insert Text
> Insert Arrow
> Insert Line
> Zoom In
> Zoom Out

Some of these options lead to other options as well. You can also change some of the properties of the axes by first clicking on them and opening up some option windows. The line size and type size and style may also be changed within some of the options. If you desire to change the width of a curve, right-click on it and a window opens that will permit the line width to be changed.

Again, experimentation is the best way to learn how to perform the various modifications. Some of them will be used at various points throughout the book.

MATLAB PROBLEMS

It is suggested that Problems 4-1 through 4-14 be solved directly in the Command Window. However, if you prefer, M-files can be used.

4-1.   An object propelled downward with an initial velocity of 50 m/s from a height above the Earth's surface has a velocity given by

$$v = 9.8t + 50$$

The actual positive direction of velocity is downward, so a positive value of the function will be considered as a downward velocity. Prepare a curve of the velocity as a function of time from $t = 0$ to $t = 10$ s and keep the default horizontal and vertical scales. Provide labeling for the two axes and provide a title.

4-2.   For the function of Problem 4-1, change the vertical scale so that the lower level of the velocity is zero and add a grid to the graph.

4-3. An object propelled directly upward with an initial velocity of 50 m/s has a velocity given by

$$v = -9.8t + 50$$

In this case, the positive direction of velocity is upward. Prepare a curve of the velocity as a function of time from $t = 0$ to $t = 8$ s and keep the default horizontal and vertical scales. Provide labeling for the two axes and provide a title.

4-4. For the function of Problem 4-3, use the **Insert Line** function within the Figure Window to add an abscissa across the graph corresponding to a velocity of zero. In addition, add a grid.

4-5. The output voltage v_2 in volts (V) for a certain circuit is a square-law function of the input voltage v_1 and is given by

$$v_2 = 0.1v_1^2$$

Using linear scales, plot the output voltage versus the input voltage for $v_1 = 0$ to $v_1 = 10$ V.

4-6. The output voltage v_3 in volts (V) for a certain circuit is a square-law function of the input voltage and is given by

$$v_3 = 0.1v_1^2 + 5$$

Using linear scales, plot the output voltage versus the input voltage for $v_1 = 0$ to $v_1 = 10$ V and keep the default horizontal and vertical scales. Provide labeling for the two axes and provide a title.

4-7. For the function of Problem 4-6, change the vertical scale so that the lower level of the voltage is zero and add a grid to the graph.

4-8. Consider the functions v_2 and v_3 of Examples 4-5 and 4-6. Plot the two functions on the same linear graph and use **gtext** to provide some labels.

4-9. Plot the following function on a **log-log** plot over the domain of $0.1 \le x \le 10$.

$$y = x^3$$

4-10. Plot the following function on a **log-log** plot over the domain of $0.01 \le x \le 100$.

$$y = \sqrt{x}$$

4-11. Plot the following function on a **log-log** plot over the domain of $0.1 \le x \le 10$.

$$y = 2x^2$$

4-12. Plot the following function on a **log-log** plot over the domain of $0.1 \le x \le 10$.

$$y = 2x^3$$

4-13. The enrollment in a particular college program since 1995 has the following data:

1995	1996	1997	1998	1999	2000	2001	2002
320	330	369	350	310	370	390	400

Prepare a **bar** graph showing the enrollment trends.

4-14. Repeat Problem 4-13 using a **stem** plot.

Common Functions and Their Properties

5

The concept of functions is a very basic part of mathematics and one that pervades the study of all forms of algebra, trigonometry, and calculus. While there are hundreds of different types of mathematical functions, certain common ones tend to occur quite often in engineering and scientific applications. In this chapter, we will explore some of these most common functions and study their behavior.

Now that the reader has progressed through the basic properties of matrix operations, MATLAB commands, and curve plotting, much of the work that follows will provide coverage of the MATLAB commands immediately after introducing the mathematical forms. This will be the norm where the commands are fairly simple and are best introduced after discussing the mathematical form. In particular, the need to plot curves of functions immediately after introducing the functions will be best achieved with appropriate MATLAB commands.

Once again, we will remind the reader that the notational form for mathematical equations will involve the usual italicized variables with both subscripts and superscripts, as required. In contrast, the MATLAB equations will utilize nonitalicized variables without subscripts or superscripts and will follow the >> prompt or the EDU>> prompt on the left-hand side of a line. The same letters and numbers will be used for both, so the meanings should be clear.

Objectives

After completing this chapter, the reader should be able to

1. Define a function and discuss various properties and classifications.
2. State the form of a *power function*.
3. State the equation of a *straight line* and discuss its properties.
4. State the form of a *polynomial function* and discuss the various properties of its roots.
5. Apply MATLAB to manipulate polynomials in various ways, including the determination of the roots, the evaluation as a function, and the construction of the polynomial from the roots.

6. State the form of an *exponential function*, sketch its form, and apply MATLAB for evaluation.
7. Define the *time constant* and the *damping constant* and convert between them.
8. State the form of a *logarithmic function*, sketch its form, and apply MATLAB for evaluation.
9. Use MATLAB to construct *semi-log* plots with the logarithmic scale applied to either variable.
10. State the definitions of the *trigonometric functions* based on a right triangle.
11. Discuss *radians* and *degrees* and convert between them.
12. Sketch the forms of *sine*, *cosine*, and *tangent* as functions and apply MATLAB for evaluation.
13. Discuss the concept of a *sinusoidal time function* and define its various parameters.
14. Convert the linear combination of a sine function and a cosine function at the same frequency to a single sine or cosine function with a phase angle.

5-2 FUNCTION CONCEPTS AND DEFINITIONS

In its simplest interpretation, a **function** is a relationship between two or more variables. For example, the air pressure in a balloon is a function of the volume of the balloon. At this point in the chapter, we will consider only two variables for a given function. For much of this chapter, we will use the traditional variables x and y. In most cases, we will consider that *x is the independent variable* and *y is the dependent variable.* This does not necessarily mean that x causes y in all cases, but it suggests that we consider x to be assigned as the first step and then determine how y varies with x.

Normally, x is assigned to the *horizontal axis* and y is assigned to the *vertical axis*. The general notation indicating that y is a function of x will take the form $y = f(x)$, and letters other than f may be used when there are several functions. Alternately, subscripts may be added to different $f(x)$ functions to give them separate identities. Sometimes, we will use the simpler notation $y = y(x)$ to mean the same thing.

A Simple Error Sometimes Seen

As elementary as the following statement may seem, this author has seen college-level students confused by the notation. The notation $f(x)$ *does not mean* that f is multiplied by x. This confusion is somewhat understandable for beginners studying functions because basic algebra does use parentheses for multiplication. Hopefully, the context of a development will assist one in distinguishing between functional notation and multiplication.

Many Applications of Functions

One cannot begin to cover all the possible variables in the real world that can be dealt with on a functional basis. To mention just a few, the output power of a motor can be considered as a function of the load, the stress in a beam can be considered as a function of the location on the beam, and the current flow in a circuit can be considered as a function of the voltage across the circuit. However, we return to the basic y and x variables to study the mathematical processes. Since they are "pure" functions, we need not provide any units at this point.

Single-Valued versus Multi-Valued Functions

A *single-valued function* is one where a single value of x results in a single value of y. A *multi-valued function* is one where there are more than one possible values of y for a given value of x. A single-valued function is shown in Figure 5-1(a), and a multi-valued function is shown in Figure 5-1(b).

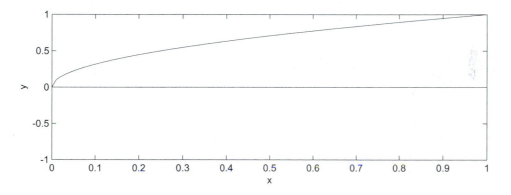

Figure 5-1(a) Example of a single-valued function

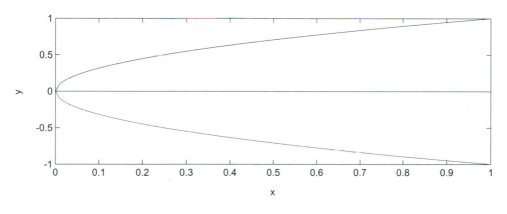

Figure 5-1(b) Example of a multi-valued function.

Continuous versus Discontinuous Functions

The formal definition of a continuous function is one in which, at any value of the independent variable, approaching the value from the left results in the same dependent value as approaching the value from the right. A simpler way to look at this is to say that a continuous function has no "sudden jumps," while a discontinuous function has one or more "jumps." An example of a continuous function is shown in Figure 5-2(a), and a function that has one finite discontinuity is shown in Figure 5-2(b). A discontinuous function can have a *finite discontinuity* such as for the case here, or it may have an *infinite discontinuity*, in which the function has an infinite jump at a certain point.

Domain and Range

Assume that a function is being evaluated over specific limits such as $x_1 \leq x \leq x_2$. This portion of the x axis is called the *domain*. All values of the dependent variable y that are

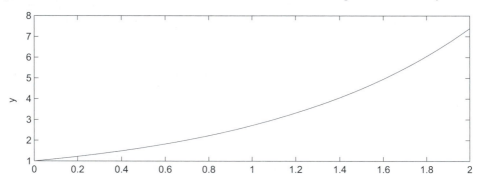

Figure 5-2(a) Example of a continuous function

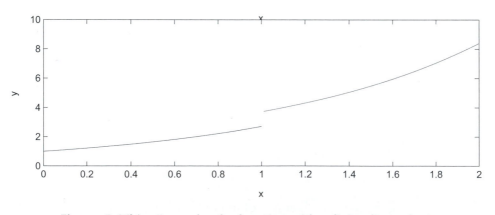

Figure 5-2(b)   Example of a function with a finite discontinuity

produced in the process are called the *range*. In casual usage, engineers and technologists (including this author) tend to refer to both as *ranges*.

Inverse Functions

If we have a function $y = f(x)$ and we can work from right to left and solve for x in terms of y, we have the *inverse function*. For the moment, we will denote the inverse simply as $x = g(y)$. As a simple example, if $y = 10x$, then $x = 0.1y$.

There are two schools of thought concerning inverse functions. Many mathematics texts reverse x and y in the inverse function so that x is always the independent variable and y is always the dependent variable. In the simple example of the preceding paragraph, the inverse function would likely be expressed as $y = 0.1x$.

The other approach is to simply keep the original notation and let y be the independent variable and x the dependent variable. In a practical engineering situation, the original variables would not change, so we will tend to use the latter approach when we are dealing with a specific problem. However, many of the functions that will be introduced in different sections will be inverses of each other, and in such cases, we will tend to return to the basic convention of x as the independent variable and y as the dependent variable.

Even and Odd Functions

An *even function* is one that satisfies the following criterion:

$$f(-x) = f(x) \qquad (5\text{-}1)$$

The even function criterion means that the function to the left of the vertical axis appears to be the mirror image of the function to the right of the vertical axis. Looking ahead momentarily, the functions of Figures 5-4 and 5-6 are even functions.

An *odd function* is one that satisfies the following criterion:

$$f(-x) = -f(x) \qquad (5\text{-}2)$$

The odd function criterion means that the function to the left of the vertical axis appears to be the *inverted* mirror image of the function to the right of the vertical axis. Looking ahead again, the functions of Figures 5-5 and 5-7 are odd functions.

Example 5-1

Consider the function

$$y = f(x) = x^2 - 1 \qquad (5\text{-}3)$$

Is the function single-valued or multi-valued?

Solution

With x as the independent variable and y as the dependent variable, there is only one value of y for a given value of x. Hence the function is single-valued.

Example 5-2

Is the function of Example 5-1 an even function, an odd function, or neither?

Solution

The test is to substitute $-x$ for x in the expression. We have

$$f(-x) = (-x)^2 - 1 = x^2 - 1 = f(x) \qquad (5\text{-}4)$$

Hence, it is an even function.

Example 5-3

Determine the inverse function for the function of Example 5-1.

Solution

The inverse function is determined by solving for x in terms of y in Equation 5-3. The equation is turned around, the term -1 is moved across the equal sign, and the square root of both sides is taken. This yields

$$x = \pm\sqrt{y+1} = g(y) \qquad (5\text{-}5)$$

Example 5-4

We now consider y as the independent variable and x as the dependent variable. Is the inverse function of Example 5-3 single-valued or multi-valued?

Solution

Since two values of x result from a given value of y, the inverse function is multi-valued. What this tells us is that a function may be single-valued yet have an inverse that is multi-valued, or vice versa. In many applications, only the positive square root would be of interest, so if the negative square root is rejected, we could interpret the result as being single-valued.

Example 5-5

Is the inverse function of Example 5-3 even, odd, or neither?

Solution

The criteria of Equations 5-1 and 5-2 must now be interpreted in terms of $g(y)$ and $g(-y)$. Substitution of $-y$ for y in Equation 5-5 results in

$$g(-y) = \pm\sqrt{-y+1} \qquad (5\text{-}6)$$

The function on the right-hand side of Equation 5-6 is neither the function of Equation 5-5 nor its negative. Hence the inverse function is neither even nor odd.

Example 5-6

Using MATLAB, plot the function of Example 5-1 and the inverse of Example 5-3 on the same page using the **subplot** command.

Solution

The **subplot** command allows more than one plot to be prepared on the same printer page. In fact, Figures 5-1 and 5-2 were both prepared using that command.

The syntax for the subplot command is as follows:

$$>> \text{subplot(m, n, k)} \qquad (5\text{-}7)$$

where **m**, **n**, and **k** are all integers. Consider the first two integers **m** and **n** almost in the same manner as a matrix, in the sense that **m** is the number of rows of subplots and **n** is the number of columns of subplots. The last integer, **k**, starts at 1 and progresses to a final value of **mn**. The sequence of progression is along a row from left to right, then back to the left for the next row, and so on. Since the size of each subplot decreases as the number increases, one would normally want to limit the number of subplots. The maximum that the author has ever employed was three rows and two columns, resulting in six subplots on the same page.

Once the command is asserted for a given value of **k**, the additional entries for the particular plot follow the same pattern as for a single plot. Just remember that there will be less space available, so the various labels will be more constrained in size. Note again that **m** and **n** remain constant for a given set of plots, but **k** varies from 1 to **mn**.

Now the author will share a secret with the reader. For most plots in the text, some experimentation was performed before the final form was created in many cases, and we expect that the reader and everyone else using this type of software will do the same thing. Thus, when we state that a plot is being created over a certain domain or range, the values involved did not necessarily emerge in their final form, but more likely evolved after some experimentation and trial-and-error. Keep that in mind throughout the text as we assume various values for presentation.

Now let us get down to business with the plots. We will choose two rows and one column, so the first two integers in the subplot command will be 2 and 1. The third integer will be 1 for the top plot and 2 for the second plot. Equation 5-3 will be evaluated over the domain $-2 \leq x \leq 2$. We will arbitrarily select 201 points. Thus, we generate the independent variable vector as

$$>> x = \text{linspace(-2,2,201)}; \qquad (5\text{-}8)$$

The vector y is then generated as

$$>> y = x.^2 - 1; \qquad (5\text{-}9)$$

As a quick review item, note that the period is required after x to ensure that the squares of the x-values are generated on a point-by-point array basis.

We now enter

$$\text{subplot}(2, 1, 1) \tag{5-10}$$

$$\text{plot}(x, y) \tag{5-11}$$

Three additional commands were entered for the **xlabel**, **ylabel**, and **title**, but they should be routine by now and will not be discussed.

For the second plot, we could use the two branches of Equation 5-6 and solve for x in terms of y, but this is not necessary since we already have the x and y vectors. The steps for the second plot now follow.

$$\text{subplot}(2, 1, 2) \tag{5-12}$$

$$\text{plot}(y, x) \tag{5-13}$$

Again, the labels and title were added. In addition, the plots were edited to provide x and y axes and the curve lines were widened. The final result is shown in Figure 5-3.

5-3   POWER FUNCTIONS

Some of the most common functions are power functions of x. We have already worked with some simple cases. In general, such functions can be described by the equation

$$y = x^n \tag{5-14}$$

Various forms may be obtained depending on the value of n. For the moment, we will consider only integer values of n for $n \geq 0$. They will be identified by placing a subscript n following each function.

For, $n = 0$, the function y_0 is given by

$$y_0 = x^0 = 1 \tag{5-15}$$

This function is shown in Figure 5-4 and is a constant for all values of the independent variable.

For, $n = 1$, the function y_1 is given by

$$y_1 = x^1 = x \tag{5-16}$$

This function is shown in Figure 5-5. This is a special case of a *straight-line* or *linear* equation with a slope of 1 and a vertical intercept of 0. More general linear equations will be considered in the next section.

For, $n = 2$, the function y_2 is given by

$$y_2 = x^2 \tag{5-17}$$

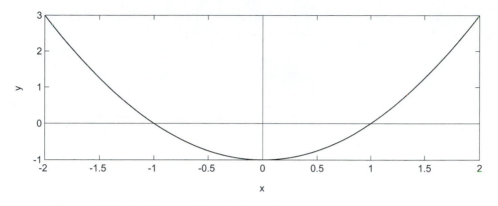

Figure 5-3(a) Curve of Examples 5-1 through 5-5 with y versus x

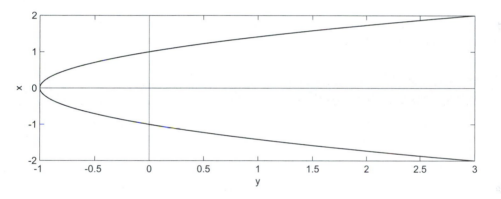

Figure 5-3(b) Curve of Examples 5-1 through 5-5 with x versus y

This function is shown in Figure 5-6. This is a special case of a *quadratic* or *second-degree* equation.

For $n = 3$, the function y_3 is given by

$$y_3 = x^3 \qquad\qquad (5\text{-}18)$$

This function is shown in Figure 5-7. This is a special case of a *cubic* or *third-degree* equation.

One could continue this process to higher-order powers of x, but this is sufficiently far to see the trend. As the power of x increases, the function tends to be more "squashed-down" for $x < 1$, and it increases more rapidly for $x > 1$.

We will not show the details, but we believe the reader who has followed all the MATLAB work up to this point can now create any of the plots in Figures 5-4 through 5-7.

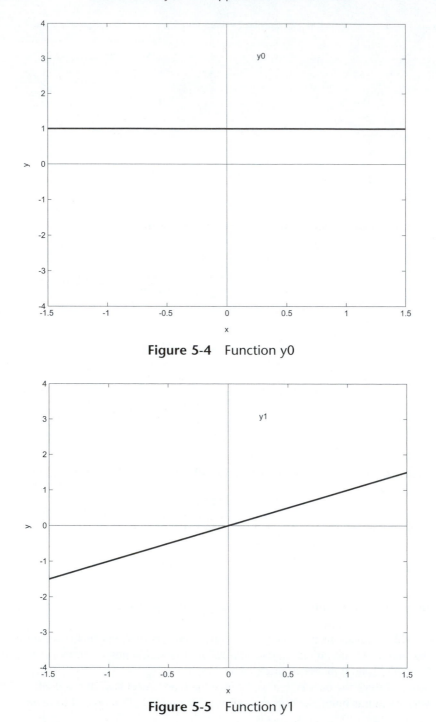

Figure 5-4   Function y0

Figure 5-5   Function y1

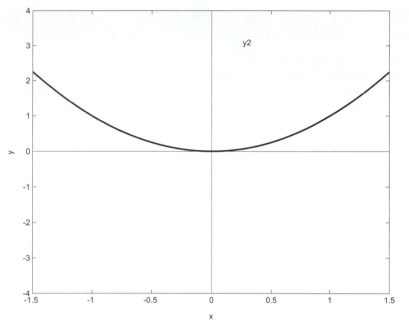

Figure 5-6 Function y2

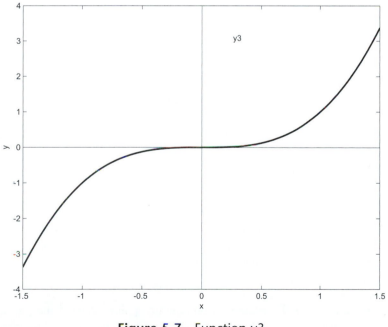

Figure 5-7 Function y3

5-4   STRAIGHT-LINE OR LINEAR EQUATIONS

Straight-line equations arise so often that they need to be emphasized within the context of mathematical functions. In basic mathematics texts, there are several different forms provided. However, we can accomplish the necessary treatment here with only one form, which will be the basis of the work that follows. A straight-line equation is also called a *linear equation*.

Slope/Vertical-Intercept Form

The slope/vertical-intercept form of the straight-line equation is

$$y = mx + b \qquad (5\text{-}19)$$

The quantity m is the *slope* of the line, and b is the vertical intercept. For $m > 0$, the slope is upward, and for $m < 0$, the slope is downward. The line crosses the vertical axis at a value $y = b$. An example of a linear equation with a positive slope will be considered in Example 5-7, and one with a negative slope will be considered in Example 5-8.

Any straight-line equation can be expressed in the form of Equation 5-19 if the slope and vertical intercept are known. Conversely, if any two points on the curve are known, the slope and intercept may be determined.

Slope

Assume that the curve passes through the points with coordinates (x_1, y_1) and (x_2, y_2). The slope may then be determined as

$$m = \frac{y_2 - y_1}{x_2 - x_1} \qquad (5\text{-}20)$$

This concept will prove useful in the study of differential calculus in Chapter 6.

Example 5-7

A straight line has a slope of 2 and a vertical intercept of -4. Write the equation and plot it.

Solution

Problems do not get much simpler that this one, since we are given the two parameters required in Equation 5-19. With one step, we have

$$y = 2x - 4 \qquad (5\text{-}21)$$

If we were plotting by hand, we would need only two points. When hand-plotting is performed, the two easiest points with which to work are usually the vertical intercept and the

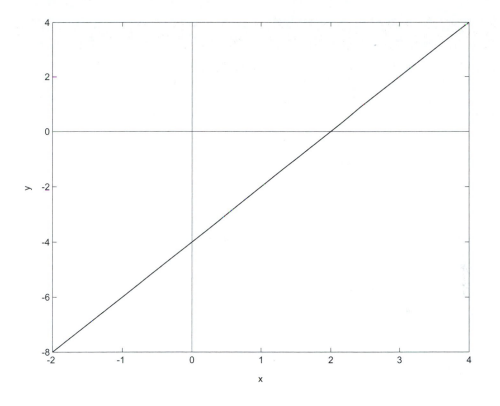

Figure 5-8   Function of Example 5-7

horizontal intercept. The former is obtained by setting $x = 0$, and we already know that $y = -4$ for that situation. The horizontal intercept is determined by setting $y = 0$, and we readily see from Equation 5-21 that $x = 2$ for that case.

While the preceding discussion is useful for a hand plot, MATLAB can also be readily employed. The resulting plot is shown in Figure 5-8.

Example 5-8

A linear equation passes through the points (3, 5) and (6, -7). Determine the equation and plot it.

Solution

The presence of two points on the equation suggests that we determine the slope. From Equation 5-20, we have

$$m = \frac{y_2 - y_1}{x_2 - x_1} = \frac{-7 - 5}{6 - 3} = \frac{-12}{3} = -4 \tag{5-22}$$

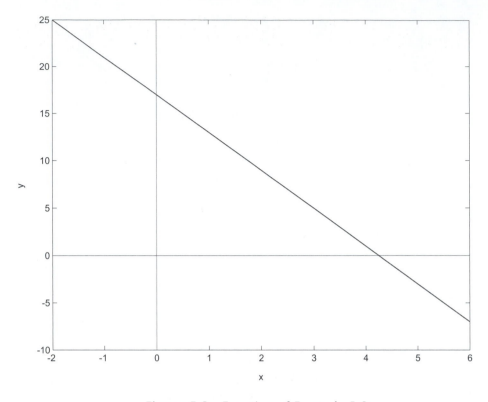

Figure 5-9   Function of Example 5-8

The slope is negative, meaning that the function y decreases as x increases.

Since we do not know the vertical intercept, let us form the equation with the one unknown left. Now we have

$$y = -4x + b \qquad (5\text{-}23)$$

Now we can pick either of the points and substitute the values of x and y to determine b. Arbitrarily selecting the first point, we substitute $x = 3$ and $y = 5$.

$$5 = -4(3) + b \qquad \text{or} \qquad b = 17 \qquad (5\text{-}24)$$

The equation is

$$y = -4x + 17 \qquad (5\text{-}25)$$

The resulting straight line utilizing MATLAB for the plot is shown in Figure 5-9.

An alternate approach would have been to begin with the form $y = mx + b$ and substitute the values of x and y at the two points, thereby yielding two equations with two unknowns. The reader is invited to work the problem in that manner.

5-5 POLYNOMIAL FUNCTIONS

A *polynomial function* is a function composed of a sum of power terms of the form of Equation 5-14, with integer values of n and arbitrary multiplicative constants. A typical polynomial function $p(x)$ of degree N can be expressed in the form

$$y = p(x) = A_N x^N + A_{N-1} x^{N-1} + \dots + A_1 x + A_0 \tag{5-26}$$

The straight-line equation considered in the last section is a special case of a polynomial function in which only the last two terms of Equation 5-26 are present.

Polynomial functions arise in many applications, and some of their basic properties need to be studied. MATLAB has some very powerful features for manipulating polynomials, but we first need to study the basic mathematical properties.

Roots of Polynomial Functions

A *root* of a polynomial equation is a value of x such that the polynomial is zero when it is evaluated for that particular value of x. This means that for any root x_k of the polynomial of Equation 5-26,

$$p(x_k) = 0 \tag{5-27}$$

In general, the roots of $p(x)$ could be determined in theory by setting the polynomial to 0 and solving for the values of x that satisfy the equation. For a first-degree or straight-line equation, the process is very simple, and for a second-degree equation, the quadratic equation can always be employed. For third and higher degree forms, however, determining the roots by a manual process can be very difficult. As we will see shortly, however, MATLAB makes it very easy.

Theorem on Roots

A fundamental theorem in algebra says that a polynomial of degree N of the form of Equation 5-26 has exactly N roots. These roots may be classified in the following four categories:

1. Real roots of first order
2. Complex roots of first order (including purely imaginary)
3. Real roots of multiple order
4. Complex roots of multiple order (including purely imaginary)

A real root of first order is a real number that occurs only once in the possible solution of Equation 5-27. A real root of multiple order is one that occurs more than once, and its *order* is the number of times it appears. In counting the roots, a multiple-order root is counted according to its order. For example, if $x = 1$ were the only value that satisfied a third-degree equation according to Equation 5-27, it would be a third-order root.

Complex Roots

We will deal with complex variables in more detail later in the text, but at this time, accept the notion that a root of an equation may involve both a real part and an imaginary part. The imaginary part is the multiplier of the purely imaginary number $i = \sqrt{-1}$, which is called j in most electrical engineering applications. A complex root would then have the form $a + ib$ (or $a + jb$ in electrical theory).

The quantity a is called the *real part* of the complex number, and b is called the *imaginary part* of the complex number. Note that b is real, but the i in front makes the product ib imaginary.

As long as the coefficients in the polynomial equation of Equation 5-26 are real, any complex roots will appear in *conjugate pairs*. This means that if $a + ib$ is a root, then $a - ib$ will also be a root. Although the real parts are the same, the imaginary parts are different and the roots are considered different. Therefore, if $a + ib$ and $a - ib$ occur only once each, they constitute a *complex pair of first-order roots*. The quantity $a - ib$ is called the *complex conjugate* of $a + ib$. If $a = 0$, the roots are $\pm ib$; these are referred to as *purely imaginary roots*.

Factored Form of Polynomial

Assume that the N roots of the polynomial are $x_1, x_2, ..., x_N$. The polynomial can be written in factored form as

$$y = p(x) = A_N (x - x_1)(x - x_2)....(x - x_N) \qquad (5\text{-}28)$$

Note that the coefficient A_N of the highest-degree term must be placed in front of the polynomial. Other than this constant multiplier, the polynomial can be determined from the values of the roots. Note also that it is quite obvious from the factored form that the value of the polynomial when x assumes any one of the root values is zero.

MATLAB Evaluation of Polynomials

Suppose we have a polynomial of the form of Equation 5-26 and we desire to evaluate it for all values of a vector x. One way this could be done would be to write out the equation in terms of the various powers. To illustrate the process without getting too unwieldy, suppose we assume a fourth-degree form for illustration. Assume that a vector x has been created that contains all domain values at which the polynomial is to be evaluated. Calling the result y, one form would be

$$\text{>> y = A4*x.^4 + A3*x.^3 + A2*x.^2 + A1*x + A0} \qquad (5\text{-}29)$$

Note that array exponentiation would be required in all terms but the last two, and the period follows x in each of the other terms.

While this form would work, there is a *much easier way*. Form a row vector consisting of the coefficients of the polynomial, starting with the highest-degree coefficient and ending with the lowest-degree coefficient. If a given exponential power is missing, a value of

0 is used for that coefficient. This means that a polynomial of degree N will require $N + 1$ terms. We will call this vector **C,** and for a fourth-degree equation, it is expressed as

$$\gg C = [A4\ A3\ A2\ A1\ A0] \tag{5-30}$$

Of course, you should assume that the vector x containing all values of the independent variable has been created. The following command will evaluate the polynomial for all values of x.

$$\gg y = \text{polyval}(C, x) \tag{5-31}$$

There is no required relationship between the sizes of C and x. In fact, x could be a simple scalar if desired.

Factoring of Polynomials

The roots of a polynomial may be easily determined by MATLAB. Assume that the coefficient vector has been formed according to the form of Equation 5-30. Let R represent a vector containing the roots. This vector, which is provided by MATLAB as a column vector, is determined by the simple command

$$\gg R = \text{roots}(C) \tag{5-32}$$

Forming the Polynomial from the Roots

Suppose the roots of a polynomial are known and it is desired to form the polynomial. Form a vector R from the roots and then apply the following command

$$\gg C = \text{poly}(R) \tag{5-33}$$

Even though the operation of Equation 5-32 provides a column vector, the reverse operation of Equation 5-33 can be applied to either a column vector or a row vector. The resulting vector will be the vector of coefficients starting with the highest-degree coefficient and ending with the constant term.

There is one possible missing element in the operation of Equation 5-33, which is a constant multiplier for all terms. The quantity is the value A_N in Equation 5-26 and 5-28. If this value is known, the final polynomial can be modified by the command

$$\gg C = AN*C \tag{5-34}$$

This type of operation, which is invalid in regular algebra, is fine in MATLAB since it is telling the program to take all the elements of vector C, multiply them by AN, and replace them in C.

Of course, the preceding two steps could be combined in the form

$$\gg C = AN*\text{poly}(R) \tag{5-35}$$

Multiplication of Polynomials

Two polynomials can be multiplied together by the use of the **conv** command. The term **conv** is a contraction of the term *convolution*, which has applications in signal processing and in both differential and difference equations. However, our immediate interest is in seeing how it can simplify the process of the multiplication of two polynomials.

We will illustrate the concept with two second-degree equations, but the procedure for arbitrary-size polynomials should be easy to infer. Assume two polynomials p_1 and p_2 with the forms

$$p_1(x) = A_2 x^2 + A_1 x + A_0 \tag{5-36}$$

and

$$p_2(x) = B_2 x^2 + B_1 x + B_0 \tag{5-37}$$

It is desired to generate a new polynomial $p_3(x)$ by the multiplicative operation

$$p_3(x) = p_1(x) p_2(x) \tag{5-38}$$

It should be clear that the degree of the product of two polynomials is the sum of the degrees of the two polynomials to be multiplied together, which in this case is $2 + 2 = 4$, meaning that the product will have 5 terms.

Switching to default MATLAB notation, we form two coefficient vectors as follows:

$$\text{>> C1 = [A2 A1 A0]} \tag{5-39}$$

$$\text{>> C2 = [B2 B1 B0]} \tag{5-40}$$

The coefficient vector C3 for the product function of Equation 5-38 is then generated by the command

$$\text{>> C3 = conv(C1, C2)} \tag{5-41}$$

The coefficients in the vector proceed from the highest-order term to the lowest-order term, as with previous coefficient vectors.

To illustrate the preceding process, we used two second-degree equations, meaning that the two vectors each had three components. Just so the reader does not get the wrong impression, the two vectors are not required to have the same size. Thus, the degrees of the two polynomials can be any arbitrary integer values.

Example 5-9

Using MATLAB, determine the roots of the equation

$$y = 3x^2 + 12x + 39 \tag{5-42}$$

Solution

We could factor this polynomial using the quadratic formula, but it is a "warmup" exercise that will be used to illustrate a few points. First, we form the coefficient vector as

$$>> C = [3\ 12\ 39]; \tag{5-43}$$

Let R represent the roots of this equation, which are determined as

$$>> R = \text{roots}(C) \tag{5-44}$$

$$R =$$

$$-2.0000 + 3.0000i$$

$$-2.0000 - 3.0000i$$

This column vector contains the two roots of the equation, which are a pair of complex conjugate roots; that is, $-2 \pm i3$. Note that MATLAB puts the i after the imaginary part value.

As a quick review of a topic covered in Chapter 1, when one is entering a complex number, there are four different ways in which the imaginary part can be entered. Assuming an imaginary 3, without the decimal points, it can be entered either as 3i or as 3j without using the normal multiplicative asterisk. One can also put the i or j in front of the 3, although the asterisk is required in that case. Thus, i*3 or j*3 would be acceptable. In all cases, however, it will print on the screen as 3i, with appropriate decimal places.

Example 5-10

Reconstruct the coefficients of the polynomial from the roots determined in the preceding example.

Solution

Assume that the roots from Example 5-9 are still in memory. Since we called the first vector of coefficients C, we will call the reconstructed vector in the first step C1. The command is

$$C1 = \text{poly}(R) \tag{5-45}$$

$$C1 =$$

$$1\ 4\ 13$$

If we compare the coefficients of C1 with those of C in Equation 5-43, we see that there is a discrepancy. Specifically, all the elements of C are exactly three times the elements of

C1. This is due to the fact that the construction process does not take into account the coefficient of the highest-degree term, which in this case is A2. Thus, the operation is equivalent to that of Equation 5-28 except for the multiplier A_N. The conclusion is that we can construct the polynomial to within a constant multiplier from the roots.

Since we know A2, we can generate the final form by the operation

$$>> C1=3*C1 \qquad (5-46)$$

$$C1 =$$

$$3 \; 12 \; 39$$

Now it can be easily seen that C1 = C and the polynomial terms can be read in order from the coefficients. We could also utilize the **polyval** function for evaluating it at specific values of x if we desired.

It should be noted here that in many applications of polynomial functions, it is a common practice to set them up so that the coefficient of the highest degree term is unity. This is particularly true in working with differential equations and Laplace transforms, subjects to be covered later in the book. Therefore, the absence of the coefficient in the construction process is usually not a serious one, but it must be understood for proper scaling.

Example 5-11

A so-called Butterworth polynomial of fifth degree is given by

$$y = x^5 + 3.2361x^4 + 5.2361x^3 + 5.2361x^2 + 3.2361x + 1 \qquad (5-47)$$

Using MATLAB, determine the five roots of the equation.

Solution

While the polynomial of the previous two examples could have been done manually, how many of us could factor this one? You need not worry, since you have MATLAB. First we form the coefficient vector C.

$$>> C=[1 \; 3.2361 \; 5.2361 \; 5.2361 \; 3.2361 \; 1]; \qquad (5-48)$$

Do not forget that, in general, we start with the term of highest degree and move downward to the term of lowest degree. One could be easily confused with this particular case because of the symmetry of the coefficients.

The roots are then easily determined by the command

$$>> R = roots(C) \qquad (5-49)$$

$$R =$$

$$-0.3090 + 0.9511i$$

-0.3090 - 0.9511i

-1.0000

-0.8090 + 0.5877i

-0.8090 - 0.5877i

This is a neat solution! In one sweep, we have determined the five roots of the fifth-degree equation. All are of first order, but there are two pairs of complex conjugate roots and one real root.

Example 5-12

Using MATLAB, reconstruct the polynomial from the five roots determined in the preceding example.

Solution

Assuming the five roots are still in memory, we enter

$$>> C1 = poly(R) \qquad (5\text{-}50)$$

$$C1 =$$

1.0000 3.2361 5.2361 5.2361 3.2361 1.0000

Since the coefficient of the highest-degree term was unity, we do not need to modify the vector, and C1 = C.

Example 5-13

For the Butterworth polynomial of Examples 5-11 and 5-12, evaluate the function at x = 0, 0.5, 1, and 2.

Solution

Assume that the polynomial coefficient vector C is still in memory. We could generate a vector x at many points and generate a detailed plot, but to be a little different in this example, we will evaluate the polynomial for only a few points. The vector x is entered as

$$>> x = [0\ 0.5\ 1\ 2]; \qquad (5\text{-}51)$$

Assuming y as the dependent function, we can generate the vector y from the command

$$>> y = \text{polyval}(C, x) \qquad (5\text{-}52)$$

$$y =$$

1.0000 4.8151 18.9444 154.0830

The values proceed in the same order as x. Thus, for x = 0, y = 1; for x = 0.5, x = 4.8151; and so on.

5-6   EXPONENTIAL FUNCTION

The basic exponential function arises in a large number of scientific and engineering problems. The "purest" form of the exponential function is as a power of the mathematical constant $e = 2.718$ to four significant digits. In terms of this constant, the basic form of the exponential function is

$$y = e^x \qquad (5\text{-}53)$$

This function is well tabulated and is available on virtually all scientific calculators.

Shape of Exponential Function

The basic shape of the exponential function for both positive and negative x is shown in Figure 5-10. As $x \to -\infty$, $e^x \to 0$, and as $x \to \infty$, $e^x \to \infty$.

Decaying Exponential Function with Time as the Independent Variable

The most common form of the exponential function in practical engineering problems is the *decaying* or *damped exponential function*. This is the portion of the function shown in Figure 5-10 for $x \leq 0$, and it usually is turned around and expressed as a function of positive time in either of the forms

$$y = e^{-t/\tau} \qquad (5\text{-}54)$$

or

$$y = e^{-\alpha t} \qquad (5\text{-}55)$$

The quantity τ is the *time constant* in seconds (assuming t is expressed in seconds) and α is the *damping constant* in seconds^{-1}. (In some applications, the latter units are referred to as *nepers*.) The two constants are obviously related simply as

$$\alpha = \frac{1}{\tau} \qquad (5\text{-}56)$$

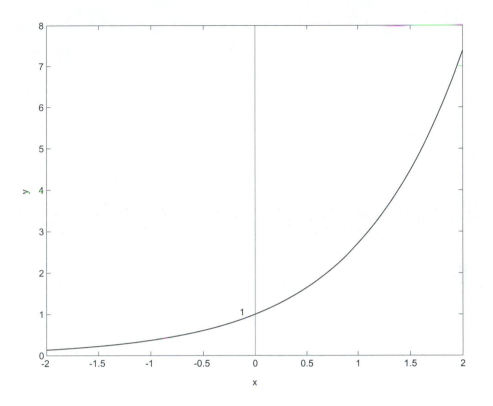

Figure 5-10 General form of the exponential function

Shape of Decaying Exponential Function

The form of the decaying exponential function for $t \geq 0$ expressed in terms of the normalized variable t / τ, which is labeled "Number of Time Constants," is shown in Figure 5-11. One interpretation of the time constant is that it would be the time when the function reached zero if it continued to decrease at its initial rate. This is illustrated for the straight line below the exponential curve. Of course, the function does not continue to decrease at its initial rate. In fact, the rate of decrease is proportional to its level, a property that will be of great interest in calculus.

Specifically, at $t = \tau$, the function has decreased to $e^{-1} = 0.368$; at $t = 2\tau$ the function has decreased to about 0.135; and so on. In theory, the decaying exponential never reaches 0, but at $t = 5\tau$, the function will have reached a level of about 0.0067, or less than 1% of its initial value. For many applications, it may be assumed to be zero at this point. Hence, the "rule of thumb" of five time constants for the decaying exponential is a practical one for many applications. Like most rules of thumb, however, this should not be considered as

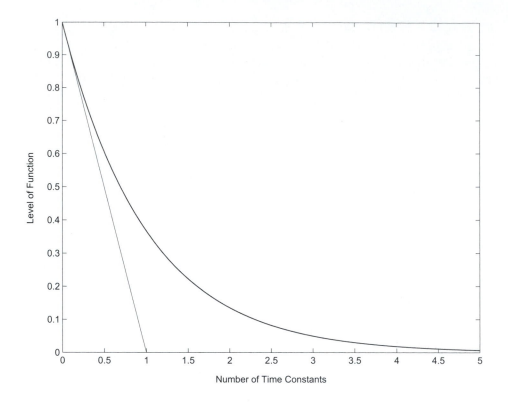

Figure 5-11   Decaying exponential function and line to one time constant

absolute. The author has worked on an instrumentation system in which about eight time constants were required to satisfy the specifications.

MATLAB Form of Exponential Function

Assuming that a vector, x, has been defined at all points at which the function is to be evaluated, the MATLAB command for generating an exponential y in terms of x is

$$>> y = \exp(x) \qquad\qquad (5\text{-}57)$$

In the form of Equation 5-57 there is no need for the array operation since x appears as a first power in the argument. Thus, MATLAB will evaluate the function for all values of x. If the argument had involved a higher degree of x, an array multiplication in the argument would have been required. Also, if the exponential function has been multiplied by some other function, it would have been necessary to utilize array multiplication for the two functions.

MATLAB Exponentiation with Other Bases

The **exp(x)** operation is a function that will produce a value for all individual values of the argument based on a vector entry. If the base of an exponential function is not e, we can use the exponentiation form introduced in Chapter 1, but it is necessary to define it as array operation. For example, if 10 is the desired base and x is a vector, we must enter the operation in the format that follows.

$$>> y = 10.^x \tag{5-58}$$

Example 5-14

An exponential function is given by

$$y = e^{-t/0.01} \tag{5-59}$$

where t is measured in seconds. Determine (a) the time constant and (b) the damping constant. Then, (c) based on the rule of thumb provided in the text, determine about how long it would take to reach a practical level of zero.

Solution

a. Since the independent variable is expressed in the form of t divided by a value, the form is that of Equation 5-54, and the value in the denominator of the argument is the time constant. Hence,

$$\tau = 0.01 \text{ s} = 10 \text{ ms} \tag{5-60}$$

b. The damping constant is the reciprocal of the time constant.

$$\alpha = \frac{1}{\tau} = \frac{1}{0.01} = 100 \text{ s}^{-1} \tag{5-61}$$

Note that an alternate way to express this exponential is

$$y = e^{-100t} \tag{5-62}$$

c. Based on the rule of thumb, the function should be sufficiently close to zero for most practical purposes at a time T given by about

$$T = 5\tau = 5 \times 10 \text{ ms} = 50 \text{ ms} \tag{5-63}$$

Example 5-15

A certain force variable f has the form of a decaying exponential function. The initial value is 20 newtons (N) and the time constant is 5 s. Write an equation for the function.

Solution

The basic exponential function has a value of 1 at $t = 0$. For an initial value of 20, we simply use a constant of that value as a multiplier for the exponential. The resulting function can be expressed as

$$f = 20e^{-t/5} = 20e^{-0.2t} \tag{5-64}$$

Example 5-16

Generate both the decaying exponential function over the domain of 5 time constants and the straight-line function shown in Figure 5-11 and plot them.

Solution

First, the vector x is generated by the following command:

$$\gg x = \text{linspace}(0,5,501); \tag{5-65}$$

where an arbitrary selection of 100 points per time constant interval was used.

The decaying exponential function is generated by the command

$$\gg y = \exp(-x); \tag{5-66}$$

The equation of this straight line in basic mathematical form for $0 \leq x \leq 1$ would be $y = 1 - x$.

However, we have already used x to cover a much larger domain and y for the exponential function. Therefore, we will use x1 and y1 for the straight line. For the straight line, only two points would be required, but we will give it 11 points just for good measure. The MATLAB commands follow.

$$\gg x1 = \text{linspace}(0,1,11); \tag{5-67}$$

$$\gg y1 = 1 - x1 \tag{5-68}$$

$$\gg \text{plot}(x, y, x1, y1) \tag{5-69}$$

The plot was provided in Figure 5-11. We will not show the labeling commands, since by now they should be fairly routine. However, it should be noted again that when the figure window appears on the screen, there are lots of options that can be added to enhance the figure. To perform any enhancements, first left-click on the **edit** option.

The one option that was added here was to increase the width of the exponential curve trace. After left-clicking on **edit** as discussed in the last paragraph, left-click on the exponential curve trace to select it. This is indicated by the appearance of a series of points. Next, right-click on the curve and a menu will appear. Move the arrow to the **line width** option. Then change the line width from 0.5 (the default value) to 1 and click **OK**. The result is a more pronounced curve for the function.

5-7 LOGARITHMIC FUNCTION

The logarithmic function is basically the inverse of the exponential function. However, because it arises in many applications, it will be represented in the usual form with x as the independent variable and y as the dependent variable. The logarithmic function to the base e is often expressed as

$$y = \ln x \tag{5-70}$$

where "ln" is the common mathematical notation for "logarithm to the base e." The form of $y = \ln x$ is shown in Figure 5-12.

Logarithms to Other Bases

Frequently, the logarithm to a base other than e is desired. In general, the logarithm to a base other than e can be determined as follows:

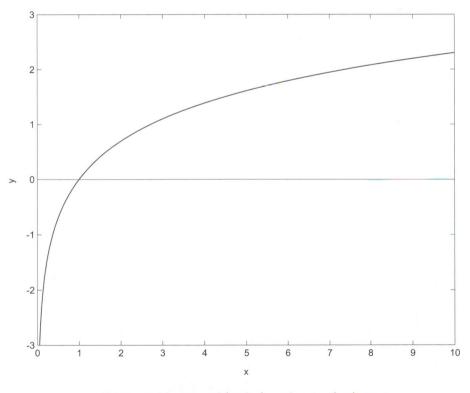

Figure 5-12 Logarithmic function to the base e

$$\log_a x = \frac{\ln x}{\ln a} \tag{5-71}$$

In digital systems and communications theory, the logarithm to the base 2 is often desired. It can be expressed as

$$\log_2 x = \frac{\ln x}{\ln 2} = 1.4427 \ln x \tag{5-72}$$

Logarithm to Base 10

The logarithm to the base 10 arises so often that it is usually available as a separate option on most scientific calculators and in MATLAB. A relationship between the logarithm to the base 10 and the logarithm to the base e is

$$\log_{10} x = \frac{\ln x}{\ln 10} = \frac{\ln x}{2.3026} = 0.4343 \ln x \tag{5-73}$$

MATLAB Logarithm Commands

Assuming a vector, x, at all points at which evaluation is desired, the logarithm to the base e in MATLAB is determined from the command

$$>> y = \log(x) \tag{5-74}$$

This could lead to a little confusion since many mathematical references use ln x to refer to the natural logarithm and log x to refer to the logarithm to the base 10. However, the logarithm to the base 10 in MATLAB is generated by the command

$$>> y = \log 10(x) \tag{5-75}$$

You must keep these differences in mind to avoid confusion. Note that most functional forms in MATLAB require the parentheses around the arguments.

Example 5-17

A unit widely employed in various forms for sound measurement and electrical signal measurement is the **decibel (dB)**. In its basic form, it is a dimensionless logarithmic ratio of two power values. However, there are many variations on the definition in which some reference power level is assumed. For example, the unit **dBm** refers to a logarithmic ratio referred to a reference standard of **1 mW**.

Assume a reference power level P_{ref} and an arbitrary power level P, with both measured in watts. Let

$$G = \frac{P}{P_{ref}} = \text{ absolute power ratio} \tag{5-76}$$

The decibel power ratio G_{dB} is defined as

$$G_{dB} = 10 \log_{10} G \qquad (5\text{-}77)$$

where $\log_{10}$ refers to the logarithm to the base 10.

Use MATLAB to develop a conversion curve in which G varies from 0.01 to 100. Use a semi-log plot with G on the horizontal logarithmic scale and G_{dB} on the vertical linear scale.

Solution

This example and the next one provide good opportunities for introducing semi-log plots with the two different combinations of one axis as linear and one axis as logarithmic. In general, a function that is already logarithmic in form will often display better if it is plotted on a linear plot, with the other variable shown on the logarithmic scale.

In this case, the independent variable is selected as the absolute power gain G. It will be displayed in logarithmic fashion. A **logspace** plot is therefore selected by the following command:

$$\gg G = logspace(\text{-}2,2,200); \qquad (5\text{-}78)$$

Recall from Chapter 4 that this means that the domain will vary from 10^{-2} (0.01) to 10^2 (100) and the row vector G will contain 200 points.

The command for generating the decibel gain is

$$\gg GdB = 10*log10(G); \qquad (5\text{-}79)$$

The command that provides a logarithmic plot on the horizontal scale and a linear plot on the vertical scale is

$$\gg semilogx(G,GdB) \qquad (5\text{-}80)$$

A grid is added by the command

$$\gg grid \qquad (5\text{-}81)$$

We will not show all the labeling commands since they have been well covered earlier in the text, but the resulting curve is shown in Figure 5-13.

Example 5-18

Take the results generated in Example 5-17 but consider that the decibel power ratio is the independent variable on a horizontal linear scale and the absolute power ratio is the dependent variable on a vertical logarithmic scale.

Solution

If we were "starting from scratch," we could solve for G in terms of G_{dB}, with the following result

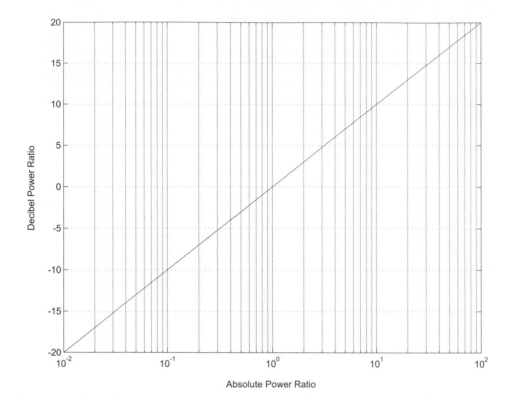

Figure 5-13   Decibel power ratio versus absolute power ratio

$$G = 10^{G_{dB}/10} \tag{5-82}$$

We could then create a linear vector for G_{dB} and use MATLAB to generate the dependent vector G. However, we already have vectors for both G and GdB, and we can switch the order of the variables in the plot command. We also want to have the logarithmic scale for G, so the command in this case is

$$\text{>> semilogy(GdB, G)} \tag{5-83}$$

The resulting plot (after adding the grid and labels) is shown in Figure 5-14.

Example 5-19

In the study of statistics later in the text, you will learn that a gaussian probability density function having a mean value of 0 and a standard deviation of 1 may be expressed by the equation

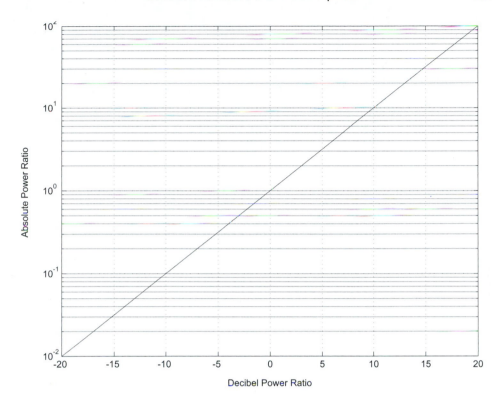

Figure 5-14 Absolute power ratio versus decibel power ratio

$$y = \frac{1}{\sqrt{2\pi}} e^{-x^2/2} \tag{5-84}$$

Use MATLAB to generate and plot the function over the domain $-3 \leq x \leq 3$.

Solution

Just to simplify the equation slightly, let us define the constant in front as **a**. Thus,

$$\gg \text{a} = 1/(\text{sqrt}(2*\text{pi})); \tag{5-85}$$

Although you probably won't be able to tell the difference if you have a fast processor, making this definition should decrease the amount of computation time, since we have now evaluated the constant and it is ready to be used with further computations. We will arbitrarily select 301 points for evaluation. Hence,

$$\gg \text{x} = \text{linspace}(-3,3,301); \tag{5-86}$$

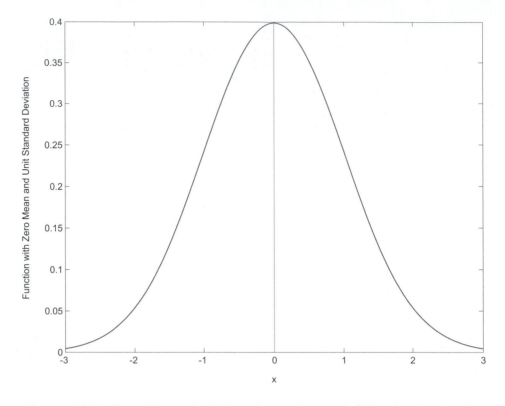

Figure 5-15   Plot of Example 5-19 with gaussian probability density function

Now since the argument of *e* involves the square of *x*, we need to use the array product form. Thus,

$$\text{>> y = a*exp(-0.5*x.^2);} \tag{5-87}$$

We then plot by the command

$$\text{>> plot(x, y)} \tag{5-88}$$

The resulting function after labeling is shown in Figure 5-15.

5-8   TRIGONOMETRIC FUNCTIONS

There are six basic trigonometric functions: (1) sine, (2) cosine, (3) tangent, (4) cotangent, (5) secant, and (6) cosecant. However, the first three tend to occur more often in practical applications than the latter three. Moreover, the latter three can be expressed as reciprocals of the first three (though not in the order listed). Therefore, we will focus on the first three, but the definitions of the latter three will be provided for reference purposes.

Angle Measurement

The most widely employed units for angles, especially in engineering applications, are *degrees*. One complete revolution around a circle corresponds to 360°. However, the *degree* is not a basic unit in terms of mathematical principles. Rather, it is an artificial unit that dates back to early usage.

The most basic mathematical unit for an angle is the *radian* (rad). It does have a mathematical basis for its form and does arise as a natural process. One complete revolution for a circle corresponds to 2π radians. To convert from radians to degrees, the following formula can be used:

$$\text{Angle (degrees)} = \frac{180}{\pi} \times \text{Angle (radians)} \qquad (5\text{-}89)$$

Conversely, to convert from degrees to radians, the reverse formula is

$$\text{Angle (radians)} = \frac{\pi}{180} \times \text{Angle (degrees)} \qquad (5\text{-}90)$$

The reader will likely have occasion to perform one or the other of these conversions many times in the future, so we have presented the formulas two ways. It can be shown that 1 radian is approximately 57.3°, so the radian is a much larger unit. Therefore, if you remember that the conversion factor is either 180 / π or π / 180, you should pick the form that will yield fewer radians or more degrees, depending on the direction in which you are going.

Why make a "big fuss" about the conversion? The fact is that the use of degrees is widely embedded in most areas of engineering and science and will probably not change anytime in the near future. However, when software such as MATLAB is employed, and when combinations of angles varying with an independent variable and a fixed angle are of interest, it is necessary to employ radians. Therefore, you must learn to think in terms of both types. In fact, surveyors even use another angle measurement, called *grads* (100 grads = 90°), but we will not deal with those units.

Right Triangle

While our emphasis is directed toward functions and their relationships, it is worthwhile to initially introduce the trigonometric definitions in terms of a triangle. Consider the right triangle shown in Figure 5-16. Since x will serve as our independent variable in the functional developments that follow, the angle will be denoted as x. Let b represent the length of the base or *adjacent side*, h represent the height or *opposite side*, and r represent the *hypotenuse*. By the Pythagorean theorem, we know that

$$r^2 = b^2 + h^2 \qquad (5\text{-}91)$$

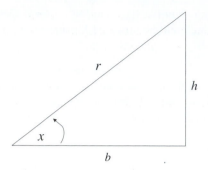

Figure 5-16   Right triangle used to define trigonometric functions

Definitions of Trigonometric Terms

The first three trigonometric terms are defined in the three equations that follow. The sine is abbreviated as sin x and is

$$\sin x = \frac{h}{r} \tag{5-92}$$

The cosine is abbreviated as cos x and is

$$\cos x = \frac{b}{r} \tag{5-93}$$

The tangent is abbreviated as tan x and is

$$\tan x = \frac{h}{b} \tag{5-94}$$

As previously mentioned, the other three terms can be expressed as reciprocals of the preceding three. They are provided here for convenience.
The cotangent is abbreviated as cot x and is

$$\cot x = \frac{b}{h} = \frac{1}{\tan x} \tag{5-95}$$

The secant is abbreviated as sec x and is

$$\sec x = \frac{r}{b} = \frac{1}{\cos x} \tag{5-96}$$

The cosecant is abbreviated as csc x and is

$$\csc x = \frac{r}{h} = \frac{1}{\sin x} \tag{5-97}$$

Numerous relationships between these various quantities, along with identities, are provided in basic algebra and trigonometry textbooks. Our focus from this point forward will be the functional forms as the angle x is allowed to vary.

Sine Function

The first function that will be considered is the sine function. Consider the following equation:

$$y = \sin x \qquad (5\text{-}98)$$

The form of this function over the domain $0 \leq x \leq 2\pi$ is shown in Figure 5-17. The function is *periodic*, meaning that it repeats the pattern shown for both positive and negative x. The domain shown constitutes *one cycle* of the periodic function; the *period,* on an angular basis, is 2π radians. If the curve were extended backward from $x = 0$, it would have the same form as at the right-hand end of the period. Therefore, the sine function is an *odd function*.

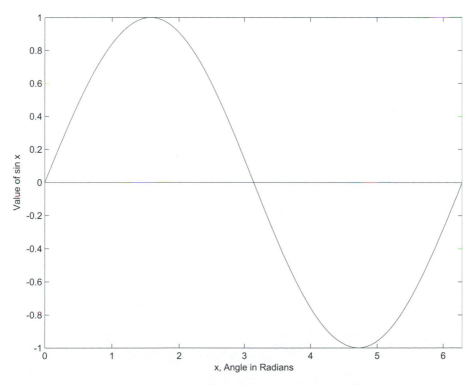

Figure 5-17   Plot of sin x over one cycle

Cosine Function

The next function that will be considered is the cosine function. It can be represented as

$$y = \cos x \tag{5-99}$$

The form of this function over the domain $0 \le x \le 2\pi$ is shown in Figure 5-18. As in the case of the sine function, the cosine function is periodic with a period of 2π radians on an angular basis. If the curve were extended backward from $x = 0$, it would have the same form as at the left-hand end of the period. In contrast to the sine function, the cosine function is an *even function*.

Tangent Function

The tangent function may be expressed as

$$y = \tan x \tag{5-100}$$

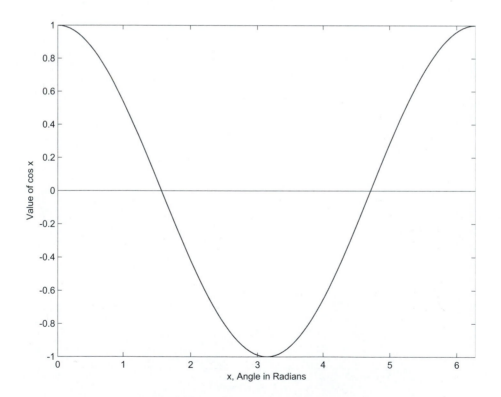

Figure 5-18   Plot of cos x over one cycle

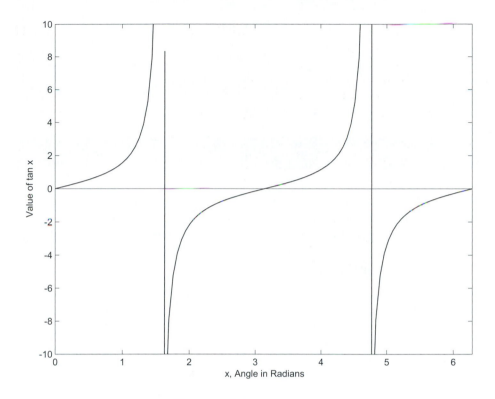

Figure 5-19   Plot of tan x

The form of this function over the domain $0 \leq x \leq 2\pi$ is shown in Figure 5-19. This function is periodic, but there are two cycles shown in the given domain. Hence, the tangent function is periodic with a period of π radians on an angular basis.

The tangent function is an odd function. Moreover, it has infinite discontinuities at odd integer multiples of $\pi / 2$.

MATLAB Trigonometric Functions

MATLAB actually allows the direct computation of all six of the trigonometric functions. Assuming all values of the independent variable have been defined as a vector x, the six functions can be generated by the following six commands:

$$\gg y = \sin(x) \tag{5-101}$$

$$\gg y = \cos(x) \tag{5-102}$$

$$\gg y = \tan(x) \tag{5-103}$$

$$>> y = \cot(x) \qquad (5\text{-}104)$$

$$>> y = \sec(x) \qquad (5\text{-}105)$$

$$>> y = \csc(x) \qquad (5\text{-}106)$$

5-9 SINUSOIDAL TIME FUNCTIONS

Probably the most common form of sine and cosine functions in engineering applications is when they are functions of the independent variable time. Such functions arise in sound waves, communication signals, pressure waves, and many other applications. Thus, their treatment warrants a special section devoted to the form and terminology involved.

What Are Sinusoidal Functions?

We need to define what we mean by sinusoidal functions. Both the sine function and the cosine function are considered sinusoidal functions. Moreover, a linear combination, which consists of the sum of sine and cosine functions with arbitrary constant multipliers, is considered a sinusoidal function. We will begin the discussion here with a cosine function of the form

$$y_c = A \cos \omega t \qquad (5\text{-}107)$$

The various parameters in the equation are defined as follows:

y_c = cosine function
A = amplitude or peak value of cosine function
ω = angular frequency or angular velocity of function in radians/second (rad/s)
t = time in seconds (s)

Next, consider the sine function

$$y_s = B \sin \omega t \qquad (5\text{-}108)$$

where

y_s = sine function
B = amplitude of sine function

The quantities ω and t have the same meanings as for the sine function.

We have studied the behavior of both of these functions when the argument was x. We have seen that the sine function begins at a level of 0 with a positive slope, and the cosine function begins at a maximum level and then starts to decrease.

Period and Cyclic Frequency

For either of the preceding functions, the quantity ω is the number of radians per second that the function undergoes in the argument. This quantity is called the *angular velocity* in mechanics and the *angular frequency* in electricity. It is related to the cyclic frequency f by the relationship

$$\omega = 2\pi f \tag{5-109}$$

For each cycle of rotation in a mechanical sense, the angle changes by 2π radians. Hence, the angular velocity is 2π times the number of rotations per second. Therefore, the cyclic frequency f can be measured in cycles per second or hertz (Hz), where 1 Hz = 1 cycle/second.

One additional quantity is the period T, which is the length of time corresponding to 1 cycle. Since there are f cycles per second, the period is

$$T = \frac{1}{f} \tag{5-110}$$

Therefore, there are several alternate forms for the argument of the sine and/or the cosine function:

$$\omega t = 2\pi f t = \frac{2\pi}{T}t \tag{5-111}$$

Combining Sine and Cosine Functions of the Same Frequency

Consider the following function:

$$y = A\cos \omega t + B\sin \omega t \tag{5-112}$$

It turns out that the sum of a cosine function and a sine function *that both have the same frequency* can be expressed as a single sinusoidal function of the same frequency, but with an additional phase angle added in the argument. This means that the function of Equation 5-112 can be expressed in either of the following forms:

$$y = C\sin(\omega t + \theta) \tag{5-113}$$

or

$$y = C\cos(\omega t + \phi) \tag{5-114}$$

The constant C is the same for both forms, but the angles are different. Without showing a proof, we will demonstrate a clever way of making the combination.

Relative Phase Sequence Diagram

Consider the rectangular coordinate system shown in Figure 5-20. The positive direction of rotation is counterclockwise. The relative phase sequence is +sin, +cos, −sin, and −cos.

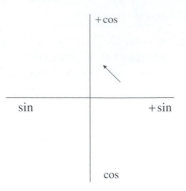

Figure 5-20   Relative phase sequence of sine and cosine functions

When a function of the form of Equation 5-112 is given, consider a point having the coordinates (B, A), which could be in any one of the four quadrants, depending on the sign pattern. The length of a line from the origin to the point is the value C and is given by

$$C = \sqrt{A^2 + B^2} \qquad (5\text{-}115)$$

The angle used will depend on whether we wish to express the result in the form of Equation 5-113 or Equation 5-114. Determine the angle with respect to the particular axis and whether it is a positive or a negative angle. This may require the use of some basic trigonometric relationships or the rectangular-to-polar conversion forms on a scientific calculator. Of course, it can be done with MATLAB, as will be demonstrated later.

Example 5-20

Use MATLAB to plot the following function over two cycles.

$$y = 20\sin(\omega t + 30^\circ) \qquad (5\text{-}116)$$

Solution

The function has been stated in the form that most engineers use (including this author), but the argument of the sine function is a mixture of "apples and oranges," as the old saying goes. For interpretation purposes it is fine, but if we need to perform computations based on the argument, the function needs to be rewritten as

$$y = 20\sin\left(\omega t + \frac{\pi}{6}\right) \qquad (5\text{-}117)$$

Since no specific period or frequency has been specified, we will consider the product ωt in the form

$$\omega t = 2\pi \frac{t}{T} = 2\pi x \qquad (5\text{-}118)$$

where

$$x = t/T \qquad (5\text{-}119)$$

The variable x is a normalized variable expressing time relative to a period. Thus, two cycles would correspond to a domain of $0 \le x \le 2$.

We will create a vector x by the following command:

$$\text{>> } x = \text{linspace}(0, 2, 201); \qquad (5\text{-}120)$$

The function y is then created by the command

$$\text{>> } y = 20\text{*sin}(2\text{*pi*x} + \text{pi/6}); \qquad (5\text{-}121)$$

A plot is then obtained by the command

$$\text{>> plot}(x, y) \qquad (5\text{-}122)$$

Additional labeling is provided by methods covered earlier, and the result is shown in Figure 5-21.

5-10   THREE-DIMENSIONAL AND POLAR PLOTS

Two other types of plots will be introduced in this section. They are the *three-dimensional* and *polar* plots.

Three-Dimensional Plots

MATLAB has the capability of creating three-dimensional plots. Assume three variables, x, y, and z. The command to generate a three-dimensional plot is

$$\text{>> plot3}(x, y, z) \qquad (5\text{-}123)$$

As in the case of two-dimensional plots, the variables involved can have any names, but the order in which they are entered into the plot command determines the order for the three variables. For three-dimensional plots, there are three label commands and they are denoted as **xlabel**, **ylabel**, and **zlabel**, respectively. The **title** label may also be used. This process will be illustrated in Example 5-21.

Polar Plots

A polar plot utilizes two variables, **theta** and **rho**. The first variable, **theta**, is the angle in a two-dimensional plane, and the second variable, **rho,** is the radius of the dependent variable from the origin to a particular point. The command is

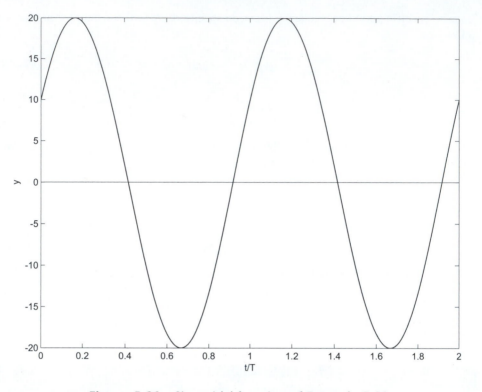

Figure 5-21 Sinusoidal function of Example 5-20

$$\gg \text{polar(theta, rho)} \qquad (5\text{-}124)$$

This process will be illustrated in Example 5-22.

Example 5-21

Consider the three-dimensional function defined as follows:

$$x = \cos t$$
$$y = \sin t \qquad (5\text{-}125)$$
$$z = t$$

Create a three-dimensional plot as the variable t varies from 0 to 100.

Solution

The time step was set somewhat arbitrarily as 0.1, and the MATLAB commands follow.

>> t = 0:0.1:100;

>> x = cos(t);

>> y = sin(t);

>> z = t;

>> plot3(x, y, z)

>> xlabel('x')

>> ylabel('y')

>> zlabel('z')

>> title('Figure 5-22. Three-dimensional plot for Example 5-21.')

The plot is shown in Figure 5-22. The figure thus created could represent a spring.

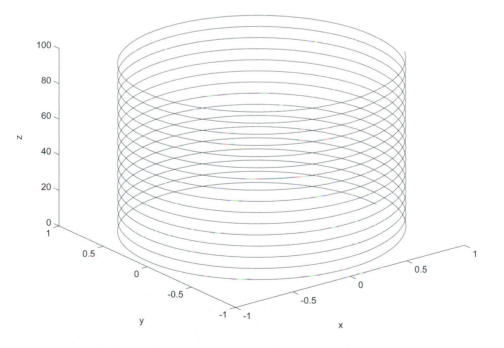

Figure 5-22 Three-dimensional plot for Example 5-21

Example 5-22

A particular type of normalized antenna power radiation pattern in two dimensions can be represented as

$$g(\theta) = \cos^2 \theta \qquad (5\text{-}126)$$

Plot the pattern using a polar plot.

Solution

An arbitrary choice was made to utilize 101 points over the domain from $\theta = 0$ to $\theta = 2\pi$. The code follows.

```
>> theta = linspace(0, 2*pi, 101);

>> g = cos(theta).^2;

>> polar(theta, g)
```

The polar plot is shown in Figure 5-23.

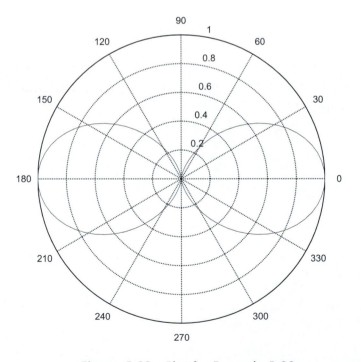

Figure 5-23   Plot for Example 5-22

GENERAL PROBLEMS

5-1. Consider the function $y = f(x) = x^2 + 4x$. Is the function single-valued or multi-valued?

5-2. Consider the function $y = x^4$. Is the function single-valued or multi-valued?

5-3. Is the function of Problem 5-1 an even function, an odd function, or neither?

5-4. Is the function of Problem 5-2 an even function, an odd function, or neither?

5-5. Determine the inverse function for the function of Problem 5-1.

5-6. Determine the inverse function for the function of Problem 5-2.

5-7. Is the inverse function of Problem 5-5 single-valued or multi-valued?

5-8. Is the inverse function of Problem 5-6 single-valued or multi-valued?

5-9. Is the inverse function of Problem 5-5 even, odd, or neither?

5-10. Is the inverse function of Problem 5-6 even, odd, or neither?

MATLAB PROBLEMS

5-11. Use vertical subplots (two rows and one column) to plot the function of Problem 5-1 and its inverse from Problem 5-5 as x varies from -5 to 5.

5-12. Use subplots (one row and two columns) to plot the function of Problem 5-2 and its inverse from Problem 5-6 as x varies from -2 to 2.

In Problems 5-13 through 5-16, use the **Insert Line** operation from the **Figure Window** to insert solid horizontal and vertical axes through the origin in each case.

5-13. A linear equation has a slope of -3 and a vertical intercept of 2. Write the equation and plot it over the domain $-3 \le x \le 3$.

5-14. A linear equation has a slope of 5 and a vertical intercept of 2. Write the equation and plot it over the domain $-3 \le x \le 3$.

5-15. A linear equation passes through the two points $(-1, -2)$ and $(3, 1)$. Write the equation and plot it over the domain $-3 \le x \le 3$.

5-16. A linear equation has a slope of 2 and a *horizontal* intercept of 1. Write the equation and plot it over the domain $-3 \leq x \leq 3$.

5-17. Determine the roots of the equation

$$y = x^3 + 1.2529x^2 + 1.5349x + 0.7157$$

5-18. Determine the roots of the equation

$$y = x^3 + 2x^2 + 2x + 1$$

5-19. Reconstruct the polynomial in Problem 5-17 from the roots.

5-20. Reconstruct the polynomial in Example 5-18 from the roots.

5-21. For the polynomial of Problems 5-17 and 5-19, evaluate the function for x = 0, 0.5, 1, and 2.

5-22. For the polynomial of Problems 5-18 and 5-20, evaluate the function for x = 0, 0.5, 1, and 2.

5-23. Determine the roots of the equation

$$y = x^5 + 1.1725x^4 + 1.9374x^3 + 1.3096x^2 + 0.7525x + 0.1789$$

5-24. Determine the roots of the equation

$$y = 2x^5 + 6.4721x^4 + 10.4721x^3 + 10.4721x^2 + 6.4721x + 2$$

5-25. Reconstruct the polynomial in Problem 5-23 from the roots.

5-26. Reconstruct the polynomial in Problem 5-24 from the roots.

5-27. For the polynomial of Problems 5-23 and 5-25, evaluate the function for x = 0, 0.5, 1, and 2.

5-28. For the polynomial of Problems 5-24 and 5-26, evaluate the function for x = 0, 0.5, 1, and 2.

5-29. A voltage in volts in a certain circuit is given by

$$v(t) = 100e^{-500t}$$

with time measured in seconds. Determine (a) the damping factor and (b) the time constant. Plot the function over the domain of five time constants.

5-30. A current in amperes in a certain circuit is given by

$$i(t) = 0.04e^{-t/0.2}$$

with time measured in seconds. Determine (a) the time constant and (b) the damping constant. Plot the function over the domain of 5 time constants.

5-31. The temperature in degrees Fahrenheit in a certain environment is changing according to the following equation:

$$T = 50 - 30e^{-t/4}$$

with time measured in hours. Plot the function over the domain of five time constants.

5-32. The temperature in degrees Fahrenheit in a certain environment is changing according to the following equation:

$$T = -10 + 80e^{-t/20}$$

with time measured in hours. Plot the function over the domain of five time constants.

5-33. Example 5-17 dealt with the development of an absolute power level to decibel conversion curve. Based on certain assumptions concerning equal electrical impedance levels, it is also possible to use voltage ratios as the basis for determining decibel forms. Assume a reference voltage level V_{ref} and an arbitrary voltage level V, with both measured in volts. Let

$$A = \frac{V}{V_{ref}} = \text{ absolute voltage ratio}$$

The decibel ratio A_{dB} is defined as

$$A_{dB} = 20\log_{10} A$$

Write a program to develop a conversion curve in which A varies from 0.1 to 10. Use a semilog plot with A on the horizontal logarithmic scale and A_{dB} on the vertical linear scale.

5-34. Repeat the process of Problem 5-33 but assume that A_{dB} is the independent variable on a horizontal linear scale and A is the dependent variable on a vertical logarithmic scale.

A set of mathematical functions that were first used in steam engine design and later used in approximation theory, including the design of electrical filters, are the **Chebyshev polynomials**. (There are several translations of the name from Russian, including some that begin with T.) The first six polynomials are defined as follows:

$$C_1(x) = x$$
$$C_2(x) = 2x^2 - 1$$

$$C_3(x) = 4x^3 - 3x$$
$$C_4(x) = 8x^4 - 8x^2 + 1$$
$$C_5(x) = 16x^5 - 20x^3 + 5x$$
$$C_6(x) = 32x^6 - 48x^4 + 18x^2 - 1$$

These functions will be used to develop a number of programming exercises. In each problem, write an M-file program to generate the figure or figures or to perform the analysis required. The domain for Problems 5-35 through 5-39 is $-1 \le x \le 1$.

5-35. Use six subplots with three rows and two columns to plot the six polynomials as separate small figures on one page. Label each x-axis as "x." Label each y-axis with the appropriate "Cn(x)" for n = 1 to 6.

5-36. Plot $C_5(x)$ versus x. Add grid. Label x-axis as "x." Label y-axis as "C5(x)." Provide title "Chebyshev Polynomial of Degree 5."

5-37. Plot $C_6(x)$ versus x. Add grid. Label x-axis as "x." Label y-axis as "C6(x)." Provide title "Chebyshev Polynomial of Degree 6."

5-38. Plot $C_3(x)$ and $C_4(x)$ versus x on the same graph. Add grid. Label x-axis as "x." Label y-axis as "Value." Provide title "Chebyshev Polynomials of Degrees 3 and 4." Use **gtext** to label curves "C3(x)" and "C4(x)."

5-39. Determine the roots of the six Chebyshev polynomials. To the extent possible, observe how these values compare with the zero crossings of the plots in Problems 5-35 through 5-38.

5-40. The domain for this problem is $0 \le x \le 2$. Plot the following function over the domain:

$$f(x) = \frac{1}{\sqrt{1 + 0.2589C_5^2(x)}}$$

Label x-axis as "x." Label y-axis as "Amplitude Response." Provide title "Amplitude Response of 1-dB Chebyshev Low-Pass Filter."

5-41. The Chebyshev functions were introduced earlier in their direct polynomial forms. There is an alternate formulation for generating these functions, which will be considered here. The function $C_m(x)$ can be generated by the formula

$$C_m(x) = \cos\left(m\cos^{-1}x\right)$$

In earlier problems, the polynomials were generated with separate equations. Now use a **for loop** to generate $C_1(x)$ through $C_6(x)$ utilizing the preceding trigonometric form over the domain $-1 \le x \le 1$.

Use a **for loop** with subplots to plot the first six Chebyshev polynomials in three rows and two columns with C_1 and C_2 along the first row, and so on. Label each *x*-axis as "x" and label each *y*-axis with the number **m**, where m varies from 1 to 6.

> *Hint 1:* The command for the inverse cosine in MATLAB is **acos(x)**.
>
> *Hint 2:* The Chebyshev polynomials may be defined with two integer arguments as C(m, n), with m being the degree and n being the index for the independent variable x. To perform an evaluation for all values of a variable, colons (:) can be used to represent "all values." Thus, an equation within the loop might read
>
>> C(m,:) = "some function of **m** and **x**." The result will be evaluated for all values of **x**.
>
> *Hint 3: A* number **m** can be converted to a string variable by the command **m1=num2str(m).**

5-42. Go back to Problem 3-32 and create a new program that will plot the earnings per day in dollars versus the day (from 1 to 30) for B on a semi-log scale with y as the logarithmic variable.

Differential Calculus

6

The two basic forms of calculus are *differential calculus* and *integral calculus*. This chapter will be devoted to the former, and Chapter 7 will be devoted to the latter. Finally, Chapter 8 will be devoted to a study of how MATLAB can be used for calculus operations.

Since calculus is a complete subject with numerous hefty books devoted to its study, what do we expect to accomplish in three chapters? First, it is anticipated that most readers of this text will have completed one or more courses in calculus, and to some extent, much of this treatment will constitute a review and strengthening of the subject area. However, we believe that even students who have little or no background in calculus, but who have a good command of basic algebra and trigonometry, will be able to absorb most of the topics in these chapters. We believe this is possible because much of the emphasis will be on the physical interpretation of the concepts rather than on a lot of equation memorization.

Both this chapter and Chapter 7 will begin with graphical approaches to the subject utilizing functions consisting of straight-line segments. This process will show the reader exactly what is happening when we apply differential or integral calculus to a function.

Objectives

After completing this chapter, the reader should be able to

1. Define the *derivative of a function* and show the various forms of notation.
2. Determine the *derivative* of a *piecewise linear continuous function* and plot it.
3. Derive the derivative of a simple function using the delta process.
4. State and apply the *chain rule* for differentiation.
5. Approximate the numerical value of a derivative by utilizing small changes.
6. Utilize tables to determine derivatives of various common functions.
7. Apply the procedure for determining the *maximum* or the *minimum* of a function using calculus.
8. State the definition of the *differential*.
9. Apply the *differential approximation* to determine small changes in a function.

6-2   THE DERIVATIVE AND DIFFERENTIATION

The study of calculus usually begins with the basic definition of a *derivative*. A derivative is obtained through the process of *differentiation,* and the study of all forms of differentiation is collectively referred to as *differential calculus*.

If we begin with a function and determine its derivative, we arrive at a new function called the *first derivative*. If we differentiate the first derivative, we arrive at a new function called the *second derivative*, and so on.

The Derivative Is the Slope

The first derivative of a function is a new function that represents the **slope** of the original function as a function of the independent variable. The slope indicates the rate at which the original function is changing. The concept of the slope is illustrated in Figure 6-1. For some value of x, consider a tangent line to the curve of $y = f(x)$, as shown. As the independent variable x changes by an amount Δx, assume that y changes by an amount Δy. The slope of the tangent line is then $\Delta y/\Delta x$. The first derivative of y versus x can be denoted in several ways. The most common forms are

$$\frac{dy}{dx} \quad \text{or} \quad y'(x) \quad \text{or} \quad \frac{df(x)}{dx} \quad \text{or} \quad f'(x) \tag{6-1}$$

Of the four forms, the first is the most common and will be used most often in our work, but we will have occasion to use the other forms in some cases.

The first derivative is defined as

$$\frac{dy}{dx} = \lim_{\Delta x \to 0} \frac{\Delta y}{\Delta x} \tag{6-2}$$

where "lim" represents the *limit* of the ratio of the changes as they become infinitesimally small. The result is the slope of the original dependent variable at a particular point.

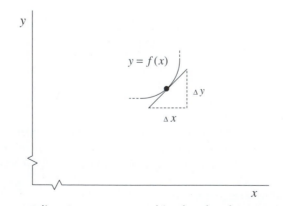

Figure 6-1   Tangent line to a curve used in the development of the derivative

Units for the Derivative

In general, the units for a derivative are the units for the dependent variable y divided by the units for the independent variable x. For example, if y is displacement in meters and x is time in seconds, the units of the derivative would be meters/second (m/s).

For much of this chapter, we will be dealing with "pure" functions involving y and x, so we will not bother with units. When we get to specific practical applications later, however, we will employ units.

Interpretation of the Derivative

The derivative is a measure of the rate of change of a function. If the function is increasing, its derivative is positive, and if it is decreasing, the derivative is negative. The more rapidly the function is changing, the greater the magnitude of the derivative. If the function is not changing at all, meaning that it is a constant value, the derivative is zero.

The next section will illustrate how differentiation of certain types of functions may be accomplished with simple arithmetic.

6-3 DERIVATIVES OF PIECEWISE LINEAR CONTINUOUS FUNCTIONS

In this section, we will show how the first derivative of a piecewise linear function can be determined by simple arithmetic calculations. A *piecewise linear function* is one that is composed of straight-line sections. An example of a piecewise linear function is shown in Figure 6-2(a). This function is also continuous, since there are no sudden jumps. An example of a piecewise linear function that has discontinuities is shown in Figure 6-2(b). It is possible to define derivatives in that type of situation utilizing *impulse functions,* but we will not consider that situation at this time. Thus, our immediate focus is on *piecewise linear functions* that are also *continuous*.

Why does one want to spend a whole section on this special class of functions? There are two good reasons: (1) First, piecewise linear functions occur often in practical engineering systems. Two good examples are electrical circuit waveforms and distributed force systems in structures. (2) Second, evaluation of the derivative (and, in the next chapter, the integral) of this type of function imparts an understanding of the process that is seldom acquired from the myriad of differentiation formulas. Many students who have studied calculus can readily write down formulas, but they often fail to understand the physical process inherent in the operations. Piecewise linear functions provide vehicles for imparting this understanding.

Slope of a Piecewise Linear Segment

A straight-line portion of some function is shown in Figure 6-3. Assuming the normal convention that positive x is to the right, the beginning of the line has the coordinates (x_1, y_1) and the ending of the line has the coordinates (x_2, y_2). The *slope* of a straight-line seg-

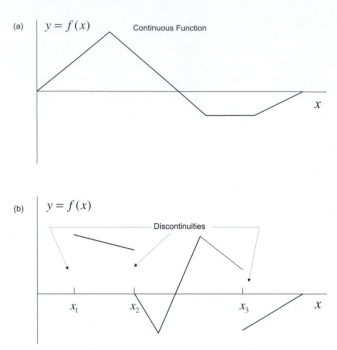

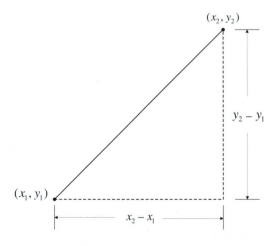

Figure 6-2   Some piecewise linear functions

Figure 6-3   Segment of a straight line leading to a simple computation of the derivative (or slope)

ment is constant over the domain from x_1 to x_2; it is the derivative of y with respect to x. Note that y changes by an amount $y_2 - y_1$ while x changes by an amount $x_2 - x_1$. The slope or derivative in this interval is then given by

$$\frac{dy}{dx} = \text{slope} = \frac{y_2 - y_1}{x_2 - x_1} \qquad (6\text{-}3)$$

This concept was utilized in dealing with straight-line functions in Chapter 5. For the particular segment shown in Figure 6-3, the derivative has a positive value since both numerator and denominator of Equation 6-3 are positive. If the orientation were downward, the change in y would be negative, but since the change in x will be positive in this case, the derivative will be negative.

Differentiation of a piecewise linear continuous function will be illustrated by the example that follows.

Example 6-1

Plot the first derivative of the function y of Figure 6-4(a).

Solution

There are four straight-line segments for the function. The derivative of each segment can be determined by the application of Equation 6-3 based on the beginning and ending points. Actually, any two points along a particular segment could be used since the slope is constant on the segment, but the end points are the most convenient. Let us consider each segment.

$0 \le x \le 4$: In this segment, $x_1 = 0$, $y_1 = 0$, $x_2 = 4$, $y_2 = 12$. The slope or derivative is

$$\frac{dy}{dx} = \frac{y_2 - y_2}{x_2 - x_1} = \frac{12 - 0}{4 - 0} = 3 \qquad (6\text{-}4)$$

$4 \le x \le 6$: In this segment, $x_1 = 4$, $y_1 = 12$, $x_2 = 6$, $y_2 = -12$. The derivative is

$$\frac{dy}{dx} = \frac{-12 - 12}{6 - 4} = -12 \qquad (6\text{-}5)$$

$6 \le x \le 8$: In this segment, y is a constant and does not change. Thus,

$$\frac{dy}{dx} = 0 \qquad (6\text{-}6)$$

$8 \le x \le 10$: In this segment, $x_1 = 8$, $y_1 = -12$, $x_2 = 10$, $y_2 = 0$. The derivative is

$$\frac{dy}{dx} = \frac{0 - (-12)}{10 - 8} = 6 \qquad (6\text{-}7)$$

The derivative is shown in Figure 6-4(b). Carefully compare the curves of the function and its derivative. Note how to determine when the derivative is positive, zero, or negative

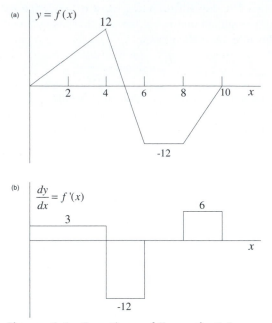

Figure 6-4   Functions of Example 6-1

according to the direction of the change of the function. Even with more complex functions, it is possible with a little practice to predict the relative size of the derivative and whether it is positive, negative, or zero.

For this example. there is an interval in which the function and its derivative are both positive, an interval in which both are negative, and intervals in which they have opposite signs. Differentiation is a process in which changes may occur instantaneously, as noted in this example. If the function changes its slope abruptly, the derivative will change instantaneously. Therefore, the derivative of a piecewise linear continuous function can have finite discontinuities, as is the case in this example. In earlier graphs showing finite discontinuities, no lines were drawn between the discontinuous points. In this example, we have chosen to draw vertical lines between the points.

6-4   DEVELOPMENT OF THE DERIVATIVES OF COMMON FUNCTIONS

There are literally hundreds of mathematical functions whose derivatives have been derived and tabulated. We would expect many readers to own or to have access to a handbook of mathematical functions in which many functions, their derivatives, and integrals (to be considered in Chapter 7) are tabulated. Moreover, as you will see in Chapter 8, MATLAB provides the capability of performing calculus operations both numerically and symbolically.

Our goal in this section is to show the thought process employed in the derivation of these mathematical formulas to aid in the understanding and interpretation of functions employed in tables. To that end, we will actually go through the derivation process for two functions in this section. The cases illustrated are some of the simplest possible functions, but there are points that will be made that can be applied to many other cases.

The Derivative of a Simple Quadratic Form

Consider the simple square-law function

$$y = x^2 \tag{6-8}$$

To derive the formula for the derivative of y with respect to x, assume that x is increased by a small amount, Δx, meaning that x is replaced by $x + \Delta x$. Assume that y changes by an amount Δy, and thus y will be replaced by $y + \Delta y$.

Before going further, note that we said that x *increased* by Δx, but we said that y *changed* by an amount Δy. The normal convention is that x increases, but in general y could either increase or decrease, depending on the function. In this case, it will increase, but ultimately the sign of Δy will automatically take care of whether it is an increase or a decrease.

Substituting the modified variables in Equation 6–8, we obtain

$$y + \Delta y = (x + \Delta x)^2 \tag{6-9}$$

Expansion of the right-hand side yields

$$y + \Delta y = x^2 + 2x\Delta x + (\Delta x)^2 \tag{6-10}$$

The first term on the left (y) is equal to the first term on the right (x^2) from the form of the given function, and these terms may be cancelled. This leaves

$$\Delta y = 2x\Delta x + (\Delta x)^2 \tag{6-11}$$

We now divide both sides by Δx, which results in

$$\frac{\Delta y}{\Delta x} = 2x + \Delta x \tag{6-12}$$

Finally, we impose the definition of the derivative as given by Equation 6–2. When this operation is applied, the single Δx standing alone on the right becomes zero. We have

$$\frac{dy}{dx} = \lim_{\Delta x \to 0} \frac{\Delta y}{\Delta x} = 2x \tag{6-13}$$

Thus, we have derived the derivative of $y = x^2$ with respect to x and have determined that it is $2x$. Most readers will likely have known the result, but most have likely forgotten the derivation process.

Some students get confused by the fact that as Δx alone approaches zero, it vanishes, but as $\Delta y / \Delta x$ approaches zero, it becomes the non-zero derivative function. The answer lies in the fact that the ratio of two very small quantities can still be a sizable quantity, even if each alone is approaching zero. This concept will be illustrated in Example 6-2.

While the function considered was quite simple in form, the process employed is more or less the same that is used in the derivations of most derivatives in which the dependent variable is directly expressed in terms of the independent variable. More complex functions differ only with respect to the manipulations involved to arrive at the final step, as given by Equation 6-13. Standard calculus books are loaded with these derivations.

Chain Rule for Differentiation

There is one more type of differentiation that we need to consider, and we will continue to use the simple quadratic as a means of illustrating the process. Assume that we have a variable u that is a function of the independent variable x. The notation $u = u(x)$ is convenient for this purpose. Then assume that y is given as a function of u. In some cases, we can merely substitute $u(x)$ in the expression for y and obtain y directly as a function of x. In other cases, that may be inconvenient or even impossible. In such cases, the use of the *chain rule* for differentiation is the way to determine the solution.

Again, we will employ a simple second-degree function to illustrate the process, but in this case it will be

$$y = u^2 \tag{6-14}$$

where $u = u(x)$. However, the quantity desired is the derivative of y with respect to x. The process begins in the same manner as the preceding development, except that $u + \Delta u$ is substituted on the right and $y + \Delta y$ is substituted on the left. After the substitution and cancellation of the original function on both sides, we obtain

$$\Delta y = 2u\Delta u + (\Delta u)^2 \tag{6-15}$$

Division of both sides by Δx yields

$$\frac{\Delta y}{\Delta x} = 2u\frac{\Delta u}{\Delta x} + \Delta u\left(\frac{\Delta u}{\Delta x}\right) = (2u + \Delta u)\left(\frac{\Delta u}{\Delta x}\right) \tag{6-16}$$

The expression on the right is a little more complex than for the preceding development, but it is quite manageable. When the limit as $\Delta x \to 0$ is formed, the left-hand side becomes dy/dx as before. The factor on the far right becomes the derivative of u with respect to x; that is, du/dx. Finally, the term $\Delta u \to 0$, and we have

$$\frac{dy}{dx} = 2u\frac{du}{dx} \tag{6-17}$$

The first part of the product on the right has the same form as for the derivative of x^2 except that it is $2u$ instead of $2x$. However, since we desire the derivative with respect to x, we must multiply the first factor by the derivative of u with respect to x.

Generalizing the Concept Somewhat

Let us now carry the process over to an arbitrary function, $y = f(u)$, in which $u = u(x)$. The derivative of y with respect to x can be expressed as

$$\frac{dy}{dx} = \frac{df(u)}{du}\frac{du}{dx} = f'(u)\frac{du}{dx} \tag{6-18}$$

where

$$f'(u) = \frac{df(u)}{du} \tag{6-19}$$

Let's see if we can make a statement to clarify the process. Consider u as the "first variable" and x as the "second variable." The preceding development says that the derivative of a dependent function y with respect to a second variable is the product of the derivative of y with respect to the first variable (u in this case) times the derivative of the first variable with respect to the second variable (x in this case).

The concept can be expanded into any number of related functional variables, and the resulting product can be loosely associated as a "chain," which explains the name of the process. For our purposes, two steps should be sufficient at this point.

Example 6-2

In this example, a derivative will be approximated by the delta process considered in this section. Based on the function $y = x^2$, develop a numerical approximation for the derivative at $x = 1$ by assuming $\Delta x = 0.01$ and determine the corresponding change Δy utilizing either MATLAB or a calculator. Then approximate the derivative at the point by the ratio $\Delta y / \Delta x$.

Solution

For $x = 1$, the value of y is

$$y(1) = (1)^2 = 1 \tag{6-20}$$

If $\Delta x = 0.01$, we need to evaluate y at $x = 1.01$; it is

$$y(1.01) = (1.01)^2 = 1.0201 \tag{6-21}$$

The value of Δy is

$$\Delta y = 1.0201 - 1 = 0.0201 \tag{6-22}$$

The approximate derivative is then

$$\frac{dy}{dx} \approx \frac{\Delta y}{\Delta x} = \frac{0.0201}{0.01} = 2.01 \tag{6-23}$$

This value compares very favorably with the exact value of 2 based on the true derivative as determined by the formula derived in this section.

Example 6-3

The derivative of the function sin u with respect to u is given by

$$\frac{d}{du}(\sin u) = \cos u \tag{6-24}$$

Use the chain rule to determine the derivative of y with respect to x, where

$$y = 4 \sin x^2 \tag{6-25}$$

Solution

In this case, we can identify u as

$$u = x^2 \tag{6-26}$$

The derivative of u with respect to x has the form developed early in this section; it is

$$\frac{du}{dx} = 2x \tag{6-27}$$

The derivative of y with respect to x can be expressed as

$$\frac{dy}{dx} = f'(u)\frac{du}{dx} = \frac{dy}{du}\frac{du}{dx} = 4(\cos u)(2x) = 8x\cos x^2 \tag{6-28}$$

6-5   TABULATION OF DERIVATIVES

In this section, the tabulation of some common derivative functions and operations will be presented. Some explanations of the meanings will be provided, and their use will be illustrated with examples at the end of the section.

First, refer to Table 6-1, at the end of the chapter. Note that each of the derivatives and operations is provided with an identifier ranging from D-1 to D-18. Note that the quantities a and n are constants and u and v are functions of x; that is, $u = u(x)$ and $v = v(x)$. The left-hand column provides the basic form of any function $f(x)$, and the next column provides the form of the derivative $f'(x)$.

D-1 and D-2

D-1 and D-2 are the "obvious" operations that are stated at the outset. D-1 states that the derivative of a constant times a function is the constant times the derivative of the function.

D-2 states that the derivative of a sum is the sum of the derivatives. Not much more can be said about these.

D-3

D-3 was stated in the last section and is an important operation that should be committed to memory for anyone using calculus. To reiterate the discussion of the last section, it states that for any $f(u)$ with $u = u(x)$, the derivative of $f(u)$ with respect to x is obtained by forming the derivative of $f(u)$ with respect to u and multiplying that result by the derivative of u with respect to x. This operation arises over and over in solving calculus problems and will be seen in most of the derivative operations that follow. We will refer to the quantity du/dx in subsequent discussions as the "chain factor."

D-4

Undoubtedly, this is the simplest of all derivatives. The derivative of a constant is 0, since no change is taking place.

D-5

D-5 is a generalization of the first derivation performed in Section 6-4 and reduces to that case when $n = 2$. In general, the derivative of x to a higher power yields the next lower power with a constant in front. The one exception is when $n = 0$. In this case, the function reduces to a constant and the derivative is zero, as already noted in D-4.

D-6

D-6 was also derived for $n = 2$ in Section 6-4 and differs from D-5 only in that the chain factor is required.

D-7

D-7 arises very often and should be carefully noted. While extensive tables provide many combinations of product functions, this operation allows the derivative of a product to be expressed in terms of the derivatives of the individual factors. Stated in words, the derivative of a product is the first function times the derivative of the second function plus the second function times the derivative of the first function.

D-8

D-is very similar to D-7 but involves the ratio of two functions. It is a bit more clumsy to carry out but arises about as often as D-7. Stated in words, the derivative of the ratio of two functions is the denominator function times the derivative of the numerator function minus the numerator function times the derivative of the denominator function, all of which is divided by the denominator function squared. (If you would like some humor to help in remembering this one, the author's daughter was provided with the following little statement from her college calculus teacher: "ho-de-hi minus hi-de-ho over ho squared.")

D-9

The exponential function e^u is remarkable in that its derivative is also the exponential function. Of course, the chain factor is required. As a special case, the derivative of e^x with respect to x is also e^x.

D-10

This situation is an extension of D-9 but with e replaced by an arbitrary value, a. The resulting derivative is a little more involved than D-9.

D-11

D-11 provides the form of the derivative of the natural logarithm of a variable. As a special case, the derivative of $\ln x$ with respect to x is $1/x$.

D-12

This situation is an extension of D-11 with the logarithm to the base e replaced by the logarithm with respect to an arbitrary base, a.

D-13 and D-14

D-13 and D-14 arise in many engineering applications. Excluding the chain factor, the derivative of the sine function is the cosine function and the derivative of the cosine function is the negative of the sine function. To remember which one has the negative sign, visualize the sine and cosine functions and their slopes.

D-15

D-15 provides the derivative of the tangent function, which is given by the secant function squared times the chain factor. Recall that $\sec u = 1/\cos u$.

D-16 through D-18

The derivatives tabulated for these three functions are those of the inverse sine, the inverse cosine, and the inverse tangent. Please note that inverse functions are *not the reciprocals* of the respective functions. For example, $\sin^{-1} u$ refers to the angle in which the sine of that angle is u.

Example 6-4

A function is given by

$$y = x^2 \sin x \qquad (6\text{-}29)$$

Determine $\dfrac{dy}{dx}$.

Solution

The function $x^2 \sin x$ is definitely not in Table 6-1. It is likely that the function can be found in a more extensive set of derivative tables. However, this is where we practice the use of D-7 and learn to determine the derivative more easily than by looking it up in a handbook.

For this first product situation, we will belabor the process somewhat as an aid in learning how to deal with more complex functions. We need to identify both u and v to utilize D-7. The obvious choice is to let $u = x^2$ and $v = \sin x$. Applying D-7, we have

$$\frac{dy}{dx} = u\frac{dv}{dx} + v\frac{du}{dx} = x^2 \frac{d(\sin x)}{dx} + \sin x \frac{d(x^2)}{dx} \tag{6-30}$$

The two derivatives required are those of D-13 and D-5, with $n = 2$ for the latter case. Applying these, we obtain

$$\frac{dy}{dx} = x^2 \cos x + \sin x(2x) = x^2 \cos x + 2x \sin x \tag{6-31}$$

Example 6-5

A function is given by

$$y = \frac{\sin x}{x} \tag{6-32}$$

Determine $\dfrac{dy}{dx}$.

Solution

Again, this is not a function given in Table 6-1. However, this function can be considered in the form of D-8, with $u = \sin x$ and $v = x$. The operation is this case is a bit more involved and is given by

$$\frac{dy}{dx} = \frac{v\dfrac{du}{dx} - u\dfrac{dv}{dx}}{v^2} = \frac{x\dfrac{d(\sin x)}{dx} - \sin x \dfrac{d(x)}{dx}}{x^2} = \frac{x\cos x - \sin x}{x^2} \tag{6-33}$$

Example 6-6

A function is given by

$$y = e^{-\frac{x^2}{2}} \tag{6-34}$$

Determine $\dfrac{dy}{dx}$.

Solution

In this case, the function can be consider as a case of D-9 provided that we define u as

$$u = -\frac{x^2}{2} \tag{6-35}$$

We will need du/dx, so we might as well form it from Equation 6-35. In the process, we will also be making use of the "obvious" form of D-1. We have

$$\frac{du}{dx} = \frac{d\left(-\dfrac{x^2}{2}\right)}{dx} = \left(-\frac{1}{2}\right)(2x) = -x \tag{6-36}$$

From D-9, we have

$$\tag{6-37} \frac{dy}{dx} = e^{-\frac{x^2}{2}}(-x) = -xe^{-\frac{x^2}{2}}$$

Using Table 6-1 to Obtain Derivatives

This exercise and the previous two demonstrate that the derivatives of many functions can be obtained from the relative few provided in Table 6-1 by utilizing various combinations and operations.

6-6   HIGHER-ORDER DERIVATIVES

Thus far, we have concentrated only on the first derivative of a function. Provided that they exist in a mathematical sense, however, one can continue with the differentiation of derivatives for an arbitrary number of times. Assuming a function of one independent variable, the derivative of the first derivative is called the second derivative and the derivative of the second derivative is called the third derivative, and so on.

Notation for Successive Derivatives

We begin again with a function

$$y = f(x) \tag{6-38}$$

As we have seen many times so far in the chapter, the first derivative can be denoted in either of the forms

$$\frac{dy}{dx} = f'(x) = \frac{df(x)}{dx} \tag{6-39}$$

The second derivative is denoted by

$$\frac{d^2y}{dx^2} = f''(x) = \frac{d^2f(x)}{dx^2} = \frac{d}{dx}\left(\frac{dy}{dx}\right) \tag{6-40}$$

Before proceeding further, a few comments are in order because the pattern for higher-order derivatives can be extrapolated from Equation 6-40. First, note that the number 2 identifying the derivative as a second-order form appears as a superscript for d in the numerator and a superscript for x in the denominator. For the second form, note that there are two primes following the f. After two or three primes, that particular form generally employs a number such as $f^{(4)}(x)$ for the fourth derivative, and so on. The third form follows the same logic as the first form, with the superscript 2 in two places. The last form is new to us, but the logic says that the second derivative is the first derivative with respect to x of the first derivative with respect to x.

The third derivative is denoted by

$$\frac{d^3y}{dx^3} = f^{(3)}(x) = \frac{d^3f(x)}{dx^3} = \frac{d}{dx}\left(\frac{d^2y}{dx^2}\right) \tag{6-41}$$

Finally, the nth derivative can be expressed as

$$\frac{d^ny}{dx^n} = f^{(n)}(x) = \frac{d^nf(x)}{dx^n} = \frac{d}{dx}\left(\frac{d^{n-1}y}{dx^{n-1}}\right) \tag{6-42}$$

Example 6-7

Determine the second derivative of the function

$$y = 5\sin 4x \tag{6-43}$$

Solution

First, we form the first derivative using D-13 of Table 6-1. Note that we can consider that $u = 4x$.

$$\frac{dy}{dx} = 5(\cos 4x) \cdot \frac{d}{dx}(4x) = 20\cos 4x \tag{6-44}$$

Next, we take the derivative of the first derivative to form the second derivative. In this case, we use D-14, but u remains the same.

$$\frac{d^2y}{dx^2} = 20(-\sin 4x) \cdot \frac{d}{dx}(4x) = -80\sin 4x \tag{6-45}$$

An interesting outcome is that the second derivative of a sine function is also a sine function, but the sign has changed.

6-7 APPLICATIONS OF DIFFERENTIAL CALCULUS

There are numerous applications of differential calculus in virtually all walks of life. We will sample a few cases in this section.

Change of Notation

This discussion may seem irrelevant and mundane to the most learned of our readers, but the author has worked with enough students over the years to believe that the discussion is worthwhile. Many students seem to comprehend the concepts of calculus in a mathematics environment where the basic variables are x and y, but a significant number seem to reach a mental block when the variables suddenly become quantities such as force, displacement, time, voltage, current, or pressure. In such cases students have difficulty transferring the mathematical processes to real physical variables.

We can not offer a panacea to completely cure the problem, but we can offer some suggestions. When you see a physical problem with completely different variables than in calculus, ask the following questions: (1) What is the independent variable? (2) What is the dependent variable? If it helps, redefine any equations involved by momentarily replacing the independent variable by x and the dependent variable by y. In going to any table to check for a derivative, this logic is certainly appropriate. However, the ultimate goal is to understand the process and adapt any of the derivative (and later, integral) formulas to the variables involved. The major way that this is achieved is by practice and more practice. In short, if you work lots of problems, the understanding will eventually come.

Maxima and Minima

One of the most important applications of differential calculus is in determining the maximum and/or the minimum points on the curve of a function. This process arises frequently in optimizing an engineering application. We will begin with the basic variables y and x and consider other variables in the examples.

Consider the function of Figure 6-5. Over the domain and range shown, there is one *maximum,* on the right, and a *minimum,* on the left. Outside of the area provided, there is no way of knowing whether there are other maxima and minima, but in most practical engineering problems, it is usually possible to identify what one might call *local maxima and/or local minima* in the region of primary interest.

Maxima and minima on a curve are characterized by one important property: *The slope or derivative is zero at either point.* Setting the derivative equal to zero and solving the resulting equation will identify points at which the derivative is zero. However, the first

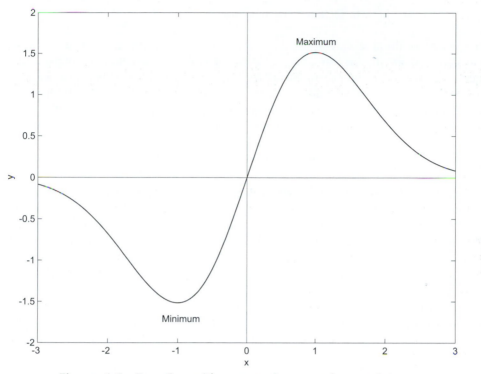

Figure 6-5 Function with one maximum and one minimum

step does not identify whether a given point is a maximum or a minimum. There are several ways to accomplish this; the overall procedure is outlined in the following steps.

Determining Maximum and Minimum Points

The procedure is as follows:

1. Determine the derivative of the function; that is, $\dfrac{dy}{dx}$

2. Set the derivative equal to 0; that is, $\dfrac{dy}{dx} = 0$, and solve for the values of x that satisfy this equation.

3. Determine the second derivative of the function, that is, $\dfrac{d^2 y}{dx^2}$. The following criteria apply:

   a. If $\dfrac{d^2 y}{dx^2} > 0$, the point is a minimum.

    b.   If $\dfrac{d^2 y}{dx^2} < 0$, the point is a maximum.

In some cases, the procedure of step 3 does not work due to the difficulty of taking the second derivative, or in some cases, the second derivative may be zero itself. A couple of practical pointers to assist in this process are appropriate. First, there are many practical problems in which one can determine from the outset whether a given point will represent a maximum or a minimum, and in such cases, it may be unnecessary to form the second derivative. However, whenever it is impractical and when nothing else works, a straightforward way to test the function is to evaluate it at two points: one just below the questionable maximum or minimum point and one just above this point. If the function is greater than at the questionable point, then the function is a minimum at the point. If the function is less than at the questionable point, then the function is a maximum at the point.

Displacement, Velocity, and Acceleration

Three physical variables that appear in a large number of applications, particularly in mechanical systems, are displacement, velocity, and acceleration. We will use these variables at a number of places in the text to illustrate various important mathematical properties. Aside from their practical value in describing many real systems, these variables are among the easiest for students from different disciplines to relate to in the "real world" For example, "My car accelerates from 0 to 60 mph in 7.6 s and yours takes 10.3 s."

Displacement

Displacement represents the position of an object relative to some reference point. Both x and y are used to represent displacement, but since displacement is usually considered as a dependent variable and since we have been using x to represent the independent variable, we will use y for displacement initially. The International System of Units (SI) unit for **displacement** is the **meter** (m). Usually, our interest is in displacement as a function of time. To that end, the independent variable will be considered as **time** in **seconds** (s) and denoted as t.

Velocity

Velocity will be denoted as v. It is the rate of change of displacement with respect to time, that is,

$$v = \frac{dy}{dt} \tag{6-46}$$

The basic SI units for velocity are meters/second (m/s).

Acceleration

Acceleration will be denoted as a. It is the rate of change of velocity with respect to time, that is,

$$a = \frac{dv}{dt} = \frac{d^2 y}{dt^2} \qquad (6\text{-}47)$$

The basic SI units for acceleration are meters per second per second, or meters/second^2 (m/s^2).

Example 6-8

A function is given by

$$y = f(x) = x^3 - 6x^2 + 9x + 2 \qquad (6\text{-}48)$$

Determine the finite values of x at which any local maxima or minima occur and determine the corresponding values of y.

Solution

The term "finite" is used here for a particular reason. An inspection of Equation 6-48 reveals that y increases without bound as x increases without bound and y decreases without bound as x decreases without bound. Mathematical forms to express these facts are $y \to \infty$ as $x \to \infty$ and $y \to -\infty$ as $x \to -\infty$. Alternately, $|y| \to \infty$ as $|x| \to \infty$, where the vertical lines represent absolute values. However, these limiting cases are not the points of interest. Rather, possible finite (or local) values represent our focus.

All of the terms in Equation 6-48 can be differentiated with D-5 and D-4 of Table 6-1. We have

$$\frac{dy}{dx} = 3x^2 - 12x + 9 \qquad (6\text{-}49)$$

Next we set the derivative to 0 and determine the values of x that satisfy it.

$$3x^2 - 12x + 9 = 0 \qquad (6\text{-}50)$$

By the use of the quadratic formula or by factoring, the roots are determined as

$$x = 1 \text{ and } x = 3 \qquad (6\text{-}51)$$

Thus, there are two finite values of x that need to be investigated as possible maxima or minima.

We will next form the second derivative. We have

$$\frac{d^2 y}{dx^2} = 6x - 12 \qquad (6\text{-}52)$$

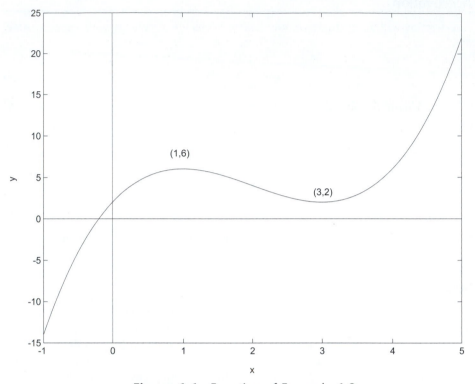

Figure 6-6   Function of Example 6-8

For $x = 1$, $f''(1) = 6(1) - 12 = -6 < 0$. Therefore, $x = 1$ is a *maximum* and the value y_{max} is

$$y_{max} = f(1) = (1)^3 - 6(1) + 9(1) + 2 = 6 \qquad (6\text{-}53)$$

For $x = 3$, $f''(3) = 6(3) - 12 = 6 > 0$. Therefore, $x = 3$ is a *minimum* and the value x_{min} is

$$y_{min} = f(3) = (3)^3 - 6(3)^2 + 9(3) + 2 = 2 \qquad (6\text{-}54)$$

The function is shown in Figure 6-6.

Example 6-9

A farmer needs to build a fence to enclose a **fixed acreage**. Based on certain practical considerations, the shape of the fence must be rectangular in form. Assume that the width is w and the length is l. Assume that there is enough land available that there is no restriction on the width and length. Determine a relationship between the width and length that **minimizes** the length of fencing that must be purchased.

Solution

Some perceptive readers may already have guessed the solution, but let us apply some calculus to determine the result. The quantity that is to be minimized is the perimeter P, which is the distance around the fence. Since a rectangle has four sides, the total perimeter is given by

$$P = 2w + 2l \tag{6-55}$$

There are two variables on the right-hand side; that is, w and l. However, a constant acreage means that the area A is fixed, which provides a second relationship, as given by

$$A = wl \tag{6-56}$$

Either w or l can be eliminated. We will arbitrarily choose to eliminate l. Solving for l in 6-56, we obtain

$$l = \frac{A}{w} \tag{6-57}$$

Substitution of this value in Equation 6-55 results in

$$P = 2w + \frac{2A}{w} \tag{6-58}$$

Since A is a constant, we now have the dependent variable P expressed in terms of the independent variable w. Thinking back to the discussion made earlier about variables, P is like y and w is like x.

Next, we determine the derivative of P with respect to w and set it equal to 0. We have

$$\frac{dP}{dw} = 2 - \frac{2A}{w^2} = 0 \tag{6-59}$$

The solution to this equation is

$$w = \sqrt{A} \tag{6-60}$$

The negative square root is rejected since it does not correspond to the real physical situation. It can be readily shown that the second derivative is positive at the critical point, so the function is a *minimum* at the given point. The corresponding value of l is determined from Equation 6-56 and is

$$l = \frac{A}{\sqrt{A}} = \sqrt{A} \tag{6-61}$$

You may have guessed it: the shape of a rectangle of fixed area that minimizes the perimeter is a *square*.

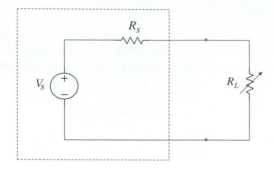

Figure 6-7   Electrical circuit of Example 6-9

Example 6-10

A classical problem in electrical circuit analysis is that of the *Maximum Power Transfer Theorem*. Consider the electrical circuit shown in Figure 6-7. A voltage of value V_s has an internal resistance of value R_s. Assume that this source resistance is part of the source and impossible to change. Assume now that an external load resistance R_L is connected across the output terminals. What value of R_L will result in maximum power being delivered to R_L?

Solution

Since readers of this text may have a diversity of backgrounds, the equation for the load power will be provided, but electrical students should be able to verify it. The power P_L delivered to the load resistance R_L is given by

$$P_L = \frac{R_L V_s^2}{\left(R_s + R_L\right)^2} \tag{6-62}$$

Assuming that R_s and V_s are fixed, the independent variable is R_L and the dependent variable is P_L. The obvious approach is to form dP_L/dR_L and set it equal to 0. While this can be done, however, it is messy because it is a u/v form and the result turns out to be somewhat unwieldy. This leads us to a process that is not often emphasized in basic calculus books and can be summarized as follows:

1. If a function y has a *maximum* at some value of x, the function $1/y$ has a *minimum* at the same point.
2. If a function y has a *minimum* at some value of x, the function $1/y$ has a *maximum* at the same point.

What the preceding statements tell us is that we can work with the reciprocal of a function when it is easier to manipulate. The only qualifier is that a maximum becomes a minimum, and vice versa.

Let us define the reciprocal of Equation 6-62 as y. It is

$$y = \frac{(R_s + R_L)^2}{R_L V_s^2} = \frac{1}{V_s^2}\frac{(R_s + R_L)^2}{R_L} = \frac{1}{V_s^2}\frac{(R_s^2 + 2R_s R_L + R_L^2)}{R_L} = \frac{1}{V_s^2}\left(\frac{R_s^2}{R_L} + 2R_s + R_L\right) \quad (6\text{-}63)$$

The derivative of y with respect to R_L is

$$\frac{dy}{dR_L} = \frac{1}{V_s^2}\left(\frac{-R_s^2}{R_L^2} + 0 + 1\right) = 0 \quad (6\text{-}64)$$

Since the resistance cannot be negative in this situation, the solution is

$$R_L = R_s \quad (6\text{-}65)$$

In theory, we would need to check whether the point is a maximum or a minimum. Returning to the original expression for load power in Equation 6-62, it can be seen that as $R_L \to 0$, $P_L \to 0$. Likewise, as $R_L \to \infty$ (an open circuit), $P_L \to 0$. Since there is only one finite point in between these extreme limits at which the slope is zero, and since power dissipated in a resistance is always positive, the conclusion is that the point represents a maximum for power delivered to the load. (Of course, the reciprocal function y has a minimum at this point.) The conclusion is that when a source has a fixed internal resistance, maximum power is delivered to an external load when the load resistance is equal to the internal resistance of the source.

Example 6-11

The displacement in meters of a certain object undergoing acceleration is given by

$$y = 50t + 100(1 - e^{-2t}) \quad (6\text{-}66)$$

Determine (a) the velocity and (b) the acceleration.

Solution

a. The velocity is the first derivative of displacement with respect to time.

$$v = \frac{dy}{dt} = 50 - 100e^{-2t}(-2) = 50 + 200e^{-2t} \quad \text{m/s} \quad (6\text{-}67)$$

b. The acceleration is the first derivative of velocity with respect to time (or the second derivative of displacement).

$$a = \frac{dv}{dt} = 200e^{-2t}(-2) = -400e^{-2t} \quad \text{m/s}^2 \quad (6\text{-}68)$$

Let us see if we can interpret these results. Going backward, the acceleration is always negative, which means that the velocity decreases with respect to time. From the expression for velocity, that fact is confirmed, since $v(0) = 50 + 200e^{-0} = 250$ m/s and $v(\infty) = 50$ m/s. The displacement contains a term $50t$, which accounts for the limiting value of velocity, but the other term approaches a constant value of 100 m as $t \to \infty$.

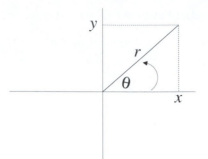

Figure 6-8   Rotating rod for Example 6-12

Example 6-12

Consider the rod with length r shown in Figure 6-8 and assume that it is rotating in a counterclockwise direction with a constant angular velocity ω. Assume that the time $t = 0$ is established at the point where it passes the positive x-axis. Determine the displacement from the origin, the velocity, and the acceleration of (a) the x-component and (b) the y-component.

Solution

Since the angular velocity is constant, the angle θ at any time t is given by

$$\theta = \omega t \tag{6-69}$$

a. The x-component of displacement is the projection of the rod on the horizontal axis and is given by

$$x = r\cos\theta = r\cos\omega t \tag{6-70}$$

The velocity v_x in the x-direction is given by

$$v_x = \frac{dx}{dt} = -r\sin\omega t(\omega) = -r\omega\sin\omega t \tag{6-71}$$

The acceleration a_x in the x-direction is

$$a_x = \frac{dv_x}{dt} = -r\omega\cos\omega t(\omega) = -r\omega^2\cos\omega t \tag{6-72}$$

b. The y-component of displacement is the projection of the rod on the vertical axis and is

$$y = r\sin\theta = r\sin\omega t \tag{6-73}$$

The velocity v_y is

$$v_y = \frac{dy}{dt} = r\cos\omega t(\omega) = r\omega\cos\omega t \tag{6-74}$$

The acceleration in the y-direction is

$$a_y = \frac{dv_y}{dt} = -r\omega\sin\omega t(\omega) = -r\omega^2\sin\omega t \tag{6-75}$$

The outcome of this problem is very interesting and should be noted. While the rod is rotating at a constant angular velocity, the components along the two axes experience both time-varying velocities and time-varying accelerations. Moreover, these functions are sinusoidal forms. Students dealing with mechanical systems and machines will likely see these relationships and their variations in various applications.

6-8 "DELTA" ALGEBRA FOR SMALL CHANGES

We have seen that the derivative is obtained by assuming a small change in the independent variable, determining the small change in the dependent variable, and taking the limit as the changes become infinitesimally small. This leads to one of the useful applications of differential calculus.

Differentials

First, we will define the classical notion of a differential. Let dy represent a very small change in the dependent variable y. We can express this infinitesimal change by forming the derivative at a point and then multiplying by the small change in x, that is, dx. This leads to the equation

$$dy = \left(\frac{dy}{dx}\right)dx \qquad (6\text{-}76)$$

The quantities dy and dx are called *differentials*. Stated in words, the differential dy is equal to the derivative dy/dx times the differential dx.

Small Changes

Instead of working with the infinitesimally small quantities dx and dy, let us take on a more realistic approach and simply say that the changes are "small." To that end we replace dx by Δx and dy by Δy. This leads to the following approximation

$$\Delta y \approx \frac{dy}{dx}\Delta x \qquad (6\text{-}77)$$

This formula states that a small change in a dependent variable may be determined by multiplying a small change in the independent variable by the derivative or slope evaluated at the point in question. As simple as this statement appears to be, it is a very powerful way of dealing with parameter variations in practical systems.

Example 6-13

The frequency f of oscillation of a certain electronic oscillator circuit is given by

$$f = \frac{1}{\sqrt{LC}} \qquad (6\text{-}78)$$

where L is the inductance in henries (H) and C is the capacitance in farads (F). Using the differential process, determine the approximate percentage change in the frequency of oscillation if the capacitance increases by 10%.

Solution

Although we have no actual numbers for the circuit parameters, we can deal with the changes on a per-unit basis, as will be demonstrated. First, we need to identify the variables. L is assumed to be a constant, C is the independent variable, and f is the dependent variable. If it helps the reader in understanding the process, feel free to use x for C and y for f. We will not go that far here, but we will rewrite the equation in the following form:

$$f = \frac{1}{\sqrt{L}}\left(C^{-\frac{1}{2}} \right) \tag{6-79}$$

This odd rearrangement allows us to better see the power relationship between C and f. Next, we form the derivative of f with respect to C.

$$\frac{df}{dC} = \frac{1}{\sqrt{L}}\left(-\frac{1}{2} C^{-3/2} \right) = -\frac{1}{2} \frac{1}{\sqrt{L}} \frac{1}{C^{3/2}} \tag{6-80}$$

Now a little "twiddling" is in order. Since $C^{3/2} = C \times C^{1/2} = C\sqrt{C}$, the latter factor may be regrouped with $\sqrt{L}$, allowing us to write

$$\frac{df}{dC} = -\frac{1}{2} \frac{1}{\sqrt{LC}} \frac{1}{C} = -\frac{1}{2} \frac{f}{C} \tag{6-81}$$

where the definition of Equation 6-78 was employed. The change in f can now be approximated as

$$\Delta f \approx \frac{df}{dC} \Delta C = -\frac{1}{2} \frac{f}{C} \Delta C \tag{6-82}$$

Rearranging, we have

$$\frac{\Delta f}{f} = -\frac{1}{2}\left(\frac{\Delta C}{C} \right) \tag{6-83}$$

The quantity in parentheses is the per-unit change in capacitance, which for 10% would correspond to 0.1. The quantity on the left is the per-unit change in frequency. Therefore,

$$\frac{\Delta f}{f} = -\frac{1}{2}(0.1) = -0.05 \tag{6-84}$$

The negative sign means that the change in frequency is in the opposite direction to the change in capacitance. The conclusion is that a 10% *increase* in capacitance will result in about a 5% *decrease* in the frequency.

GENERAL PROBLEMS

6-1. Plot the first derivative of the function *y* below.

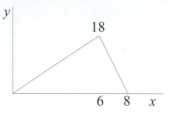

6-2. Plot the first derivative of the function *y* below.

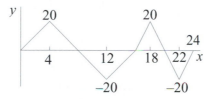

6-3. Plot the first derivative of the function *y* below.

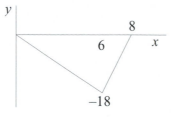

6-4. Plot the first derivative of the function *y* below.

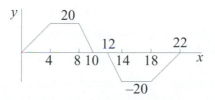

6-5. Approximate the derivative of $y = x^3$ at $x = 2$ by assuming $\Delta x = 0.001$ and determining the corresponding change, Δy. Compare the approximate value with the exact value.

6-6. Approximate the derivative of $y = \sqrt{x}$ at $x = 4$ by assuming $\Delta x = 0.01$ and determining the corresponding change, Δy. Compare the approximate value with the exact value.

In Problems 6-7 through 6-14 , you may use Table 6-1 (at the end of the chapter) as necessary, along with required modifications. In each case, determine dy/dx.

6-7. $y = x \sin x$

6-8. $y = x \cos x$

6-9. $y = x \cos(2x^2)$

6-10. $y = x \sin(2x^2)$

6-11. $y = \dfrac{\sin x}{x^2}$

6-12. $y = \dfrac{\cos x}{x}$

6-13. $y = xe^{-x}$

6-14. $y = xe^{-x^2}$

6-15. For the function of Problem 6-7, determine $\dfrac{d^2 y}{dx^2}$.

6-16. For the function of Problem 6-8, determine $\dfrac{d^2 y}{dx^2}$.

6-17. A function is given by

$$y = 2x^3 + 3x^2 - 12x + 5$$

Determine the finite values of x at which any local maxima or minima occur and determine the corresponding values of y.

6-18. A function is given by

$$y = -2x^3 - 3x^2 + 36x - 12$$

Determine the finite values of x at which any local maxima or minima occur and determine the corresponding values of y.

6-19. A farmer needs to build a fence to enclose a fixed acreage, but because of a very long existing fence at the back of the property, only three new sides will be needed to form a rectangle. Let l represent the one side added that is parallel to the existing back, and let w represent the width, which will require two sides. Determine the relationship between l and w that minimizes the length of new fence required.

6-20. Assume that we are back at the farmer's need to provide four sides, as indicated in Example 6-9. However, because of certain aesthetic requirements, the fence along one side on the front costs twice as much per linear unit of measurement as that on the other three sides. Determine the relationship between l and w that minimizes the cost.

6-21. The circuit of Figure P6-21 contains an ac source and an ideal transformer connected between a source with internal resistance on the left and a fixed load on the right. The so-called turns ratio of the transformer is denoted by n. This is the ratio of turns on the secondary side (right side) of the transformer to turns on the primary (left side). It can be shown by circuit analysis that the voltage gain, that is, $A = V_2/V_s$, is given by

$$A = \frac{nR_L}{R_L + n^2 R_s}$$

Determine the value of n that maximizes the voltage gain and determine the maximum voltage gain.

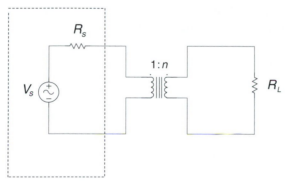

6-22. Consider Example 6-10 dealing with the maximum power transfer theorem for the circuit shown in Figure 6-7. However, assume that R_L is fixed and R_s is variable. Determine the value of R_s that maximizes the power delivered to R_L. *Hint:* Application of calculus in this case leads to a negative value of resistance, which is impossible (except in some specialized electronic circuits). Instead, observe the value of the load power with the understanding that none of the quantities can be negative and use reason to determine the value of source resistance that maximizes the load power.

6-23. The displacement of a certain object undergoing acceleration is given by

$$y = 10t^2 + 100e^{-t}$$

Determine (a) the velocity and (b) the acceleration.

6-24. The displacement of a certain object undergoing acceleration is given by

$$y = -10t + 100(1 - e^{-2t})$$

Determine (a) the velocity and (b) the acceleration.

6-25. Consider the rod of Example 6-12 and Figure 6-8 but assume that it rotates in a **clockwise** direction, starting from the same point at $t = 0$. Since a positive angle is measured in a counterclockwise direction, the angle is now given by $\theta = -\omega t$. Perform the computations of Example 6-12 based on this assumption.

6-26. Consider the rod of Example 6-12 and Figure 6-8 but assume the counterclockwise direction of that example. However, assume that at $t = 0$, the rod starts from a positive vertical position instead of a horizontal position and consider the angle to be 0 at the starting point. Perform the computations of Example 6-12 based on this assumption.

6-27. The output voltage v_2 of a certain square-law device is related to the input voltage v_1 by the relationship $v_2 = Kv_1^2$. Assume that v_1 increases by 10%. Then, using the differential process, compute the approximate per-unit change in the output, that is, $\Delta v_2 / v_2$.

6-28. The output, y, of a certain mechanical device is related to the input, x, by the relationship $y = K\sqrt{x}$. Assume that x increases by 10%. Then, using the differential process, compute the approximate per-unit change in the output, that is, $\Delta y/y$.

Note: Since the total emphasis in this chapter was analytical in nature, MATLAB problems dealing with differential calculus will be delayed to the end of Chapter 8.

Table 6-1 Common Derivatives of Functions and Operations

$f(x)$	$f'(x)$	Derivative Number
$af(x)$	$af'(x)$	D-1
$u(x)+v(x)$	$u'(x)+v'(x)$	D-2
$f(u)$	$f'(u)\dfrac{du}{dx}=\dfrac{df(u)}{du}\dfrac{du}{dx}$	D-3
a	0	D-4
$x^n \qquad (n\neq 0)$	nx^{n-1}	D-5
$u^n \qquad (n\neq 0)$	$nu^{n-1}\dfrac{du}{dx}$	D-6
uv	$u\dfrac{dv}{dx}+v\dfrac{du}{dx}$	D-7
$\dfrac{u}{v}$	$\dfrac{v\dfrac{du}{dx}-u\dfrac{dv}{dx}}{v^2}$	D-8
e^u	$e^u\dfrac{du}{dx}$	D-9
a^u	$(\ln a)a^u\dfrac{du}{dx}$	D-10
$\ln u$	$\dfrac{1}{u}\dfrac{du}{dx}$	D-11
$\log_a u$	$(\log_a e)\dfrac{1}{u}\dfrac{du}{dx}$	D-12
$\sin u$	$\cos u\left(\dfrac{du}{dx}\right)$	D-13
$\cos u$	$-\sin u\dfrac{du}{dx}$	D-14
$\tan u$	$\sec^2 u\dfrac{du}{dx}$	D-15
$\sin^{-1} u$	$\dfrac{1}{\sqrt{1-u^2}}\dfrac{du}{dx}\qquad\left(-\dfrac{\pi}{2}\leq\sin^{-1}u\leq\dfrac{\pi}{2}\right)$	D-16

$\cos^{-1} u$	$\dfrac{-1}{\sqrt{1-u^2}} \dfrac{du}{dx}$	$\left(0 \leq \cos^{-1} u \leq \pi\right)$	D-17
$\tan^{-1} u$	$\dfrac{1}{1+u^2} \dfrac{du}{dx}$	$\left(-\dfrac{\pi}{2} < \tan^{-1} u < \dfrac{\pi}{2}\right)$	D-18

Note: a nd *n* are constants and *u* and *v* are functions of *x;* that is, $u = u(x)$ and $v = v(x)$

Integral Calculus 7

The basic concepts of *differential calculus* were covered in the preceding chapter. This chapter will be devoted to *integral calculus*, which is the other broad area of calculus. Chapter 8 will be devoted to how both differential and integral calculus manipulations can be performed with MATLAB.

As in the case of Chapter 6, the early emphasis in this chapter will be based on graphical approaches to the subject utilizing functions consisting of straight line segments. This process will show the reader exactly what is happening when we integrate a function.

Objectives

After completing this chapter, the reader should be able to

1. Define the *integral* or *anti-derivative* of a function and show the various forms of notation.
2. Explain the difference between an *indefinite integral* and a *definite integral*.
3. Explain the interpretation of the definite integral as the area underneath the curve.
4. Determine the definite integral of a *piecewise* linear function over a specified domain.
5. Plot the definite integral of a *piecewise linear function when the upper limit is the independent variable.*
6. Utilize tables to determine both indefinite and definite integrals of common functions.
7. State and apply the integral definitions among acceleration, velocity, and displacement.
8. Apply integration techniques to solve practical problems based on given delta or differential forms.

7-2 ANTI-DERIVATIVES

A concept often used to introduce integral calculus is the idea of the *anti-derivative*. Basically, an anti-derivative of a function $f(x)$ is a new function, $F(x)$, such that

$$\frac{dF(x)}{dx} = f(x) \tag{7-1}$$

Stated in words, the anti-derivative is a new function such that when it is differentiated, it will yield the given function. This concept forms the basis of one approach to integral calculus. The approach is that we are working in the opposite direction to that of differential calculus, that is, instead of finding the slope, we determine a new function whose slope is the given function.

The idea that integration is just the opposite process of differentiation is a very useful one, and it can be utilized to develop many tables of integrals. However, it does not provide the interpretation of the physical significance of the process. Therefore, our initial approach will be to concentrate on the physical significance of the integral, and then we will use the anti-derivative approach to assist in the development of integration pairs.

Integrals

The mathematical operation of *integration* is the process of determining the *integral* (or anti-derivative) of a given function. The study of all forms of integration is collectively referred to as *integral calculus*.

Indefinite and Definite Integrals

While the forms differ only in their application, an integral may be classified as either *indefinite* or *definite*. An *indefinite integral* is simply a function that satisfies the anti-derivative definition of Equation 7-1. Most mathematical tables of integrals display the indefinite forms since they may be applied in different ways. An *indefinite integral* is usually identified by the integration symbol *without* any limits placed on the integral. A typical case is as follows:

$$\int f(x)dx \tag{7-2}$$

The meaning of the terminology following the integral sign will be explained shortly.

A *definite integral* is characterized by the presence of *limits* on the integral sign. A typical case is

$$\int_{x_1}^{x_2} f(x)dx \tag{7-3}$$

It is this latter form that will be used in the next section to develop the physical basis for integration. It will be stated at the outset that *a definite integral* of the form of Equation 7-3 may be interpreted as the *area under the curve* of $f(x)$ from $x = x_1$ to $x = x_2$. This concept will serve as the basis for the development of the next section.

7-3 DEVELOPMENT OF THE DEFINITE INTEGRAL

Refer to Figure 7-1 for the development that follows. The definite integral of the function $y = f(x)$ from $x = a$ to $x = b$ is the area under the curve over that domain. This area concept

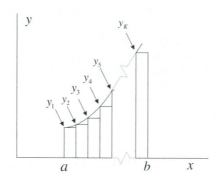

Figure 7-1   Sum of areas used in developing the definite integral

is established in calculus by considering first the approximate area as a finite summation of the areas of a number of rectangular slabs as shown. The width of a given slab is Δx, the height at the left-hand side of the kth slab is $y_k = f(x_k)$.

Strictly speaking, the area of a given slab could be better approximated by treating it as a trapezoid. That is the approach with the trapezoidal algorithm for integration that will be considered in the next chapter. However, because we will soon allow the slabs to become infinitesimally narrow in width, we will use the height at the beginning of the slab as the basis for the area. Hence, a given slab will have an area given by $y_k \Delta x$. The net approximate area can then be determined by summing the areas of all the slabs. This can be denoted by the terminology

$$\text{Approximate Area} = \sum_k y_k \Delta x \qquad (7\text{-}4)$$

where the summation is performed over all values of k. The definite integral from $x = a$ to $x = b$ can then be defined as

$$\int_a^b y\,dx = \lim_{\Delta x \to dx} \sum_k y_k \Delta x \qquad (7\text{-}5)$$

The quantity on the left of Equation 7-5 is the definite integral of y with respect to x from $x = a$ to $x = b$. The quantity $y\,dx$ is defined as the *integrand*; and a and b, which are specific values of x, are called the *limits* of the integral. The integrand can also be expressed as $f(x)dx$.

When the form of the definite integral is established, the area is *exact* rather than an approximation. Thus, the definite integral of a function from $x = a$ to $x = b$ is the area under the curve between the limits.

Running Integral

The development thus far has assumed fixed values for the lower and upper limits of a definite integral, thereby yielding a fixed value for the complete area under the curve between the limits. This is a common situation, but there is another form that needs to be considered.

Many physical processes are characterized by what could be loosely termed a *running integral*. In this form, the lower limit is fixed but the upper limit assumes the value of the independent variable (or sometimes, a function of the independent variable). A typical form is illustrated by the following integral:

$$\int_a^x y\,dx \qquad\qquad (7\text{-}6)$$

The process underway here is the goal of determining how the area under the curve varies as the upper limit moves from $x = a$ to any arbitrary value x, where it is assumed that $x > a$. The result will be a new function that describes the variation of the area under the curve as we move along the x-axis. Area above the x-axis is considered as positive and area below the x-axis is considered as negative. Whenever the function being integrated is 0, the area remains at the level previously established.

Formal Notation in Mathematics Texts

The notation of Equation 7-6 is not quite in line with advanced mathematics texts. In that setting, most likely a so-called dummy variable would be introduced and y would be expressed in terms of the dummy variable, say u. The integral might then be expressed as

$$\int_a^x y(u)\,du \qquad\qquad (7\text{-}7)$$

The reason for introducing the dummy variable is to avoid confusion between the integration variable and the upper limit of the integral. However, we believe that for the purposes of this treatment, the notation expressed by Equation 7-6 will be acceptable, and we will retain that approach in the work that follows.

Summing the Areas

When the upper limit is a variable, x, the process can be visualized by considering that the area is divided up into a series of narrow segments of area in the same fashion as in defining the integral. As x increases in the positive direction from left to right, the area will either increase, decrease, or remain at the same value. The integral function at any point is the net algebraic sum of the areas of all preceding segments. The integral of a constant level for a function either increases or decreases at a linear rate, depending on the sign of the area. The integral of a first-degree function will increase or decrease at a second-degree level, and so on.

Derivative versus Integral

We will now make a few comparisons about the differentiation and integration processes. The derivative of a function indicates the rate of change of the function. The derivative can change quickly, and the first derivative of a continuous function may have discontinuities.

The integral of a function cannot usually change too quickly except under impulsive conditions, which will not be considered at this point. As the algebraic sum of the area is accumulated, an integral either increases or decreases, but never with a sudden jump unless impulse functions are used. If the area is positive, the integral increases; but if the area is negative, the integral decreases. The change is always from the level previously reached. During a segment in which the function being integrated is zero, the integral remains at the level previously established.

7-4 INTEGRATION OF PIECEWISE LINEAR FUNCTIONS

As was done in the previous chapter for differentiation, we will now consider the graphical integration involved with piecewise linear functions. For integration purposes, we can relax the requirement that the function be continuous. Provided that the discontinuities are finite, we can integrate them without difficulty.

Integral of Piecewise Linear Function

The area between two limits on a line segment with a constant slope may be determined by a basic geometric analysis. Consider the line segment, L, of Figure 7-2. This segment starts at $x = x_1$ and ends at $x = x_2$. The area enclosed is the definite integral, and it may be determined from evaluating the area of the trapezoid involved. It can be expressed as

$$\int_{x_1}^{x_2} y\,dx = \frac{1}{2}(y_2 + y_1)(x_2 - x_1) \tag{7-8}$$

The first two factors, $(1/2)(y_2 + y_1)$, represent the average height of the trapezoid, and $x_2 - x_1$ is the width of the base. This formula reduces to the area of a triangle (either y_1 or y_2 is zero) or a rectangle ($y_1 = y_2$). These shapes may be considered as limiting cases of a trapezoid.

The net integral of a piecewise linear function over the entire domain of x may be determined by first evaluating the individual areas under the curve for different segments using

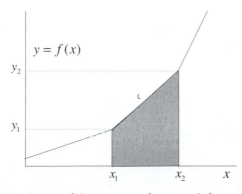

Figure 7-2 Definite integral (or area under curve) for a straight line segment.

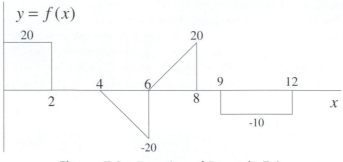

Figure 7-3   Function of Example 7-1

the concept of Equation 7-8. The algebraic sign of each segment must be noted. The net area is the algebraic sum of the individual areas. For example, consider three segments based on the domains from x_1 to x_2, x_2 to x_3, and x_3 to x_4. The net area or integral from x_1 to x_4 is

$$\int_{x_1}^{x_4} y\,dx = \int_{x_1}^{x_2} y\,dx + \int_{x_2}^{x_3} y\,dx + \int_{x_3}^{x_4} y\,dx \tag{7-9}$$

Example 7-1

A piecewise linear function is shown in Figure 7-3. Determine the integral $\int_{0}^{12} y\,dx$.

Solution

The integral over the domain from $x = 0$ to $x = 12$ is the area underneath the curve over that interval. We can divide the total area into six segments as follows:

Segment 1: $0 \le x \le 2$
Segment 2: $2 \le x \le 4$
Segment 3: $4 \le x \le 6$
Segment 4: $6 \le x \le 8$
Segment 5: $8 \le x \le 9$
Segment 6: $9 \le x \le 12$

The division into separate segments is based on the same form and the same sign within a segment.

The next step is to determine the area of each segment. As explained earlier, the area underneath a segment of a piecewise linear function is the area of either a trapezoid, a rectangle or a triangle. On the assumption that the reader can readily determine the area of each segment, the values will be tabulated in the table that follows. Note that Segments 1 and 6 are rectangles, with a positive value for Segment 1 and a negative value for Segment 6. Segments 3 and 4 are triangles, with a positive value for Segment 3 and a negative value for Segment 4. Finally, Segments 2 and 5 have zero area.

Segment	1	2	3	4	5	6
Area	40	0	−20	20	0	−30

From the principle established in Equation 7-9, the net integral from $x = 0$ to $x = 12$ is given by

$$\int_0^{12} y\,dx = 40 + 0 - 20 + 20 + 0 - 30 = 10 \tag{7-10}$$

Example 7-2

For the function of Example 7-1, plot $\int_0^x y\,dx$ as x varies from 0 to a value just greater than 12.

Solution

Determining the total area under the curve in Example 7-1 was pretty straightforward. However, displaying the variation as x changes from 0 to 12 and above takes a little more insight into the process. Several points should be noted:

1. If y is a constant, the integral is either increasing linearly or decreasing linearly, depending on whether y is positive or negative.
2. If the given line segment is triangular in form, imagine little slabs of area. Note how the successive slabs are changing in area as a guide to the parabolic nature of the integral function.
3. During any interval in which $y = 0$, the integral will remain at the value established at the end of the preceding interval.
4. For any interval in which $y \neq 0$, the integral always moves up or down from the level reached at the end of the last interval; that is, there are no sudden jumps (except for impulsive conditions).
5. The beginning and ending points for different segments represent good places to establish levels.

Refer now to Figure 7-4, in which the function of Example 7-1 is shown again in part (a). Refer to Figure 7-4(b) in the discussion that follows. You may also want to refer back to the table in Example 7-1 as an aid in determining critical values.

Segment 1: In this interval, y is a positive constant, so the integral increases linearly. A critical point is $x = 2$ and the area reached at that point is 40.

Segment 2: In this interval, $y = 0$, so the integral remains at the initial value of 40 established in the first interval.

Segment 3: This interval and the next one are a bit tricky, so note the logic that follows. The function in this interval is negative, so that means that the integral must be decreasing. Thinking in terms of slabs, each subsequent slab will have a little more negative area than the previous one, so the area decreases at an increasing rate. Thus, it will have a parabolic shape of the form shown. A critical point is $x = 6$, and the net

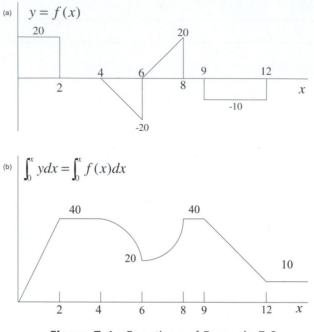

Figure 7-4   Functions of Example 7-2

area at that point will be the area at the beginning of the interval (40) plus the algebraic area accumulated in the interval (–20), so the integral function decreases to a level of 20.

Segment 4: Once again, we have a triangular function, but in this case, it is positive. Hence, the integral function increases in this interval. The areas of the initial slabs are small at first and become increasingly greater as x increases. The integral function increases slowly at first, and then more rapidly later as shown. The net area of this triangle is 20, so the integral function increases back to 40.

Segment 5: In this interval, $y = 0$ again, so the integral remains at 40.

Segment 6: In this interval y is a negative constant, so the integral decreases in a linear manner. A critical point is at $x = 12$ at which the integral function has the value $40 - 30 = 10$. Since there is no additional area, the integral remains at the level of 10 thereafter.

7-5   TABULATION OF INTEGRALS

In this section, the tabulation of some common integral functions and operations will be provided. Some explanations of the meanings will be provided, and their use will be illustrated with examples at the end of the section. First, refer to Table 7-1, at the end of this chapter. Note that each of the derivatives and operations is provided with an identifier ranging from I-1 to I-17. Note that the quantities a, n, and C are constants. Any function whose integral is to be determined is denoted as $f(x)$ in the left-hand column. The second column is labeled as $F(x)$ and is the *indefinite integral*, which is defined as

$$F(x) = \int f(x)dx \qquad (7\text{-}11)$$

Arbitrary Constant for Indefinite Integral

In using the table to determine an indefinite integral, an arbitrary constant, C, should always be added to the integral function. The reason is that the derivative of a constant is zero, and any integral that might be determined through this process could have had an additional constant in its expression. Hence, you must always add C to the integral when it is an indefinite form. In many practical problems, there will be conditions provided that will allow the value of C to be determined.

Definite Integrals

With definite integrals, it is not necessary to add an arbitrary constant, but the integral must be evaluated at the two end points. Assume that the definite integral I is expressed as

$$I = \int_a^b f(x)dx \qquad (7\text{-}12)$$

To determine I, the indefinite integral $F(x)$ is first determined, and I is evaluated as

$$I = F(x)\big]_a^b = F(b) - F(a) \qquad (7\text{-}13)$$

Observe the notation associated with $F(x)$ in the middle term. The upper limit (b) is placed at the top of the bracket, and the lower limit (a) is placed at the bottom. The function $F(x)$ is then evaluated at the two limits, and the value at the lower limit is subtracted from the value at the upper limit.

We will now discuss some of the various integration pairs.

I-1 and I-2

As in the case of the derivatives, I-1 and I-2 are the simple linear operations that are stated at the outset. I-1 states that the integral of a constant times a function is the constant times the integral of the function. I-2 states that the integral of a sum is the sum of the integrals.

I-3

I-3 states that the integral of a constant is the constant times the independent variable. This is equivalent to saying that the area under the curve of a constant increases or decreases at a linear rate.

I-4

I-4 provides the basis for integrating any power function (except for $n = -1$). For each level of integration, the power is increased by 1. This is just the opposite pattern as compared to differentiation.

I-5

I-5 is the integration of the exponential function, and the result is an exponential function.

I-6

I-6 takes care of the one power of x that was excluded by I-4.

I-7 and I-8

I-7 and I-8 involve integration of the sine and cosine functions.

I-9 and I-10

I-9 and I-10 involve integration of the square of sine and cosine functions. These integrals could also be obtained by expanding the square of sine and cosine with the double angle trigonometric identities.

I-11 and I-12

These two integrals are derived through a procedure known as "integration by parts." Anyone who has taken a complete course in integral calculus will likely have derived these forms.

I-13 and I-14

These two integrals involve products of sine and cosine functions and could be derived by expanding with the double angle formulas.

I-15

This integral can also be derived by the "integration by parts" procedure.

I-16 and I-17

No special comments will be made here, but these two arise in certain engineering applications.

Examples 7-3 through 7-5

A function, z, in Examples 7-3 through 7-5 is given by

$$z = \int y dx \tag{7-14}$$

Determine the function z in each case.

Example 7-3

$$y = 12e^{4x} \tag{7-15}$$

Solution

The primary integral is that of I-5, although I-1 also applies.

$$z = \int 12e^{4x} dx = 12\frac{e^{4x}}{4} + C = 3e^{4x} + C \tag{7-16}$$

Example 7-4

$$y = 12x \sin 2x \tag{7-17}$$

Solution

The primary integral is that of I-11, but as in the previous example, I-1 also applies (as it often does).

$$z = \int 12x \sin 2x dx = 12\left(\frac{1}{(2)^2} \sin 2x - \frac{x}{2}\cos 2x\right) + C \tag{7-18}$$

$$= 3\sin 2x - 6x\cos 2x + C$$

Example 7-5

$$y = 6x^2 + \frac{3}{x} \tag{7-19}$$

Solution

The two primary integrals involved are I-4 and I-6, although both I-1 and I-2 are also used.

$$z = \int\left(6x^2 + \frac{3}{x}\right) dx = \int 6x^2 dx + \int \frac{3}{x} dx = \frac{6x^3}{3} + 3\ln x + C \tag{7-20}$$

$$= 2x^3 + 3\ln x + C$$

A value, *I*, in Examples 7-6 and 7-7 is given by the definite integral

$$I = \int_a^b y dx \qquad (7\text{-}21)$$

In each case, the constants *a* and *b* are specified. Determine *I* for each example.

Example 7-6

$$I = \int_0^\pi \sin x dx \qquad (7\text{-}22)$$

Solution

The integral involved is I-7.

$$I = \int_0^\pi \sin x dx = -\cos x \big]_0^\pi = -\cos \pi - (-\cos 0) = -(-1) - (-1) = 2 \qquad (7\text{-}23)$$

Example 7-7

$$I = \int_0^1 8xe^{-2x} dx \qquad (7\text{-}24)$$

Solution

The integral involved is I-15.

$$I = \int_0^1 8xe^{-2x} dx = 8\frac{e^{-2x}}{(-2)^2}\left[-2x - 1\right]_0^1 = 2e^{-2}\left[-2(1) - 1\right] - 2e^{-0}\left[0 - 1\right] \qquad (7\text{-}25)$$

$$= -6e^{-2} + 2 = 1.1822$$

7-6   APPLICATIONS OF INTEGRAL CALCULUS

There are numerous applications of integral calculus in virtually all engineering and scientific disciplines. Some of the common ones stressed in basic calculus texts involve determining areas under various types of curves and the volumes of many different types of solids. While these applications are important, especially in learning the operations of basic calculus, the emphasis here will be on some of the physical properties of systems that require the integration process for full analysis and comprehension.

Displacement, Velocity, and Acceleration

In Chapter 6, the derivative relationships among displacement, velocity, and acceleration were provided. In this chapter, we will reverse the process and begin with acceleration and develop the formulas for velocity and displacement. To that end, the following quantities are reviewed:

$a = a(t) =$ acceleration in meters/second^2 (m/s^2)

$v = v(t) =$ velocity in meters/second (m/s)

$y = y(t) =$ displacement in meters (m)

We begin with

$$\frac{dv}{dt} = a(t) \tag{7-26}$$

Solve for the differential dv.

$$dv = \left(\frac{dv}{dt}\right)dt = a(t)dt \tag{7-27}$$

Integrate the outside terms of Equation 7-27.

$$\int dv = \int a(t)dt \tag{7-28}$$

Ignoring the constant of integration momentarily, which will be taken care of on the right, the first term of the integration of Equation 7-28 becomes simply

$$\int dv = v \tag{7-29}$$

We have belabored the process here so the reader can carefully note what is happening. In all cases, the function must be manipulated so that an integration of a differential occurs on both sides of the equation. In particular, note the simple result of Equation 7-29. In subsequent developments, we will shorten the process somewhat.

Returning to Equation 7-28 with the definition of Equation 7-29 that we employed, we have

$$v = \int a(t)dt + C_1 \tag{7-30}$$

The constant C_1 can be determined from some additional given parameter, such as the initial velocity.

Next, the displacement is formulated.

$$\frac{dy}{dt} = v(t) \tag{7-31}$$

The differential dy is formed

$$dy = \left(\frac{dy}{dt}\right)dt = v(t)dt \tag{7-32}$$

Integration of the outer terms yields

$$y = \int v(t)dt + C_2 \tag{7-33}$$

The constant C_2 is determined from some physical property such as the initial displacement.

Alternate Formulation in Terms of Definite Integrals

When we turn to numerical approximations for integration in Chapter 8, we will see that the definite integral is easier to work with than the indefinite integral. For that purpose, it is more convenient at the outset to state the initial value of the quantity being sought. We will assume that the time scale is established such that $t = 0$ is the beginning point for analysis, although the results that follow could be adapted to any reference time value.

Assume that the initial value of the velocity is $v(0)$ and the initial value of the displacement is $y(0)$. As has been stated earlier in the text, we are assuming that there are no impulsive conditions. With this assumption, any definite integral cannot establish any measurable area in an infinitesimally short time. This means that a *definite integral starting at t = 0 plus a constant must equal the constant at t = 0.*

The expressions for velocity and displacement can be reformulated as follows:

$$v(t) = \int_0^t a(t)dt + v(0) \tag{7-34}$$

$$y(t) = \int_0^t v(t)dt + y(0) \tag{7-35}$$

While we are concentrating here on acceleration, velocity, and displacement, the general ideas apply to virtually all physical systems having similar relationships. When an analytical approach is being used, one can choose to use either constants of integration or definite integrals with initial values. For numerical integration, however, the definite integral approach is the better choice, as we will see in the next chapter.

Example 7-8

A certain object subjected to both external forces and resistance experiences acceleration in meters/second^2 given by

$$a(t) = 20e^{-2t} \tag{7-36}$$

The initial conditions are $v(0) = 0$ and $y(0) = 0$. Determine expressions for (a) the velocity and (b) the displacement using indefinite integrals and constants determined from the initial conditions.

Solution

The first step involves determining the velocity from the acceleration.

$$\frac{dv}{dt} = a(t) = 20e^{-2t} \tag{7-37}$$

From this expression, we determine the velocity as

$$v(t) = \int 20e^{-2t}dt = \frac{20}{-2}e^{-2t} + C_1 = -10e^{-2t} + C_1 \tag{7-38}$$

Since $v(0) = 0$, we evaluate C_1 as follows:

$$v(0) = -10e^{-0} + C_1 = -10 + C_1 = 0 \tag{7-39}$$

and

$$C_1 = 10 \tag{7-40}$$

The velocity is then

$$v(t) = 10 - 10e^{-2t} \tag{7-41}$$

Next, we determine the displacement from the velocity.

$$y(t) = \int \left(10 - 10e^{-2t}\right)dt = 10t - \frac{10}{-2}e^{-2t} + C_2 = 10t + 5e^{-2t} + C_2 \tag{7-42}$$

From the condition that $y(0) = 0$, we have

$$y(0) = 0 + 5e^{-0} + C_2 = 5 + C_2 = 0 \tag{7-43}$$

and

$$C_2 = -5 \tag{7-44}$$

The displacement is then

$$y(t) = 10t + 5e^{-2t} - 5 \tag{7-45}$$

These functions will be shown in Example 7-13 after an alternate approach has been employed.

Example 7-9

For the acceleration function of Example 7-8, determine expressions for (a) the velocity and (b) the displacement using the **definite integral** forms of Equations 7-34 and 7-35.

Solution

The velocity is given by

$$v(t) = \int_0^t a(t)dt + v(0) = \int_0^t 20e^{-2t}dt + 0 = \frac{-20}{2}e^{-2t}\Big]_0^t \tag{7-46}$$

$$= -10e^{-2t} - \left(-10e^{-0}\right) = 10 - 10e^{-2t}$$

The displacement is given by

$$y(t) = \int_0^t v(t)dt + y(0) = \int_0^t (10 - 10e^{-2t})dt + 0 = 10t + 5e^{-2t}\Big]_0^t \tag{7-47}$$

$$= 10t + 5e^{-2t} - 0 - 5e^{-0} = 10t + 5e^{-2t} - 5$$

Comparing Examples 7-8 and 7-9, it can be argued that both approaches are about equal as far as the amount of work is concerned. However, the reader should strive to understand both perspectives since there is widespread usage of both approaches in solving applied calculus problems. As stated earlier, when numerical integration is to be employed, this latter approach using the definite integral is preferred.

Example 7-10

Use MATLAB to plot the acceleration, velocity, and displacement in Examples 7-8 and 7-9 over the domain $0 \le t \le 2s$.

Solution

A review of the three functions follows.

$$a(t) = 20e^{-2t} \tag{7-48}$$

$$v(t) = 10 - 10e^{-2t} \tag{7-49}$$

$$y(t) = 10t + 5e^{-2t} - 5 \tag{7-50}$$

While the procedure for plotting has been used extensively up to this point, a quick review for these functions is worthwhile. Switching notation to the MATLAB default form, we will arbitrarily select a time step of 0.02 s. The command is

$$\gg t = 0{:}0.02{:}2; \tag{7-51}$$

The next three commands generate the acceleration, velocity, and displacement, respectively.

$$\gg a = 20*\exp(-2*t); \tag{7-52}$$

$$\gg v = 10 - 10*\exp(-2*t) \tag{7-53}$$

$$\gg y = 10*t + 5*\exp(-2*t)-5; \tag{7-54}$$

The functions can now be plotted by the command

$$\gg \text{plot}(t, a, t, v, t, y) \tag{7-55}$$

Additional labeling was provided, and the final results are shown in Figure 7-5. Each function is expressed in terms of its basic units. Thus, acceleration is given in meters/second^2, velocity is given in meters/second, and displacement is given in meters. Fortunately, the ranges of the dependent variables are such that they all display prominently with one scale. (They were planned that way!)

An important phase of any engineering analysis is to make an interpretation of the results. This particular object accelerates very rapidly at first but tapers off gradually. This causes a rapid increase of velocity from its initial value of zero at first, but then the rate of increase of velocity tapers off eventually and the limiting form is a constant velocity and zero acceleration. The displacement is always increasing, but it increases more rapidly as

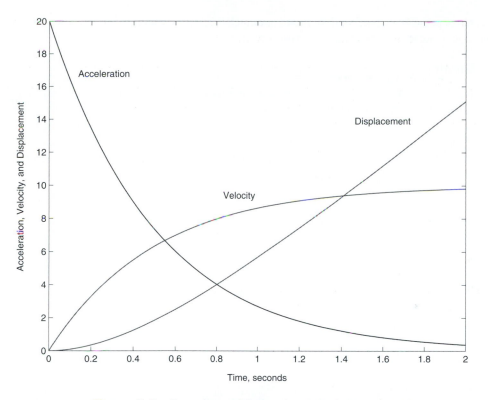

Figure 7-5 Functions of Examples 7-8, 7-9, and 7-10

the velocity increases. Eventually, it becomes asymptotic to a straight line, corresponding to the limiting constant velocity.

Example 7-11

An ideal spring is characterized by the fact that the force, f, required to stretch the spring from its initial equilibrium position is directly proportional to the distance, x, that the spring is stretched. This can be stated as

$$f = Kx \qquad (7\text{-}56)$$

where f can be measured in newtons (N), x is measured in meters (m) and K is the spring constant measured in newtons/mcter (N/m).

The increment of work ΔW in joules (J) generated by a force, f, moving through a distance, Δx, is given by

$$\Delta W = f \Delta x \qquad (7\text{-}57)$$

Determine the amount of work required to stretch the spring from its initial position, which will be denoted as $x = 0$, to a final position $x = L$.

Solution

This is a good example that will demonstrate how integral calculus must be used in a cumulative process in which the variable is constantly changing. We cannot simply multiply the force by the net distance to get the work done, since the force is a function of the position.

To solve this problem, we substitute the value of f as a function of x from Equation 7-56 into Equation 7-57, and we have

$$\Delta W = Kx\Delta x \tag{7-58}$$

Changing the deltas to differential forms, we have

$$dW = Kxdx \tag{7-59}$$

We then integrate both sides and obtain

$$W = \int_0^L Kxdx = \frac{Kx^2}{2}\Bigg]_0^L = \frac{KL^2}{2} \tag{7-60}$$

Assuming no losses in the stretching process, the work performed on the spring is now stored in the spring as potential energy.

GENERAL PROBLEMS

7-1. A piecewise linear function is shown in the figure below. Determine the integral $\int_0^8 ydx$.

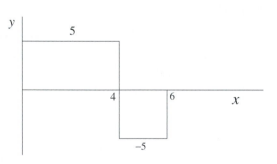

7-2. A piecewise linear function is shown in the figure at the top of the next page. Determine the integral $\int_0^8 ydx$

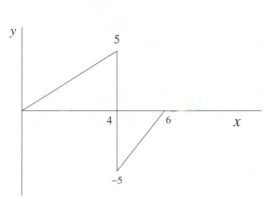

7-3. For the function of Problem 7-1, plot $\int_0^x ydx$ as x varies from 0 to an arbitrary value greater than 8.

7-4. For the function of Problem 7-2, plot $\int_0^x ydx$ as x varies from 0 to an arbitrary value greater than 8.

7-5. A piecewise linear function is shown below. Determine the integral $\int_0^{10} ydx$.

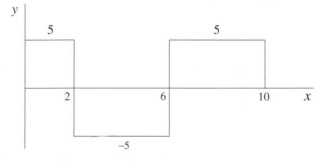

7-6. A piecewise linear function is shown below. Determine the integral $\int_0^{12} ydx$.

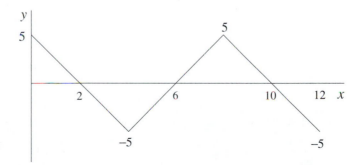

7-7. For the function of Problem 7-5, plot $\int_0^x y\,dx$ as x varies from 0 to an arbitrary value greater than 10.

7-8. For the function of Problem 7-6, plot $\int_0^x y\,dx$ as x varies from 0 to 12.

Problems 7-9 through 7-16 determine the **indefinite integral** in each case. You may use Table 7-1, at the end of the chapter, as necessary, along with required modifications.

7-9. $z = \int 12x^3\,dx$

7-10. $z = \int 20x^4\,dx$

7-11. $z = \int 10e^{-2x}\,dx$

7-12. $z = \int 400e^{-200x}\,dx$

7-13. $z = \int 20x\cos 4x\,dx$

7-14. $z = \int 20x\sin 4x\,dx$

7-15. $z = \int xe^{-x}\,dx$

7-16. $z = \int 20xe^{-4x}\,dx$

Problems 7-17 through 7-26 involve the same integrals as in Problems 7-9 through 7-16, but they will be established as **definite integrals** over specific limits.

7-17. $I = \int_0^2 12x^3\,dx$

7-18. $I = \int_{-1}^2 20x^4\,dx$

7-19. $I = \int_0^2 10e^{-2x}\,dx$

7-20. $I = \int_0^{0.01} 400e^{-200x}\,dx$

7-21. $I = \int_0^\pi 20x\cos 4x\,dx$

7-22. $I = \int_0^\pi 20x \sin 4x\, dx$

7-23. $I = \int_0^2 xe^{-x}\, dx$

7-24. $I = \int_0^{0.5} 20xe^{-4x}\, dx$

7-25. A certain object experiences acceleration in meters/second^2 according to the equation

$$a(t) = 20(1 - e^{-2t})$$

The initial conditions are $v(0) = 0$ and $y(0) = 0$. Determine expressions for (a) the velocity and (b) the displacement using indefinite integrals and constants determined from the initial conditions.

7-26. A certain object experiences acceleration in meters/second^2 according to the equation

$$a(t) = 10 \sin 2t$$

The initial conditions are $v(0) = 0$ and $y(0) = 0$. Determine expressions for (a) the velocity and (b) the displacement using indefinite integrals and constants determined from the initial conditions.

7-27. Rework Problem 7-25 using definite integral forms.

7-28. Rework Problem 7-26 using definite integral forms.

Note: As in the case of Chapter 6, the total emphasis in this chapter was analytical in nature. Hence, MATLAB problems dealing with integral calculus will be delayed to the end of Chapter 8.

Table 7-1 Common Integrals of Functions and Operations

$f(x)$	$F(x) = \int f(x)dx$	Integral Number
$af(x)$	$aF(x)$	I-1
$u(x) + v(x)$	$\int u(x)dx + \int v(x)dx$	I-2
a	ax	I-3
$x^n \qquad (n \neq -1)$	$\dfrac{x^{n+1}}{n+1}$	I-4
e^{ax}	$\dfrac{e^{ax}}{a}$	I-5
$\dfrac{1}{x}$	$\ln x$	I-6
$\sin ax$	$-\dfrac{1}{a}\cos ax$	I-7
$\cos ax$	$\dfrac{1}{a}\sin ax$	I-8
$\sin^2 ax$	$\dfrac{1}{2}x - \dfrac{1}{4a}\sin 2ax$	I-9
$\cos^2 ax$	$\dfrac{1}{2}x + \dfrac{1}{4a}\sin 2ax$	I-10
$x\sin ax$	$\dfrac{1}{a^2}\sin ax - \dfrac{x}{a}\cos ax$	I-11
$x\cos ax$	$\dfrac{1}{a^2}\cos ax + \dfrac{x}{a}\sin ax$	I-12
$\sin ax \cos ax$	$\dfrac{1}{2a}\sin^2 ax$	I-13
$\sin ax \cos bx$ for $a^2 \neq b^2$	$-\dfrac{\cos(a-b)x}{2(a-b)} - \dfrac{\cos(a+b)x}{2(a+b)}$	I-14

xe^{ax}	$\dfrac{e^{ax}}{a^2}(ax-1)$	I-15
$\ln x$	$x(\ln x - 1)$	I-16
$\dfrac{1}{ax^2 + b}$	$\dfrac{1}{\sqrt{ab}}\tan^{-1}\left(x\sqrt{\dfrac{a}{b}}\right)$	I-17

Note : a and n are arbitrary constants.

For indefinite integrals, an arbitrary constant, C, should be added.

For definite integrals of the form $\int_a^b f(x)dx$, the final result should be $F(b) - F(a)$.

Calculus Operations with MATLAB

8

Now that we have developed the analytical concepts of differential and integral calculus, we are ready to see how such operations can be performed with MATLAB. First, it should be emphasized that a digital computer operates on a numerical basis employing the basic operations of addition, subtraction, multiplication, and division, along with memory storage and logic. Therefore, true differentiation and integration can never be achieved *exactly* with numerical processes. However, there are two ways that differentiation and integration can be achieved on a practical level with MATLAB.

First, MATLAB can be viewed as a very comprehensive "look-up" table of a sort in that numerous derivatives and integrals as character strings can be manipulated with software. The **Symbolic Toolbox** permits the exact formulas for derivatives and integrals to be extracted, manipulated, and plotted.

The second way that MATLAB can be used for calculus operations is through **numerical approximations** to differentiation and integration. While such approximations are not exact, they can be used to provide extremely close approximations to the derivatives and integrals of experimental data. When used correctly, they can be applied to practical situations for which closed-form derivative and integral functions do not exist, and the results can be achieved to sufficient accuracy for virtually all practical applications.

Objectives

After completing this chapter, the reader should be able to perform the following operations within the MATLAB Command Window:

1. Establish a *symbolic variable*.
2. Employ *symbolic differentiation* and plot the function with **ezplot**.
3. Employ *symbolic integration*, determine the arbitrary constant, and plot the function with **ezplot**.
4. Apply *numerical differentiation*.
5. Apply *zero-order numerical integration*.
6. Apply *first-order* or *trapezoidal numerical integration*.

7. Apply the numerical methods to solve practical problems such as those involving acceleration, velocity, and displacement.

8-2   SYMBOLIC DIFFERENTIATION

MATLAB provides a number of very useful software supplements based on particular subject and application areas. These supplements are referred to by MathWorks, Inc., as "toolboxes." If you are using the **Professional Version**, the operations referred to in this section and some others require that the **Symbolic Toolbox** be installed. At the time of this writing, the **Symbolic Toolbox** is included with the **Student Version**. Other toolboxes may be added to the **Student Version** for a modest cost, but they are not required to support this text.

Symbolic Variables

A symbolic variable is one that can be manipulated in the same manner as in an equation, and it may or may not ever take on any numerical values. Rather, the first step may simply be to obtain an expression involving a functional relationship between two or more variables as so-called *string variables*.

In general, there are several ways of defining symbolic variables. A common way, and the one we will emphasize, is through the command **syms**, which is followed by a list of the variable or variables that are to be symbolic. Strictly speaking, you only need to list the independent variables and parameters, since any dependent variables defined in terms of the independent variables will automatically assume the form of symbolic variables. For example, suppose x and a are to constitute symbolic variables. The appropriate command would be

$$\gg \text{syms x a} \qquad (8\text{-}1)$$

If y is now expressed as a function of x and the parameter a, it will automatically become a symbolic variable.

Another way to accomplish the same result with some of the operations is to enclose the symbolic portion of the expression in apostrophes (' '). This latter form will be demonstrated later.

Differentiation

The command for differentiation is **diff( )**, with the quantity to be differentiated in parentheses. In the case of symbolic differentiation, the derivative will be listed if it can be manipulated in the proper form from the MATLAB library. In the case of numerical differentiation, a different format for the argument is used, which will be considered later.

Example 8-1

Assume that y is related to x by the fifth-degree function

$$y = 4x^5 \qquad (8\text{-}2)$$

Using MATLAB symbolic operations, create the function and the first derivative with respect to x.

Solution

From Table 6-1, we could easily determine the derivative as $4 \times 5 \times x^4 = 20x^4$, but like so many places in the text, we will use easily identifiable functions to study how MATLAB performs the operations. First, we need to define x as a symbolic variable.

$$\gg \text{syms x} \qquad (8\text{-}3)$$

Note that even though there is only one independent variable, it is necessary to put the "s" at the end of **syms**. Just for clarification, it would be fine to define y as a symbolic variable at this point, but it will be taken care of automatically in the next step. We now define y according to the standard MATLAB form.

$$\gg \text{y = 4*x\textasciicircum5} \qquad (8\text{-}4)$$

$$y =$$

$$4\text{*x\textasciicircum5}$$

It is not necessary to put the array identification period for this operation even if we evaluate it later over a domain of x. We will define the derivative of y with respect to x as yprime. It is then evaluated by the command

$$\gg \text{yprime = diff(y)} \qquad (8\text{-}5)$$

$$\text{yprime} =$$

$$20\text{*x\textasciicircum4}$$

It is obvious that the result is what we expected.

There are a couple of alternate ways that the process could be performed. The shortest way is to form the definition of y and the derivative in one step. The command to achieve that operation is

$$\gg \text{yprime = diff(4*x\textasciicircum5)} \qquad (8\text{-}6)$$

$$\text{yprime} =$$

$$20\text{*x\textasciicircum4}$$

This procedure saves one step in the process, but it has the disadvantage that we do not have a separate expression for the variable y.

A final form is to skip the step of defining the symbolic variable or variables in advance, but to do it by using apostrophes. Going back first to the two-step process, the steps are as follows:

$$>> y = \text{'}4*x\char`\^5\text{'} \tag{8-7}$$

$$y =$$

$$4*x\char`\^5$$

The value of yprime is then determined by the command

$$>> \text{yprime} = \text{diff}(y) \tag{8-8}$$

$$\text{yprime} =$$

$$20*x\char`\^4$$

Finally, this approach can also be performed in one step.

$$>> \text{yprime} = \text{diff}('4*x\char`\^5') \tag{8-9}$$

$$\text{yprime} =$$

$$20*x\char`\^4$$

In this latter form, the apostrophes are placed around the function within the **diff** argument.

Example 8-2

Use the command **ezplot** to plot the functions y and yprime of Example 8-1 over the domain from x = −1 to x = 1.

Solution

The **ezplot** operation is one that can be used to make a "quick and easy" plot. While it does not offer all the options employed with the more general plots considered earlier in the text, it is particularly convenient for plotting symbolic variables.

There are several variations on **ezplot** that can be found in the MATLAB help file, but we will employ only a few here. First, if there is a symbolic variable y that is a function of a symbolic variable x, the simple command **ezplot(y)** will actually generate a plot of y versus x. In that simple form, however, the default domain used by MATLAB is $-2\pi \le x \le 2\pi$. One can then use the **axis** command to change the domain and range. However, the form

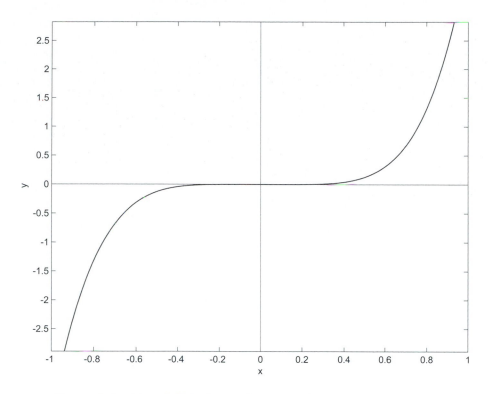

Figure 8-1   Plot of fifth-degree function of Examples 8-1 and 8-2

that we will use in this example is one where we can establish the domain within the argument of the **ezplot** command. It reads as follows:

$$\gg \text{ezplot(y, [x1 x2])} \tag{8-10}$$

where x1 and x2 define the beginning and ending values of the domain, respectively.

First, we will plot the function y over the stated domain. It is generated by

$$\gg \text{ezplot(y, [-1 1])} \tag{8-11}$$

Additional labeling can be provided using the operations considered earlier for regular plots; the final result is shown in Figure 8-1.

Next, we will plot the function yprime over the same domain. It is generated by

$$\gg \text{ezplot(yprime, [-1 1])} \tag{8-12}$$

Again, additional labeling has been provided, and the result is shown in Figure 8-2.

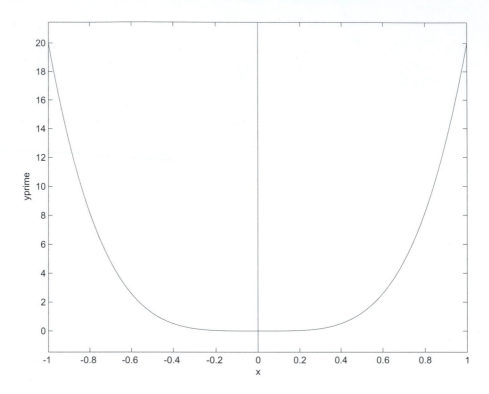

Figure 8-2   Plot of derivative of function of Figure 8-1

8-3   SYMBOLIC INTEGRATION

While symbolic operations are fresh in our minds, let us explore the reverse process, that is, symbolic integration. The process of defining the variables is the same as with differentiation, but the command for integration is **int( )**, with the symbolic function and possible additional parameters listed in parentheses.

Indefinite Symbolic Integration

As in symbolic differentiation, we must use the command **syms** followed by the independent variables, or the function can be enclosed by apostrophes within the **int** argument. An important fact to consider is that the arbitrary constant of integration is not included in the result. Therefore, that constant must be added and evaluated before the final expression can be used.

Assume that a **syms** command has been used to define the independent symbolic variable or variables. Let y represent the symbolic function that is to be integrated, and let yint represent the symbolic integrated function. This would be generated by the command

$$\gg \text{yint} = \text{int(y)} \tag{8-13}$$

Once the expression has been generated, the constant of integration, C, can be determined from some given condition. The value can then be added to the symbolic function by a simple algebraic addition.

Definite Symbolic Integration

If a net integration over a domain $a \le x \le b$ is desired, the format for symbolic integration is as follows:

$$\gg \text{yint} = \text{int(y, a, b)} \tag{8-14}$$

The result will be the area under the curve of y from a to b.

Example 8-3

Assume that a function z is to be determined from a function y by integration, as follows:

$$z = \int y dx = \int x^2 e^x dx \tag{8-15}$$

Using MATLAB symbolic operations, determine z. Evaluate the unknown constant from the condition that $z(0) = 0$.

Solution

It should be noted that this integral is not in Table 7-1. You will likely find it in a comprehensive set of integral tables, and it could be evaluated using the procedure called "integration by parts." However, we are oriented toward MATLAB here, and we will perform the integration symbolically.

We begin by defining x as a symbolic variable.

$$\gg \text{syms x} \tag{8-16}$$

Then we define y as

$$\gg \text{y} = \text{x\textasciicircum2*exp(x)} \tag{8-17}$$

$$\text{y} =$$

$$\text{x\textasciicircum2*exp(x)}$$

Next we evaluate z by the command

$$\gg \text{z} = \text{int(y)} \tag{8-18}$$

$$\text{z} =$$

$$\text{x\textasciicircum2*exp(x)-2*x*exp(x)+2*exp(x)} \tag{8-19}$$

Recall that the constant of integration is not provided with the symbolic result, so it must be added manually. We will then make use of the fact that $z(0) = 0$. For $x = 0$, the first two terms of Equation 8-19 are zero, but the third term has the value 2. Hence, the additive constant must have the value -2 to force the net function to be 0 at $x = 0$. A simple way to modify the function is the simple command

$$\text{>> } z = z - 2 \tag{8-20}$$

$$z =$$

$$\text{x\^2*exp(x)-2*x*exp(x)+2*exp(x)-2}$$

While we should be able to convert this function back to the standard form, there is an interesting MATLAB command that attempts to do so. It is called **pretty( )**, with the function placed within the parentheses. Let us try it.

$$\text{>> pretty(z)} \tag{8-21}$$

$$x^2 \exp(x) - 2 \, x \exp(x) + 2 \exp(x) - 2$$

We will leave it to the reader to decide if the result is "prettier" than the original! The actual function in proper mathematical form is

$$z = x^2 e^x - 2xe^x + 2e^x - 2 = e^x(x^2 - 2x + 2) - 2 \tag{8-22}$$

Next, let us use ezplot to plot the two functions over the domain $0 \le x \le 1$. The commands are

$$\text{>> ezplot(y, [0 1])} \tag{8-23}$$

$$\text{>> ezplot(z, [0 1])} \tag{8-24}$$

These two functions, with additional labeling, are shown in Figures 8-3 and 8-4, respectively.

8-4   NUMERICAL DIFFERENTIATION

We will now turn our attention to the process of numerical differentiation. Numerical differentiation is the process of utilizing arithmetic operations to approximate the analytical derivative.

As you might expect, the next section will be devoted to numerical integration, but since analytical differentiation was studied first, the same order will be used with the numerical operations. However, it should be stressed that numerical differentiation generally is more prone to error than numerical integration. The reason is that differentiation is a "sudden" process in which rapidly changing functions can experience rapid variations in the slope.

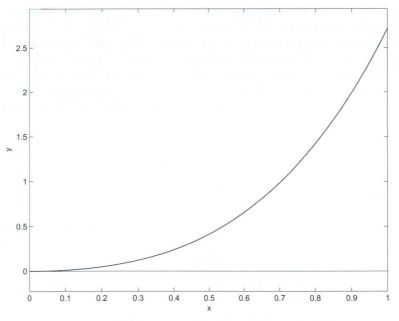

Figure 8-3 Plot of function of Example 8-3

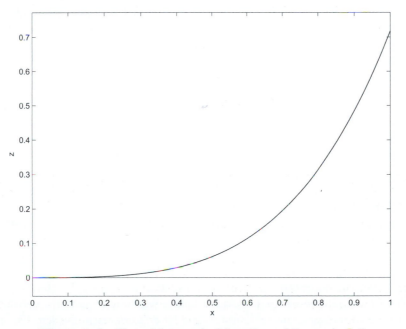

Figure 8-4 Plot of integral of function of Example 8-3

However, if the function is "reasonably well-behaved," numerical approximations to the derivative function can be reasonably accurate.

Since differentiation tends to be a simpler analytical process than integration, why would we want to use numerical differentiation? The fact is that a great deal of practical data obtained from engineering applications is somewhat random in nature and cannot be described exactly by an equation. Moreover, many engineering systems utilize a computer as a controller for the system, and when derivatives are required in the process, it is necessary to resort to the numerical processes.

Differentiation by Differences

Recall from Chapter 6 that the derivative is obtained as a limiting process of $\Delta y/\Delta x$ as the changes become infinitesimally small. Therefore, a reasonable estimate of the derivative is the ratio of the two small changes, that is,

$$\frac{dy}{dx} \approx \frac{\Delta y}{\Delta x} \tag{8-25}$$

The degree to which this approximation is "good" or "bad" depends on the nature of the function and the size of the changes used in making the approximation. While there are variations on this theme making use of so-called "backward differences" versus "forward differences," we will limit our consideration here to a simple form.

The **diff** Command for Numerical Data

We have seen that the **diff** command determines the actual derivative of a symbolic function. For a numerical sequence, however, the operation is quite different.

Assume that vectors x and y have been defined at $N + 1$ points. Assume that the values of y are denoted as $y_1, y_2, y_3,....y_N, y_{N+1}$. We will define a vector u whose discrete values are defined as follows;

$$u_1 = y_2 - y_1$$
$$u_2 = y_3 - y_2$$
$$u_3 = y_4 - y_3 \tag{8-26}$$
$$.\qquad.\qquad.$$
$$u_N = y_{N+1} - y_N$$

Note that for $N + 1$ points in y, there are N points for u.

Changing notation according to the MATLAB default form, the vector u is generated by the command

$$>> u = \text{diff}(y) \tag{8-27}$$

Now each of the components of the vector u represents a Δy change. Therefore, if we divide the u-vector by Δx, the result might be a reasonable approximation to the derivative. Let yprime represent the approximate derivative and let delx represent the step size between successive values of x. We have

$$\text{>> yprime = diff(y)/delx} \tag{8-28}$$

There is a minor bugaboo that must be reconciled before yprime can be plotted versus x. Recall that the vector x had N+1 points while yprime only has N points. They must be adjusted to the same length. Assuming that mysterious "well-behaved" condition and assuming that the step size is sufficiently small, the author finds the following command to be adequate for most practical purposes:

$$\text{>> x = x(1:N)} \tag{8-29}$$

This operation amounts to discarding the last value of x so that the vectors x and yprime have the same length. One could also argue that discarding the first point would be just as logical. The fact is that we are dealing with approximations anyway, so care must be taken to ensure that the results are reasonable.

We may now plot the derivative function by the command

$$\text{>> plot(x, yprime)} \tag{8-30}$$

If we desire to plot y and yprime both on the same scales, we need to "trim" y also by one value.

$$\text{>> y = y(1:N)} \tag{8-31}$$

We could then use the plot command

$$\text{>> plot(x, y, x, yprime)} \tag{8-32}$$

Example 8-5

Consider the sinusoidal function

$$y = \sin x \qquad 0 \le x \le 2\pi \tag{8-33}$$

Investigate the validity of the numerical differentiation process by considering two different values for the number of points in the domain: (a) 11 and (b) 101.

Solution

First, the exact derivative is readily determined as

$$\frac{dy}{dx} = \cos x \tag{8-34}$$

Therefore, any values determined through the numerical process can be readily checked with the exact values.

a. It does not take a "rocket scientist" to predict that the results for the case of 11 points will be pretty bad. After all, 11 points in a full cycle of a sine wave do not provide much in the way of accuracy, especially for differentiation. Anyway, we will proceed by first generating the values of x.

$$\text{>> x = linspace(0, 2*pi, 11);} \tag{8-35}$$

The vector y is then generated by the command

$$\gg y = \sin(x); \tag{8-36}$$

The value of Δx will be computed once and used in the remainder of this part. Be careful here. Even though there are 11 points, there are only 10 segments. Hence, delx is

$$\gg delx = 2*pi/10; \tag{8-37}$$

The approximate derivative yprime is then given by

$$\gg yprime = diff(y)/delx; \tag{8-38}$$

Now we need to shorten x by 1 point.

$$\gg x = x(1:10); \tag{8-39}$$

We are now ready to plot the approximate derivative, but we wish to compare it with the exact value. The command that follows will simultaneously plot the approximation yprime and 10 values of the exact value. The "o" label following the second function, which is $\cos(x)$, tells MATLAB to represent those values only at the values of x and to label them o.

$$\gg plot(x, yprime, x, \cos(x), 'o') \tag{8-40}$$

The result is shown in Figure 8-5. About the best we can say is that the approximation is in the "ballpark" of the exact values. Clearly the step size is too great for an accurate representation.

b. The procedure in this case follows the same pattern as part (a) except we begin with 101 points, corresponding to 100 intervals, and obtain a 100-point approximation for yprime. Again, the exact values are compared with the approximate values and the results are shown in Figure 8-6. While there is a small shift in the approximate function, the result is far more accurate than in part (a).

8-5 NUMERICAL INTEGRATION

The emphasis in this section is directed toward numerical techniques for approximating integration. In general, the techniques for numerical integration have been required in more applications than for numerical differentiation. Moreover, depending on the function and the method employed, the results are often as accurate as analytical procedures for most practical purposes.

While MATLAB offers a number of different methods for numerical integration, some of which will be considered later in the text in conjunction with differential equations, the treatment at this point will be limited to two of the simpler procedures: zero-order integration and first-order or trapezoidal integration. The second procedure is generally more accurate than the first, but there are circumstances in which the first method is used due to its simplicity.

For either method, assume that vectors x and y have been defined at n points. Assume that the values of x are $x_1, x_2, x_3, ... x_N$ and the corresponding values of y are $y_1, y_2,$

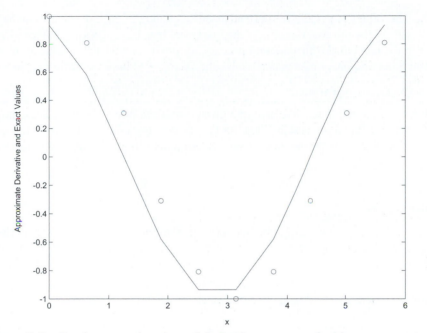

Figure 8-5   Crude approximation of derivative compared with some exact points

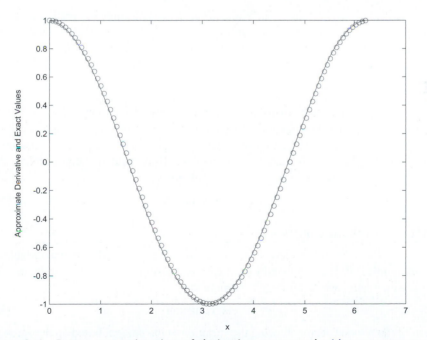

Figure 8-6   Better approximation of derivative compared with some exact points

$y_3, ..., y_N$. The values of x and y are defined at the beginning of each of the area slabs and all slabs have the width Δx, where Δx is the increment between successive values of x. In contrast to the numerical differentiation process, we don't need the extra value at the end, and when a running integration is performed, the vectors will have the same lengths.

Zero-Order Integration

With zero-order integration, the area of a given slab is the value of y at the beginning of the slab times the width of the slab. Thus, for the kth slab, the area is $y_k \Delta x$. Let z_k represent the approximate area under the curve from the beginning to the kth slab. The recursion process follows.

$$z_1 = y_1 \Delta x \qquad \text{8-41)}$$

The value of z_2 is

$$z_2 = z_1 + y_2 \Delta x = (y_1 + y_2)\Delta x \qquad \text{(8-42)}$$

The value of z_3 is

$$z_3 = z_2 + y_3 \Delta x = (y_1 + y_2 + y_3)\Delta x \qquad \text{(8-43)}$$

The pattern is definitely emerging. The last value z_k is

$$z_k = z_{k-1} + y_k \Delta x = \left(\sum_{n=1}^{k} y_n \right) \Delta x \qquad \text{(8-44)}$$

The conclusion from Equation 8-44 is that the zero-order definite integration is achieved merely by summing all the values of y up to the desired end point and multiplying by the independent variable increment. This simple algorithm is particularly useful in dealing with programmable logic systems where summing is easily achieved.

MATLAB **sum** Command

If a definite integral between two points is desired using this approximation, the command **sum** is the choice. Assuming a vector y, this command provides the sum of all the y values over the vector. However, this result must be multiplied by the distance between successive x values, which will be denoted here by delx. Thus, a suitable command to obtain the definite integral, which will be denoted by "area," is

$$\text{>> area = delx*sum(y)} \qquad \text{(8-45)}$$

MATLAB **cumsum** Command

Assume that it is desired to show the area as a function of the independent variable, that is, a running integral. The command in that case is **cumsum**. Let z represent the resulting function, which will have N points. The command is

$$\text{>> z = delx*cumsum(y)} \qquad \text{(8-46)}$$

The result is an N-point vector showing the variation of the integral based on the zero-order approximation over the domain defined by the independent variable.

First-Order Integration

The first-order integration technique is called the *trapezoidal rule*. Basically, it assumes a first-degree variation between successive points. It is widely used as a compromise between the less accurate zero-order algorithm and the more sophisticated, but complex, higher order methods. We will develop the procedure in the steps that follow.

The way the algorithm is implemented with MATLAB, you may think of the process as adding an extra value of 0 prior to the first value of the integral. This provides a degree of smoothing and it allows the result of a running integral to have the same number of points as the independent variable. Assuming again a function z for the integral, the algorithm proceeds as follows:

$$z_1 = 0 \tag{8-47}$$

$$z_2 = z_1 + \left(\frac{y_1 + y_2}{2} \right) \Delta x \tag{8-48}$$

$$z_3 = z_2 + \left(\frac{y_2 + y_3}{2} \right) \Delta x \tag{8-49}$$

The value of z_k is

$$z_k = z_{k-1} + \left(\frac{y_{k-1} + y_k}{2} \right) \Delta x \tag{8-50}$$

MATLAB **trapz** Command

If a definite integral between two points is desired using this approximation, the command **trapz** is the choice. Assuming a vector y, this command provides the operations of the preceding equations, but it assumes a unit width. Therefore, the result must be multiplied by delx. An appropriate command to determine the area between two points is

$$\gg \text{area} = \text{delx*trapz(y)} \tag{8-51}$$

MATLAB **cumtrapz** Command

A running integral can be generated by the command **cumtrapz,** as follows:

$$\gg z = \text{delx*cumtrapz(y)} \tag{8-52}$$

The result is an N-point vector showing the variation of the integral based on the trapezoidal rule over the domain defined by the independent variable.

At the end of this section, some examples will be provided to compare the results of the zero-order and first-order algorithms. Once the comparison has been made, the tendency in later work will be to use the first-order or trapezoidal rule due its inherent greater accuracy.

Example 8-6

Consider the definite integral to determine the area A as follows:

$$A = \int_0^2 4x^3 dx \qquad (8\text{-}53)$$

Determine (a) the exact area using a basic integral from Table 7-1, (b) a zero-order approximation using MATLAB, and (c) a first-order approximation using MATLAB. Use a step size of 0.05 in (b) and (c).

Solution

a. The exact integral is readily determined from I-4 of Table 7-1. We have

$$A = \frac{4x^4}{4} = x^4 \Big]_0^2 = (2)^4 - (0)^4 = 16 \qquad (8\text{-}54)$$

b. Using MATLAB, a vector x must be defined over the given domain, and the step size is 0.05. Since we will need the step size later, we first define it and then define x.

$$>> \text{delx} = 0.05; \qquad (8\text{-}55)$$

$$>> \text{x} = 0:\text{delx}:2; \qquad (8\text{-}56)$$

The result of this operation is a 41-point vector.

The function to be integrated is now defined as

$$>> \text{y} = 4*\text{x.}^3; \qquad (8\text{-}57)$$

Since the value of the approximation will likely be different than the exact value, A1 will be used to indicate the result. It is given by

$$>> \text{A1} = \text{delx*sum(y)} \qquad (8\text{-}58)$$

A1 =

16.8100

Before commenting on this result, let us first evaluate the next part.

c. We have already defined delx, x, and y in part (b), so all we need to do now is to apply the **trapz** command. Let A2 represent the result of this step and we have

$$>> \text{A2} = \text{delx*trapz(y)} \qquad (8\text{-}59)$$

A2 =

16.0100

Clearly, the first-order approximation has provided a result that is far better than the zero-order form, and it will be our choice for much of the later work. In all fairness to the zero-order form, we could have done better by selecting a smaller step size, but the intent was to show how the methods compare for a moderate step size and a function having a third-degree variation.

Example 8-7

Consider the running integral

$$z = \int_0^x \sin x\, dx \tag{8-60}$$

Determine (a) the exact value and (b) the first-order approximation based on 100 points per cycle. (c) Plot the approximation as a continuous curve and the exact value as a sequence of o's.

Solution

a. The exact integral is determined as follows.

$$z = -\cos x\Big]_0^x = -\cos x - (-\cos 0) = 1 - \cos x \tag{8-61}$$

b. To employ numerical integration with MATLAB, we first define the increment delx as

$$\text{>> delx = 2*pi/100;} \tag{8-62}$$

Next, the vector x is defined as

$$\text{>> x = 0:delx:2*pi;} \tag{8-63}$$

The latter command actually generates 101 points since the last point could be considered as the first point of a new cycle. This is back to that "extra point" concept that arises often in linear spacing.

The function y being integrated is now defined as

$$\text{>> y = sin(x)} \tag{8-64}$$

Let z1 represent the first-order approximation to the integration. The command is

$$\text{>> z1 = delx*cumtrapz(y);} \tag{8-65}$$

The result is a 101-point vector providing a running approximation to the integral.

c. We could take the exact value and evaluate it separately for all values of the vector x. However, it can be done within the plot statement, and we will take that approach. These values will be indicated as o's. The command follows.

$$\text{>> plot(x, z1, x, 1-cos(x), 'o')} \tag{8-66}$$

The plotting results are shown in Figure 8-7. The points corresponding to the exact solution are right on top of the first-order approximation. Clearly, it does not get much better than this! Of course, sinusoidal functions are "well-behaved," and numerical results do not always turn out this well.

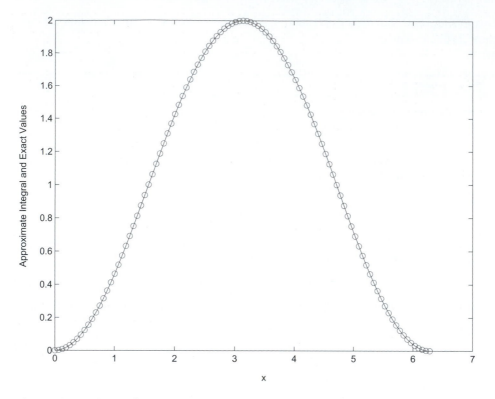

Figure 8-7 First-order approximation compared with exact values in Example 8-7

Judging Approximations

This chapter was developed to illustrate both "bad" and "good" approximations for numerical differentiation and integration. In all cases, we had exact results for comparison. When we don't have exact values, how do we know when to trust an answer? A simple but candid answer is that we don't know. This is where experience and judgment come into play, and no software can ever take the place of that process. When results are critical, it is customary to try different methods and different step sizes to see if the results converge to the same final values. If not, then alternate methods must be sought.

Example 8-8

A test is performed on a mechanical part to determine its response to certain stresses. An accelerometer connected to the device provides a reading of the acceleration a in m/s^2 at 0.1-second intervals over the time interval from $t = 0$ to $t = 2$ s. The readings are recorded as follows:

t,s	0	0.1	0.2	0.3	0.4	0.5	0.6	0.7	0.8
a, m/s^2	9	9.05	16.37	22.22	26.81	30.33	32.93	34.76	35.95

0.9	1.0	1.1	1.2	1.3	1.4	1.5	1.6	1.7	1.8	1.9	2.0
36.59	36.79	36.62	36.14	35.43	34.52	33.47	32.30	31.06	29.75	28.42	27.07

Use the MATLAB **cumtrapz** command to determine (a) the approximate velocity in m/s as a function of time and (b) the approximate displacement in meters as a function of time. Assume that the initial values of both velocity and displacement are zero. (c) Plot the acceleration, velocity, and displacement.

Solution

In a modern instrumentation system, these values would be digitized and stored in memory, but we will have to enter part of them the old-fashioned way. Fortunately, the time vector t can be entered in MATLAB with one command.

$$>> t = 0:0.1:2; \tag{8-67}$$

Instead of entering all 21 values with one long command, it is time we show the reader how to combine smaller matrices together to form larger matrices, providing that the combination makes sense. We will form four row vectors of length 5 each and then add the remaining point as a scalar. To that end, we define a1, a2, a3, and a4, each of which will consist of 5 values in successive order. The steps follow.

$$>> a1 = [0\ 9.05\ 16.37\ 22.22\ 26.81]; \tag{8-68}$$

$$>> a2 = [30.33\ 32.93\ 34.76\ 35.95\ 36.59]; \tag{8-69}$$

$$>> a3 = [36.79\ 36.62\ 36.14\ 35.43\ 34.52]; \tag{8-70}$$

$$>> a4 = [33.47\ 32.30\ 31.06\ 29.75\ 28.42]; \tag{8-71}$$

We could have easily placed the 21st point in the last vector, which would have given it a length of 6 points. However, we want to show how smaller matrices and possible scalars can be combined. Consider the following operation:

$$>> a = [a1\ a2\ a3\ a4\ 27.07]; \tag{8-72}$$

The vector now consists of the 21 points in the proper order.

To generalize the preceding concept somewhat, it is acceptable to place smaller matrices in a larger matrix if the combination is a valid matrix form. In the present case, all but the last point were row vectors and even it could be considered as a "trivial" row matrix, so the combination of placing them in row order is valid.

a. We are now ready to determine the velocity using the MATLAB trapezoidal approximation. Since the initial velocity is zero, we do not need to add a term to the definite integral form. However, we will designate the time increment Δt as delt and enter it in memory

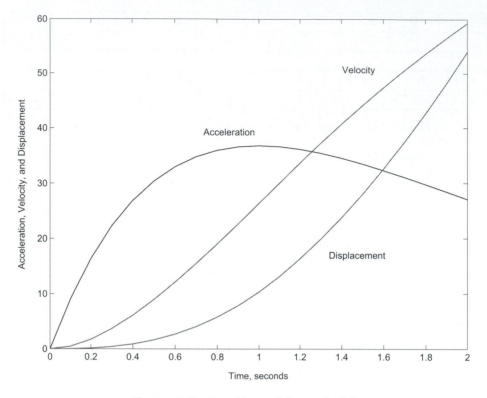

Figure 8-8 Functions of Example 8-8

$$>> \text{delt} = 0.1 \qquad (8\text{-}73)$$

The velocity function v is then estimated as

$$>> v = \text{delt*cumtrapz(a)}; \qquad (8\text{-}74)$$

b. We will now employ the same algorithm on the velocity to determine the displacement. This means that we will be performing two levels of numerical integration, so we must assume more care in the process. The displacement y is estimated as

$$>> y = \text{delt*cumtrapz(v)} \qquad (8\text{-}75)$$

c. The plots are generated by the command

$$>> \text{plot(t, a, t, v, t, y)} \qquad (8\text{-}76)$$

The results are shown in Figure 8-8. The author has checked these results with more sophisticated integration methods, and the results shown here are very accurate. In general, the trapezoidal algorithm is a very good approximation for relatively well-behaved functions and can be easily applied to a variety of situations. Some readers may have heard of digital filters. They are numerical algorithms that may be used with digital processors to separate undesired signals and noise from a desired signal or data set. One of the most widely used digital filter techniques makes use of the trapezoidal algorithm to achieve integration with a digital system.

MATLAB PROBLEMS

Many of the problems that follow are the same as or similar to problems at the ends of Chapters 6 and 7. In those chapters, you were expected to solve the problems by hand. In the present situation, you may use the MATLAB Command Window to the extent possible.

In Problems 8-1 through 8-10, use symbolic differentiation to determine the derivatives involved.

8-1. $y = x\sin x$

8-2. $y = x\cos x$

8-3. $y = x\cos(2x^2)$

8-4. $y = x\sin(2x^2)$

8-5. $y = \dfrac{\sin x}{x^2}$

8-6. $y = \dfrac{\cos x}{x}$

8-7. $y = xe^{-x}$

8-8. $y = xe^{-x^2}$

8-9. For the function of Problem 8-1, determine $\dfrac{d^2 y}{dx^2}$.
   Hint: Consider the command **diff(diff(y))**.

8-10. For the function of Problem 8-2, determine $\dfrac{d^2 y}{dx^2}$.
   (See *Hint* in Problem 8-9.)

In Problems 8-11 through 8-18, use symbolic integration to determine the integrals involved.

8-11. $z = \int 12x^3 dx$

8-12. $z = \int 20x^4 dx$

8-13. $z = \int 10e^{-2x} dx$

8-14. $z = \int 400 e^{-200x} dx$

8-15. $z = \int 20x \cos 4x dx$

8-16. $z = \int 20x \sin 4x dx$

8-17. $z = \int xe^{-x} dx$

8-18. $z = \int 20xe^{-4x} dx$

Note: You may prefer to write M-files to solve some of the problems that follow.

8-19. Consider the function $y = e^{-x}$ $0 \le x \le 5$.

Apply numerical differentiation to determine an approximation with 101 points. Plot the approximation as a continuous function and the exact derivative as a series of o's.

8-20. Consider the function $y = \cos x$ $0 \le x \le 2\pi$.

Apply numerical differentiation to determine an approximation with 101 points. Plot the approximation as a continuous function and the exact derivative as a series of o's.

8-21. Consider the definite integral to determine an area as follows:

$$A = \int_0^{\pi/2} \cos x dx$$

Determine (a) the exact area using a tabulated integral, (b) a zero-order approximation with MATLAB, and (c) a first-order approximation using MATLAB. Use a step size of 0.01π in (b) and (c).

8-22. Consider the definite integral to determine the area as follows:

$$A = \int_0^2 5x^4 dx$$

Determine (a) the exact area using a tabulated integral, (b) a zero-order approximation with MATLAB, and (c) a first-order approximation using MATLAB. Use a step size of 0.02 in (b) and (c).

8-23. Consider the running integral

$$z = \int_0^x \cos x dx$$

Determine (a) the exact value and (b) the first-order approximation over one cycle based on a step size of 0.01π. (c) Plot the approximation as a continuous curve and the exact value as a sequence of o's.

8-24. Consider the running integral

$$z = \int_0^x xe^{-x^2} dx$$

Determine (a) the exact value and (b) the first-order approximation over the domain $0 \leq x \leq 2$ based on 101 points. (c) Plot the approximation as a continuous curve and the exact value as a sequence of o's.

8-25. Write a program that will perform a numerical approximation of the derivative with respect to time of an array of experimental data and plot both functions as two separate subplots in a column format.

> *Inputs*: A row vector **y** consisting of an arbitrary number of input data points. The time step **delt.**

> *Outputs:* Upper plot of **y** as a function of time **t.**
> Lower plot of the derivative **z** as a function of time **t.**

Appropriate labeling should be provided for both curves.

> *Hint 1:* Since the length of the function is arbitrary, you may prefer to have the program determine its length rather than having to enter it as a separate value. Consider the simple command **n=size(y)**. The quantity **n** in general will be a row vector having two values in the following order: (1) the number of rows in **y** and (2) the number of columns in **y**. The size of **y** will be n(2) or, alternately, **length(y)**.

> *Hint 2:* To suppress the printing on the screen of input data values that you have entered, place a semicolon to the right of the right-hand parenthesis at the end of an **Input** statement.

a. Run the program, and when polled for the input row vector, type **0:100** and enter it. This will generate a 101-point array with values from 0 to 100 and, in effect, create a linearly increasing input function (a **ramp)**. When polled for **delt**, enter a value of 0.01.

b. Run the program, and when polled for the input row vector, type **randn(1,101)** and enter it. This will generate a 101-point row vector with a random pattern of numbers having a gaussian normal distribution (to be studied later). Statistically, the mean value of the array is 0 and the standard deviation is 1. Again, for **delt**, enter a value of 0.01.

8-26. Write a program that will perform a **first-order** numerical approximation of the running integral with respect to time of an array of experimental data and plot both functions as two separate subplots in a column format.

> *Inputs:* A row matrix **y** consisting of an arbitrary number of input data points. The time step **delt.**

> *Outputs:* Upper plot of **y** as a function of time **t.**
> Lower plot of the integral **z** as a function of time **t.**

Appropriate labeling should be provided for both curves. See the *Hints* in Problem 8-25.

a. Run the program for the same conditions of part (a) of Problem 8-25.

b. Run the program for the same conditions of part (b) of Problem 8-25.

c. Compare the results for the random input in Problem 8-25 and 8-26 and comment on the relative size and behavior of the derivative versus the integral.

Differential Equations: Classical Methods

9

Differential equations arise in virtually all areas of engineering and most areas of science. In general, a differential equation (DE) may be defined *as an equation involving one or more derivatives of an unknown dependent variable or several variables with respect to one or more independent variable or variables.* The objective in studying differential equations is to be able to solve for the desired dependent variables in terms of the independent variables. The outcome may be one or more closed-form solutions or it may be data obtained through some approximation technique.

There are many types of differential equations, but the most common types encountered in engineering applications are those of the *constant coefficient linear ordinary differential equations.* Such equations will constitute the bulk of this chapter. However, we will provide some initial classification schemes to assist the reader in recognizing the various types.

Objectives

After completing this chapter, the reader should be able to

1. Classify differential equations as either *linear* or *non-linear* and as either *ordinary* or *partial.*
2. Determine the *order* of a differential equation.
3. Discuss *boundary conditions* or *initial conditions* and their role in solving differential equations.
4. State the form of a *constant coefficient linear ordinary differential equation* (CCLODE) and recognize this form when it appears.
5. State the form of a *homogeneous* CCLODE and recognize this form when it appears.
6. Determine the solution of a differential equation that can be solved by simply integrating both sides.
7. Determine the form of the *characteristic equation* arising from a CCLODE.
8. Determine the form of the *homogeneous* or *complementary solution* from the roots of the characteristic equation.

9. Determine the *particular solution* for a non-homogeneous equation based on common forcing functions.

10. Apply *boundary* or *initial conditions* to determine a complete solution for a CCLODE.

11. Define *natural response, forced response, transient response,* and *steady-state* response and relate these terms to *homogeneous solution, particular solution,* and the properties of both the system and the forcing function.

12. For a system having complex (or purely imaginary) roots in the characteristic equation, state the form of the homogeneous solution based on real functions.

13. For a system characterized by the condition of 12, define *overdamped, critically damped, underdamped,* and *undamped.*

14. Apply CCLODE forms to deal with various applications such as trajectories, heat transfer, mechanical vibrations, and electrical circuits.

9-2   DIFFERENTIAL EQUATION CLASSIFICATIONS

Prior to studying the solutions of any differential equations, it is important to understand the classification scheme. A significant part of the history of applied mathematics has been the study of the different forms of differential equations and their possible solutions. Not all differential equations can be solved by conventional means, and some can only be solved by numerical approximations.

In the development that follows, we will assume the variable y as the dependent variable and the variable t as the primary independent variable. The choice of t is based on the fact that a large number of practical applications of differential equations involve various variables that change with *time*. When a second independent variable is needed, we will use x. We will frequently use DE as an abbreviation for differential equation.

Linear versus Non-Linear

At the broadest level, a differential equation may be classified as either *linear* or *non-linear*. A *linear differential equation* is one in which the dependent variable and its derivatives with respect to the independent variable are of the first degree and all multiplicative factors are either constants or functions only of the independent variable. Do not confuse *degree* with *order*. The *order* of the DE is the order of the highest derivative. A *higher degree* involves squaring or a higher power of the dependent variable or one of its derivatives.

An example of a linear differential equation is

$$2\frac{d^2y}{dt^2} + 5\frac{dy}{dt} + 3y = \sin 4t \tag{9-1}$$

This equation is a *second-order* DE since the highest derivative is a second derivative. It is *linear* since the factors of the dependent variable and its derivative are constants (2, 5, and 3) and all terms of the dependent variable are of the *first degree*.

The DE given by Equation 9-1 is a *constant coefficient linear differential equation* since all the coefficients are constants. There is a second type of linear differential equation in which the coefficients on the left can be functions of the independent variable *t*, but we will not consider those here.

A *non-linear differential equation* is one in which the coefficients are functions of the dependent variable *y* or there are higher powers of the dependent variable and its derivatives. An example of a non-linear differential equation is

$$y^2 \frac{dy}{dt} + y = 10 \tag{9-2}$$

This equation is non-linear because the coefficient of the first derivative (y^2) is a function of the dependent variable *y*.

Another example of a non-linear differential equation is

$$\left(\frac{dy}{dt}\right)^2 + 5y = 20 \tag{9-3}$$

This equation is non-linear because there is a higher-degree term (second degree) for the first derivative.

Ordinary versus Partial

An *ordinary differential equation* is one in which there is only one independent variable. The preceding three differential equations are ordinary because the derivatives are all formulated in terms of a single independent variable, which is *t* in all cases.

A *partial differential equation* is one in which the dependent variable (or variables) is a function of more than one independent variable. Assume, for example, that *y* is a function of two variables, *t* and *x*, that is, $y = y(x,t)$. The following is an example of a partial differential equation:

$$\frac{\partial^2 y}{\partial x^2} + a \frac{\partial^2 y}{\partial t^2} = b \tag{9-4}$$

The symbol ∂ is used for partial derivatives. This means that differentiation is performed only with respect to the variable in the denominator and other independent variables are treated as constants.

Continuous-Time versus Discrete-Time

The preceding definitions relate to *continuous-time* or *"analog"* systems. However, the same forms may be adapted to *discrete-time* or *"digital"* systems. In such cases, the equations are generally known as *difference equations* rather than *differential equations*. In fact, the numerical or computer solutions of differential equations are often performed by approximating the differential equations by difference equations.

Boundary Conditions

When differential equations are solved, the general solutions often contain arbitrary constants. The values of these constants are determined from the physical conditions surrounding the problem in terms of known constraints. These conditions are known as the *boundary conditions*. In general, if there are N arbitrary constants, there must be N independent conditions specified in order that the constants can be determined.

Initial Conditions

In many cases, particularly those that involve time as the independent variable, the *initial values* of the functions and various derivatives may be used in the solution. Therefore, the term *initial conditions* will be used interchangeably with *boundary conditions* in many situations. In general, if the DE is of order N, the function y and its first $N - 1$ derivatives will need to be specified for initial conditions. These will be represented by notation of the form $y(0)$, $y'(0)$, $y''(0)$, and so on.

Example 9-1

Classify the following differential equation as (a) linear or non-linear and (b) ordinary or partial:

$$t^2 \frac{d^2 y}{dt^2} + 2t \frac{dy}{dt} + 5y = e^{-2t} \tag{9-5}$$

Solution

a. The equation is *linear* since none of the coefficients of y and its derivatives are functions of the dependent variable y and there are no higher degree terms of y and its derivatives. Although the coefficients are a function of the independent variable t, the equation is still classified as linear.

b. The equation is *ordinary* because no partial derivatives are involved, that is, as far as the equation is concerned, y is a function only of t.

9-3 CONSTANT COEFFICIENT LINEAR ORDINARY DIFFERENTIAL EQUATIONS

The most common type of differential equation encountered in "routine" engineering and science applications is the *constant coefficient linear ordinary differential equation*. For simplicity, we will refer to such a DE in the future by the abbreviation *CCLODE*.

There is a good reason that the CCLODE forms dominate the engineering and science worlds. Many systems can be described over a reasonable operating range by such equations. Even in non-linear systems, a CCLODE form may be used to approximate the behavior over a limited region. Moreover, their solutions are well known and readily determined. Finally, equations that do not fit that form may be extremely difficult or impossible to solve and the linear model may be the only one that can be employed.

Until approximation methods using a digital computer were readily available, many solutions more involved than the CCLODE forms could not be solved by any reasonable methods. Therefore, the major emphasis for most of the chapter will be on the CCLODE forms.

Form of CCLODE

The basic form of a *CCLODE* of order m is given by

$$b_m \frac{d^m y}{dt^m} + b_{m-1} \frac{d^{m-1} y}{dt^{m-1}} + \dots + b_1 \frac{dy}{dt} + b_0 y = f(t) \tag{9-6}$$

Before proceeding further, please pause to study how all the adjectives fit in the definition. It is obviously a *differential equation* because it has terms involving derivatives of a dependent variable y in terms of an independent variable t. It is a *constant coefficient* type since all the factors of y and its derivatives are constants. It is *linear* since all the coefficients of y and its derivatives are not functions of y and there are no higher degrees of y or its derivatives. Since the highest order derivative is of order m, it is an mth order DE. Finally, it is an *ordinary* DE since the dependent variable y is a function only of the independent variable t and there are no partial derivatives.

Example 9-2

A certain differential equation is given by

$$3\frac{d^4 y}{dt^4} + 5\frac{d^3 y}{dt^3} + 7\frac{d^2 y}{dt^2} + 8\frac{dy}{dt} + 4y = \cos 5t + t^2 \tag{9-7}$$

(a) Is this equation a CCLODE form? (b) What is its order?

Solution

a. All the coefficients of y and its derivatives are constants and there are no higher-degree terms for y and its derivatives. Moreover, there are no partial derivatives, so it meets all the criteria and is a CCLODE form.

b. The highest derivative is a fourth derivative, so the order of the differential equation is $m = 4$.

In case the reader is beginning to feel intimidated, fear not. We will not be solving differential equations with this level of complexity by hand. The example was chosen simply to illustrate the terminology.

9-4   SIMPLE INTEGRABLE FORMS

Probably the simplest form of a CCLODE is one in which there is only one term on the left-hand side of Equation 9-6. A kth order differential equation of this form can be expressed as

$$b_k \frac{d^k y}{dt^k} = f(t) \tag{9-8}$$

In theory, this equation can be readily solved by integrating both sides k times. It may be convenient to introduce intermediate derivative variables so that at each stage, the equation is of first order. This process will be illustrated in Examples 9-3 and 9-4. For each step of integration, an arbitrary constant can be added. If the boundary conditions are the function and its first $k - 1$ derivatives at $t = 0$, the constant can actually be evaluated prior to the next integration step. Alternately, a definite integral form may be used.

Indefinite Integral Formulation

Consider, for example, a first-order differential equation of this form as given by

$$b_1 \frac{dy}{dt} = f(t) \tag{9-9}$$

In differential form, this equation can be rewritten as

$$dy = \frac{1}{b_1} f(t)dt \tag{9-10}$$

Integrating both sides, we obtain

$$y = \frac{1}{b_1} \int f(t)dt + C \tag{9-11}$$

Assume that we are given the initial value of y, that is, $y(0)$. The value of C may be determined from this value by substituting $t = 0$ in the indefinite form of Equation 9-11 and solving for the constant.

Definite Integral Formulation

As was noted in Chapter 7, a definite integral form may also be used. Provided there are no impulse conditions present, we simply add the initial value to the definite integral. The result is

$$y = \frac{1}{b_1} \int_0^t f(t)dt + y(0) \tag{9-12}$$

Example 9-3

Assume an object is dropped from some height h at $t = 0$ and assume negligible air resistance. The acceleration due to gravity is indicated as g. Near the surface of the Earth, the value of this constant is about $g = 9.80$ m/s^2 or 32.2 ft/s^2. The variable y will represent the distance in the *downward* direction from the point at which the object is dropped. Determine equations for the velocity v and displacement y, with the positive direction downward as indicated.

Solution

The velocity v is related to the displacement y by

$$v = \frac{dy}{dt} \tag{9-13}$$

The acceleration a in general is the derivative of velocity with respect to time. Thus,

$$a = \frac{dv}{dt} \tag{9-14}$$

If the velocity from Equation 9-13 is substituted in Equation 9-14, the acceleration can be expressed as

$$a = \frac{d^2 y}{dt^2} \tag{9-15}$$

However, it is easier to work with first derivatives, so by starting with velocity and working back to displacement, each equation is a first-order form. Therefore, we begin with Equation 9-14, but substitute the constant value g for the acceleration, resulting in

$$\frac{dv}{dt} = g \tag{9-16}$$

or

$$dv = gdt \tag{9-17}$$

Integrating both sides of Equation 9-17 we obtain

$$v = gt + C_1 \tag{9-18}$$

where C_1 is the constant of integration. Assuming that the ball is dropped from rest, we assume the boundary condition $v(0) = 0$. Substituting this value on the left and $t = 0$ on the right, we obtain

$$C_1 = 0 \tag{9-19}$$

The velocity is then given by

$$v = gt \tag{9-20}$$

Now that we have performed the first level of integration from acceleration to velocity, we can substitute the relationship for velocity in terms of displacement from Equation 9-13 and write

$$\frac{dy}{dt} = gt \tag{9-21}$$

or

$$dy = gtdt \tag{9-22}$$

Integrating both sides of Equation 9-22, we obtain

$$y = \frac{1}{2} gt^2 + C_2 \tag{9-23}$$

where C_2 is the constant of integration in this case.

The boundary condition in this case is based on the assumption that $y = 0$ when $t = 0$. Substitution of these values in Equation 9-23 leads to the result

$$C_2 = 0 \tag{9-24}$$

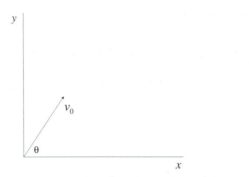

Figure 9-1   Projectile orientation of Example 9-4

Thus, the displacement is given by

$$y = \frac{1}{2} g t^2 \qquad (9\text{-}25)$$

Strictly speaking, this result is completely valid only in a vacuum where the air resistance is zero. However, it is a very good approximation for objects having reasonably large masses and at reasonably low altitudes. (It is obviously invalid for a feather!)

Example 9-4

Continuing with the simple integrable forms and with acceleration and velocity functions, consider a projectile fired from a ground point as shown in Figure 9-1. Assume that the projectile is fired at an angle θ from the ground and assume that the initial velocity is v_0. Assume that the point at which the trajectory begins is considered as $x = 0$ and $y = 0$. Neglecting air resistance and assuming a constant value of g, derive equations for the velocity and the displacement as a function of time. (Neglect any curvature of the Earth.)

Solution

The initial velocity has both x and y components. Let v_y represent the component of velocity in the y direction and let v_x represent the component of velocity in the x direction. Let $v_y(0)$ and $v_x(0)$ represent their respective initial velocities. From the trigonometry of the figure, we have

$$v_y(0) = v_0 \sin \theta \qquad (9\text{-}26)$$

and

$$v_x(0) = v_0 \cos \theta \qquad (9\text{-}27)$$

In contrast to Example 9-3, it is better in this case to choose y as positive in the upward direction, as depicted in Figure 9-1. The DE in the y direction is then given by

$$\frac{dv_y}{dt} = -g \tag{9-28}$$

Note the negative sign in this case since the acceleration of gravity is in opposition to the positive direction of velocity and displacement.

We will skip the step of showing the differential form in this case, but when it is formed and both sides are integrated, the result is

$$v_y = -gt + C_1 \tag{9-29}$$

Substitution of the value of $v_y(0)$ from Equation 9-26 at $t = 0$ yields

$$C_1 = v_0 \sin\theta \tag{9-30}$$

The velocity in the y direction is then given by

$$v_y = -gt + v_0 \sin\theta \tag{9-31}$$

We can now express an equation governing the displacement in the y direction as

$$\frac{dy}{dt} = v_y = -gt + v_0 \sin\theta \tag{9-32}$$

Converting to differential form and integrating both sides, we obtain

$$y = -\frac{1}{2}gt^2 + (v_0 \sin\theta)t + C_2 \tag{9-33}$$

Assuming $y(0) = 0$, the constant is readily determined as

$$C_2 = 0 \tag{9-34}$$

Thus, the vertical displacement as a function of time is given by

$$y = -\frac{1}{2}gt^2 + (v_0 \sin\theta)t \tag{9-35}$$

Since the effect of gravity is completely in the y direction, we will assume no external forces of acceleration in the x direction. Hence, the differential equation for the x component of velocity is simply

$$\frac{dv_x}{dt} = 0 \tag{9-36}$$

This leads to

$$v_x = C_3 \tag{9-37}$$

where C_3 is the constant of integration. Substitution of the initial condition of the x component of velocity leads to

$$C_3 = v_0 \cos\theta \tag{9-38}$$

This means that the x component of velocity is a constant given by

$$v_x = v_0 \cos\theta \tag{9-39}$$

We can now substitute the expression for the derivative of x with respect to time and obtain

$$\frac{dx}{dt} = v_0 \cos\theta \qquad (9\text{-}40)$$

Changing to differential form and integrating both sides, we obtain

$$x = (v_0 \cos\theta)t + C_4 \qquad (9\text{-}41)$$

Substituting $x = 0$ when $t = 0$, the value of C_4 is

$$C_4 = 0 \qquad (9\text{-}42)$$

Hence, the value of the x coordinate

$$x = (v_0 \cos\theta)t \qquad (9\text{-}43)$$

Equations 9-35 and 9-43 describe the values of the coordinates x and y in parametric form. In several exercises at the end of the chapter, some of the properties of the trajectory will be explored. It will be shown that the trajectory is a parabola and other properties will be determined.

9-5   DIFFERENTIAL EQUATION SOLUTIONS

In the previous section, we considered one of the simplest forms of a differential equation, namely one that can be solved by successive integration steps. We now turn our attention to those constant coefficient linear ordinary differential equation (CCLODE) forms that cannot be solved in that manner. Fortunately, the general forms of the solutions are well known and have been thoroughly investigated over the years.

Let us begin by repeating the CCLODE form, which is given by

$$b_m \frac{d^m y}{dt^m} + b_{m-1} \frac{d^{m-1} y}{dt^{m-1}} + \ldots + b_1 \frac{dy}{dt} + b_0 y = f(t) \qquad (9\text{-}44)$$

The function $f(t)$ is called the *forcing function*. This term is suggestive of the fact that it often represents some external force or other physical variable that is producing some action in the system. In some disciplines, it is referred to as the *excitation* and the solution y is referred to as the *response*.

If $f(t) = 0$, the differential equation is said to be *homogeneous*. In this case, the solution or response is determined by the system parameters, and the boundary conditions determine the level of the response terms.

In general, the solution to a CCLODE can be represented in two parts: (1) the solution to the homogeneous equation, indicated by y_h, and (2) a particular solution y_p dictated by the forcing function. Thus, y can be expressed as

$$y = y_h + y_p \qquad (9\text{-}45)$$

The *homogeneous solution* y_h is the solution of the *homogeneous* form of the differential equation, which is obtained from Equation 9-44 by setting the right-hand side to 0. (The homogeneous solution is also referred to in many references as the *complementary solution*.)

The *particular solution* y_p will have a form similar to that of $f(t)$ and is determined as a separate step in the process. The procedures for determining each of the solutions follow.

Homogeneous Solution

By setting $f(t) = 0$ in Equation 9-44, the *homogeneous differential equation* assumes the form

$$b_m \frac{d^m y}{dt^m} + b_{m-1} \frac{d^{m-1} y}{dt^{m-1}} + \dots + b_1 \frac{dy}{dt} + b_0 y = 0 \qquad (9\text{-}46)$$

To simplify the notation until the last step in this section, we will employ y without a subscript, but it should be understood that it is the solution to the homogeneous solution that is being considered. It has been determined that the basic form of a solution to Equation 9-46 can be expressed as

$$y = Ce^{pt} \qquad (9\text{-}47)$$

where C is an arbitrary constant and the form or forms for p will be determined shortly.

In order for Equation 9-47 to be a solution for Equation 9-46, it is necessary to substitute y and all derivatives up to the mth derivative in the equation. The various derivatives are easily obtained from the properties of the exponential function as

$$\frac{dy}{dt} = pCe^{pt} \qquad (9\text{-}48)$$

$$\frac{d^2 y}{dt^2} = p^2 Ce^{pt} \qquad (9\text{-}49)$$

$$\bullet$$
$$\bullet$$

$$\frac{d^m y}{dt^m} = p^m Ce^{pt} \qquad (9\text{-}50)$$

Substitution of y and the various derivatives in Equation 9-46 yields

$$b_m p^m Ce^{pt} + b_{m-1} p^{m-1} Ce^{pt} + \dots + b_1 pCe^{pt} + b_0 Ce^{pt} = 0 \qquad (9\text{-}51)$$

Note that Ce^{pt} is a common factor in all the terms of Equation 9-51 and can be cancelled. This results in

$$b_m p^m + b_{m-1} p^{m-1} + \dots + b_1 p + b_0 = 0 \qquad (9\text{-}52)$$

Characteristic Equation

Equation 9-52 is called the *characteristic equation,* and its roots determine the possible values of p that appear in the solution. Note that with a little practice, the characteristic equation may be written down from a simple inspection of the DE. The coefficients in the characteristic equation are the coefficients of the various derivatives and the exponent of each p term is the order of the corresponding derivative.

Equation 9-52 is a polynomial of order m and will, therefore, have m roots. Each of the roots expressed as a factor of t in Equation 9-47 will satisfy the homogeneous equation.

Let $p_1, p_2, \ldots, p_m$ represent the m roots. For the moment, *assume that there are no repeated roots, that is, each root has a different value.*

Since any one of the roots, when expressed in exponential form, results in zero for the homogeneous equation, it follows that the sum of the exponential terms utilizing the different roots will also satisfy the homogeneous equation. Therefore, the *homogeneous solution* y_h will be of the form

$$y_h = C_1 e^{p_1 t} + C_2 e^{p_2 t} + \ldots + C_m e^{p_m t} \tag{9-53}$$

where $C_1, C_2, \ldots, C_m$ are arbitrary constants. These constants will eventually be determined from the boundary conditions. However, it is first necessary to determine the particular solution.

It should be noted that *when the original differential equation is of homogeneous form, the homogeneous solution will be the complete solution and there will be no particular solution.*

Particular Solution

The particular solution depends directly on the form of $f(t)$. The classical approach is to first inspect the form of $f(t)$ and then deduce the form of the particular solution. We will restrict our considerations here to the common forms most often encountered in applied engineering and science problems. This procedure is known as the *method of undetermined coefficients*.

With some exceptions (to be discussed later), the table that follows provides the general guidelines for selecting the form of the particular solution y_p. Initially, there will also be some arbitrary constants that must be determined prior to the determination of the constants in the homogeneous solution. To distinguish the constants in the particular solution from those in the homogeneous solution, we will employ different symbols, depending on the form.

Form of $f(t)$	Form assumed for y_p
K	A
Kt	$A_1 t + A_0$
Kt^2	$A_2 t^2 + A_1 t + A_0$
$K_1 \cos \omega t$ and/or $K_2 \sin \omega t$	$A_1 \sin \omega t + A_2 \cos \omega t$
$Ke^{-\alpha t}$	$Ae^{-\alpha t}$

Along with the previous assumption of no repeated roots in the characteristic equation, we need to add one more restriction for the preceding table to be completely valid at this point. It will be assumed for the moment that *none of the roots of the characteristic equa-*

tion will produce a term exactly equivalent to any of the forms in the table. The equivalency for a sinusoidal function corresponds to purely imaginary roots $\pm i\omega$, where the same ω appears in a sinusoidal forcing function. The equivalency for an exponential function corresponds to a real root $-\alpha$, where the same value appears as an argument of an exponential forcing function. It should be noted that either the sine function or the cosine function or both in $f(t)$ require that both cosine and sine functions be assumed in the particular solution.

The particular solution is formulated according to the rules provided in the table, with the unknown constants assumed. This solution form is then substituted in the original differential equation (not the homogeneous form), and by equating coefficients of various functions on the two sides of the equation, the constants of the particular solution are then determined. However, there still remains the chore of determining the constants assumed in the homogeneous solution. At that point, the complete solution is formulated as the sum of the homogeneous solution and the particular solution. The boundary conditions are then used to determine the constants that were assumed in the homogeneous solution.

The preceding concepts will be illustrated in the examples that follow.

Example 9-5

A differential equation is given by

$$\frac{dy}{dt} + 2y = 0 \tag{9-54}$$

with a boundary condition given by

$$y(0) = 10 \tag{9-55}$$

Determine the solution.

Solution

Aside from the simple integrable forms considered earlier, this first-order form is about as simple as they can get and still be called a differential equation. Yet such forms arise frequently in many systems. Since $f(t) = 0$, the DE is of homogeneous form and will thus have only a homogeneous solution.

We could start with the assumed exponential form of Equation 9-47, differentiate and substitute in the DE. However, it is simpler to go directly to the characteristic equation whose form is that of Equation 9-52, and we then have

$$p + 2 = 0 \tag{9-56}$$

and

$$p = -2 \tag{9-57}$$

Thus,

$$y = Ce^{-2t} \tag{9-58}$$

We now employ the boundary condition by substituting the value of $y(0)$ on the left and $t = 0$ on the right. This leads to

$$10 = Ce^{-0} = C \qquad (9\text{-}59)$$

or

$$C = 10 \qquad (9\text{-}60)$$

The complete solution is then

$$y = 10e^{-2t} \qquad (9\text{-}61)$$

The reader is invited to verify that this function satisfies both the differential equation and the initial condition.

The reader is reminded that if there were a particular solution, it would have to be determined before the constant C can be determined. In this example, there was no particular solution and so the constant could be immediately determined. The next example will be more representative of the first-order situation.

Example 9-6

A differential equation is given by

$$\frac{dy}{dt} + 2y = 12 \qquad (9\text{-}62)$$

with a boundary condition given by

$$y(0) = 10 \qquad (9\text{-}63)$$

Determine the solution.

Solution

This DE differs from that of Example 9-5 only in the fact that there is a non-zero constant on the right-hand side, that is, we can say that the forcing function is $f(t) = 12$ (even though it really is not varying with time in this case). In general, we should form the homogeneous equation first. However, we can save a few steps by recognizing that the DE of Example 9-5 is, in fact, the homogeneous form of Equation 9-62. Therefore, the homogeneous solution can be expressed as

$$y_h = Ce^{-2t} \qquad (9\text{-}64)$$

However, we cannot determine C until we determine the particular solution. From the table provided earlier, it was shown that for a constant value of $f(t)$, the assumed value of the particular solution is

$$y_p = A \qquad (9\text{-}65)$$

Now the particular solution must be substituted into the DE in order to determine A. Since the derivative of a constant A with respect to t is 0, substitution yields

$$0 + 2A = 12 \tag{9-66}$$

or

$$A = 6 \tag{9-67}$$

The particular solution is then

$$y_p = 6 \tag{9-68}$$

The total solution is

$$y = y_h + y_p = Ce^{-2t} + 6 \tag{9-69}$$

We can now evaluate C by substituting $y = 10$ when $t = 0$. This yields

$$10 = Ce^{-0} + 6 = C + 6 \tag{9-70}$$

This results in

$$C = 4 \tag{9-71}$$

The complete solution is then

$$y = 4e^{-2t} + 6 \tag{9-72}$$

Example 9-7

A differential equation is given by

$$\frac{dy}{dt} + 2y = 12\sin 4t \tag{9-73}$$

with a boundary condition given by

$$y(0) = 10 \tag{9-74}$$

Determine the solution.

Solution

As in Example 9-6, the homogeneous form is the same as that of Example 9-5, and we can say again that

$$y_h = Ce^{-2t} \tag{9-75}$$

The particular solution is this case is a bit more involved. From the table provided earlier, in the "Particular Solution" section, the presence of a sine function or a cosine function on the right-hand side means that we must assume both a sine function and a cosine function for the particular solution. Thus, we have

$$y_p = A_1 \sin 4t + A_2 \cos 4t \tag{9-76}$$

In order to substitute this function in the DE, we must also determine the derivative with respect to t. This function is

$$\frac{dy_p}{dt} = 4A_1 \cos 4t - 4A_2 \sin 4t \qquad (9\text{-}77)$$

Substituting the function and its derivative in the DE, we have

$$4A_1 \cos 4t - 4A_2 \sin 4t + 2\left(A_1 \sin 4t + A_2 \cos 4t\right) = 12 \sin 4t \qquad (9\text{-}78)$$

The next step is to group the coefficients of sin $4t$ and cos $4t$ on both sides of the equation.

$$\left(4A_1 + 2A_2\right)\cos 4t + \left(2A_1 - 4A_2\right)\sin 4t = 12 \sin 4t \qquad (9\text{-}79)$$

The "trick" here is to recognize that since Equation 9-79 is an equality, the coefficient of sin $4t$ on the left must equal the coefficient of sin $4t$ on the right, and likewise for cos $4t$. This leads to the two simultaneous equations

$$4A_1 + 2A_2 = 0 \qquad (9\text{-}80)$$

and

$$2A_1 - 4A_2 = 12 \qquad (9\text{-}81)$$

We will leave it to the reader to verify that the simultaneous solution of the two preceding equations yields

$$A_1 = 1.2 \qquad (9\text{-}82)$$

and

$$A_2 = -2.4 \qquad (9\text{-}83)$$

The particular solution is then

$$y_p = 1.2 \sin 4t - 2.4 \cos 4t \qquad (9\text{-}84)$$

The complete solution can then be expressed as

$$y = Ce^{-2t} + 1.2 \sin 4t - 2.4 \cos 4t \qquad (9\text{-}85)$$

We still must evaluate C. Substituting $y = 10$ when $t = 0$, we have

$$10 = Ce^{-0} + 1.2 \sin(0) - 2.4 \cos(0) = C + 0 - 2.4(1) \qquad (9\text{-}86)$$

This results in

$$C = 12.4 \qquad (9\text{-}87)$$

The complete solution is then given by

$$y = 12.4e^{-2t} + 1.2 \sin 4t - 2.4 \cos 4t \qquad (9\text{-}88)$$

It is clear that even for a first-order DE with simple numbers, the process starts to get messy when the driving function is anything more than a constant. Be patient, and wait to see what MATLAB can do.

Example 9-8

A differential equation is given by

$$\frac{d^2 y}{dt^2} + 3\frac{dy}{dt} + 2y = 0 \tag{9-89}$$

with boundary conditions given by

$$y(0) = 10 \text{ and } y'(0) = 0 \tag{9-90}$$

Determine the solution.

Solution

A few points will be made at the beginning. In the previous three examples, the DEs were of the first order, but this one is a second-order form. However, it is a homogeneous form, so there will be no particular solution. Note that there are two boundary (initial) conditions given, which is necessary for a second-order form.

As has been done on several occasions, the form of the characteristic equation will be written down by inspection. It is

$$p^2 + 3p + 2 = 0 \tag{9-91}$$

The roots of this equation are $p_1 = -1$ and $p_2 = -2$. Therefore, the homogeneous solution, which is the complete solution, can be expressed as

$$y = C_1 e^{-t} + C_2 e^{-2t} \tag{9-92}$$

To determine the constants, it will be necessary to apply boundary conditions to both y and its derivative. The derivative is

$$\frac{dy}{dt} = -C_1 e^{-t} - 2C_2 e^{-2t} \tag{9-93}$$

We now substitute $y = 10$ and $t = 0$ in Equation 9-92 and $y' = 0$ and $t = 0$ in Equation 9-93. The resulting two equations are

$$10 = C_1 + C_2 \tag{9-94}$$

and

$$0 = -C_1 - 2C_2 \tag{9-95}$$

Simultaneous solution yields

$$C_1 = 20 \tag{9-96}$$

and

$$C_2 = -10 \tag{9-97}$$

The solution is then

$$y = 20e^{-t} - 10e^{-2t} \tag{9-98}$$

Example 9-9

A differential equation is given by

$$\frac{d^2 y}{dt^2} + 3\frac{dy}{dt} + 2y = 24 \tag{9-99}$$

with boundary conditions given by

$$y(0) = 10 \text{ and } y'(0) = 0 \tag{9-100}$$

Determine the solution.

Solution

The homogeneous equation has the same form as the DE Example 9-8, so the homogeneous solution is

$$y_h = C_1 e^{-t} + C_2 e^{-2t} \tag{9-101}$$

Since the forcing function is simply a constant, the particular solution will be of the form

$$y_p = A \tag{9-102}$$

All derivatives will be zero, so substitution of the particular solution form in Equation 9-99 results in

$$0 + 0 + 2A = 24 \tag{9-103}$$

or

$$A = 12 \tag{9-104}$$

The general solution is then

$$y = C_1 e^{-t} + C_2 e^{-2t} + 12 \tag{9-105}$$

We now proceed to determine the two constants. First we differentiate to obtain

$$\frac{dy}{dt} = -C_1 e^{-t} - 2C_2 e^{-2t} \tag{9-106}$$

Substituting $y = 10$ when $t = 0$ and $y' = 0$ when $t = 0$, we have

$$10 = C_1 + C_2 + 12 \tag{9-107}$$

and

$$0 = -C_1 - 2C_2 \tag{9-108}$$

Simultaneous solution of these two equations yields

$$C_1 = -4 \tag{9-109}$$

and

$$C_2 = 2 \tag{9-110}$$

The complete solution can then be expressed as

$$y = -4e^{-t} + 2e^{-2t} + 12 \qquad (9\text{-}111)$$

9-6 SOME PROPERTIES OF DIFFERENTIAL EQUATION SOLUTIONS

In the last section, we began studying the properties of CCLODEs and how the solutions are formulated. At the end of the section, several first-order and second-order DEs were solved. For the examples given, the homogeneous solution was a simple exponential for the first-order form and two exponentials for the second-order examples. For a first-order CCLODE having both a derivative term and a function term, the homogeneous solution will always be an exponential function. As we will see later, however, second- and higher-order equations *may* exhibit terms that are best expressed in a different form than exponentials.

For the DEs studied in the last section, the particular solutions, when present, assumed the general form of the forcing function. We are now ready to generalize these properties somewhat and indicate how they relate to real physical systems.

Homogeneous Solution and Natural Response

It turns out that the homogeneous solution generally arises from the physical properties of the system being analyzed. The form could display time constants of electrical, thermal, or mechanical systems, or it could display natural vibration frequencies under some conditions. While the term *homogeneous solution* is a common term in mathematics, a term more common to engineers is that of the *natural response*. Thus, for most practical purposes, we will consider the following terms as synonymous:

homogeneous solution ⇔ natural response

Particular Solution and Forced Response

As indicated earlier, the particular solution exhibits the form of the forcing function for the system. Its magnitude, however, is a function of the system parameters. Again, we have different terminology between mathematicians and engineers. The mathematical form denoted as the *particular solution* is usually called the *forced response* by engineers. For our purposes, we will consider the following terms as synonymous:

particular solution ⇔ forced response

Stability

The *stability* of an engineering system is a very important consideration and can be related in many cases to the nature of the describing DE. In simple terms, a stable system is one in which the response is bounded and does not "run away" or have disastrous levels in its response functions. Stability can be predicted from the natural response (homogeneous solution). *A system is said to be stable if its natural response approaches zero as the time increases without limit.* If this condition is met, the system will be stable for any finite forcing function.

Some systems allow a natural response that assumes a continued bounded form such as a sinusoidal natural response or a constant. Such systems are said to be *marginally stable.* This condition would be totally undesirable in a position control system, for example, but is the actual design goal in an electrical oscillator circuit. Finally, an unstable system is one in which the natural response grows without bound. In any real system, there will be certain bounds at which the system will break down, so the model employing a CCLODE usually becomes invalid after a certain point in an unstable system.

Transient Response and Steady-State Response

There are two other terms that are used often in describing an engineering system: the *transient response* and the *steady-state response.* Strictly speaking, these terms apply only to a stable system in which the forced response (particular solution) assumes some repetitive or constant form after the natural response vanishes. In this qualified case, the following terms are often considered synonymous:

$$\text{natural response} \Leftrightarrow \text{transient response}$$

$$\text{forced response} \Leftrightarrow \text{steady-state response}$$

Again, we emphasize that in this context, the terms *transient* and *steady-state* are meaningful only when the system is stable and with certain types of input forcing functions.

Classification of Roots of the Characteristic Equation

Thus far, all the roots of characteristic equations considered have been real roots of first-order. The second-order DE examples had two roots, but since they were different, they were first-order roots. In general, the roots can be classified in four ways:

1. first-order and real
2. first-order and complex (including purely imaginary)
3. multiple-order and real
4. multiple-order and complex (including purely imaginary)

Classical mathematics texts tend to cover all the various types. This author, who is an engineer with many years of experience working with applied mathematics, has never encountered a system that fell into category (4). Moreover, the cases that fell into category (3) all involved second-order roots. Therefore, on the basis of relative likelihood of practical application, we will consider in this text only cases (1) and (2) and the second-order form involving case (3). (If one of the ignored cases ever arises, just use MATLAB!)

The major case of practical importance involving case 4 is where the roots are purely imaginary and the forcing function has a frequency component corresponding to a resonant frequency, which would be represented by a pair of purely imaginary roots. The resulting response in this case would be unstable, and some of the natural disasters that have occurred can be partially explained by this phenomenon. However, it is unlikely that a linear model would hold up in such a situation, so we will not consider it for our purposes.

Example 9-10

The differential equation describing the displacement y of a certain machine part as a function of time t is given by

$$2\frac{d^2y}{dt^2} + 10\frac{dy}{dt} + 8y = 80 \tag{9-112}$$

(a) State the form of the natural response. (b) State the form of the forced response. (c) Is the system stable? (d) If the system is stable, indicate the transient and steady-state response forms.

Solution

a. The homogenous equation is obtained by setting the right-hand side of Equation 9-112 to zero. With this step understood, the characteristic equation can be written by inspection as

$$2p^2 + 10p + 8 = 0 \tag{9-113}$$

The roots of this equation are readily determined as $p_1 = -1$ and $p_2 = -4$. The natural response (or homogeneous solution) will then be of the form

$$y_h = C_1e^{-t} + C_2e^{-4t} \tag{9-114}$$

b. The forced response (particular solution) will be of the form

$$y_p = A \tag{9-115}$$

It is easily shown that by substituting this function in the DE, the forced response is simply

$$y_p = 10 \tag{9-116}$$

The complete solution then has the form

$$y = C_1e^{-t} + C_2e^{-4t} + 10 \tag{9-117}$$

We are unable to determine the two constants C_1 and C_2 without any boundary conditions, but we are aware of the form.

c. The system is stable since both of the exponential terms approach zero as time increases without limit.

d. Since the system is stable and the forced response approaches a constant value (10), we can say that the *natural response is a transient response and the forced response is a steady-state response.*

9-7 SECOND-ORDER SYSTEMS

This section will be devoted to a study of second-order systems, that is, those that can be described by a second-order DE. Many physical systems can be described by second-order

CCLODEs. Moreover, second-order systems exhibit the forms that arise in higher-order systems. In other words, once the forms of second-order systems are studied, one is poised to see the forms that arise in higher-order systems.

In general, the response y of a second-order CCLODE system can be described by

$$b_2 \frac{d^2 y}{dt^2} + b_1 \frac{dy}{dt} + b_0 y = f(t) \tag{9-118}$$

The characteristic equation will be of the form

$$b_2 p^2 + b_1 p + b_0 = 0 \tag{9-119}$$

Let p_1 and p_2 represent the two roots of this equation. In general, they can be (1) real and different, (2) real and equal, or (3) complex (which will include purely imaginary as a special case). Each form will now be considered.

Roots Are Real and Different

The homogeneous solution in this case will have the form

$$y_h = C_1 e^{p_1 t} + C_2 e^{p_2 t} \tag{9-120}$$

In order for the system to be stable, both p_1 and p_2 must be negative. It can be shown that a necessary and sufficient condition for this to be the case is that the three constants b_2, b_1, and b_0 be non-zero and have the same sign. In order to delineate the importance of a stable system, let $p_1 = -\alpha_1$ and $p_2 = -\alpha_2$. The stable form usually encountered will have a natural response of the form

$$y_h = C_1 e^{-\alpha_1 t} + C_2 e^{-\alpha_2 t} \tag{9-121}$$

Roots Are Real and Equal

This is the first point in the chapter where we have considered the situation of roots that are real and equal. In this case, we will assume that p is a second-order root, which would arise from Equation 9-119 if both roots are the same. Assuming again a stable situation, we let $p = -\alpha$. It turns out in this case that the homogeneous solution must assume a different form:

$$y_h = (C_0 + C_1 t) e^{-\alpha t} \tag{9-122}$$

Without showing it, because of its rarity, an astute reader might guess that for a higher-order root, which would naturally require a higher-order DE, the form of Equation 9-122 would require higher-order degrees of t in the parentheses.

Roots Are Complex (Including Purely Imaginary Roots)

In this case, if the coefficients of the DE are real (the usual case), any complex roots of the characteristic equation occur as a complex conjugate pair. Assume that the pair of complex conjugate roots is given by $p_1, p_2 = -\alpha \pm i\omega$, where $\alpha \geq 0$ and $i = \sqrt{-1}$. Thus, we are assuming that the roots will either have a negative real part or a real part of zero. The first case corresponds to a stable system, and the second corresponds to a marginally stable case. If the real part were positive, the system would be unstable, and we will not consider that situation.

One form of the homogeneous solution can be expressed as

$$y_h = B_1 e^{(-\alpha + i\omega)t} + B_2 e^{(-\alpha - i\omega)t} = e^{-\alpha t}\left(B_1 e^{j\omega t} + B_2 e^{-i\omega t}\right) \tag{9-123}$$

where the notation has been changed momentarily because of the change in form to follow.

While the form of Equation 9-123 is perfectly correct mathematically, it is not very satisfying because it contains exponential terms with imaginary arguments. By Euler's equation for an exponential with an imaginary argument, it is possible to reformulate the preceding equation in the form

$$y_h = C_1 e^{-\alpha t} \sin \omega t + C_2 e^{-\alpha t} \cos \omega t \tag{9-124}$$

This is the form that is easier to apply in practical problems and will be used in subsequent developments. Note that when $\alpha = 0$, corresponding to purely imaginary roots, the homogeneous solution will be of the form

$$y_h = C_1 \sin \omega t + C_2 \cos \omega t \tag{9-125}$$

In this case, the natural response will not approach zero, and there will be continued oscillations with constant amplitude. Thus, the system will be marginally stable.

Relative Damping

Second-order systems falling in the preceding three categories may be classified as follows:

1. If the roots are *real and unequal*, the system is said to be *overdamped*.
2. If the roots are *real and equal*, the system is said to be *critically damped*.
3. If the roots are *complex*, the system is said to be *underdamped*.
4. A special case of an *underdamped* system occurs if $\alpha = 0$. In this case, the system is said to be *undamped*.

Example 9-11

A differential equation is given by

$$\frac{d^2 y}{dt^2} + 2\frac{dy}{dt} + 5y = 0 \tag{9-126}$$

with initial conditions given by

$$y(0) = 0 \text{ and } y'(0) = 10 \tag{9-127}$$

Determine the solution.

Solution

The DE is a homogeneous form, so there will be no particular solution. The characteristic equation is

$$p^2 + 2p + 5 = 0 \tag{9-128}$$

The roots of this equation are determined as $p_1, p_2 = -1 \pm 2i$. From the discussion in this section, we can deduce that the response will be *underdamped* and of the form

$$y = C_1 e^{-t} \sin 2t + C_2 e^{-t} \cos 2t \tag{9-129}$$

It should be noted that the real part of the roots represents a damping constant for the exponential function and the imaginary part represents a radian oscillatory frequency.

In order to determine the constants, we need to first form the first derivative of the function. Each of the terms in Equation 9-129 is the product of two functions, which means that the derivative will initially have four terms. The function is

$$\frac{dy}{dt} = 2C_1 e^{-t} \cos 2t - C_1 e^{-t} \sin 2t - 2C_2 e^{-t} \sin 2t - C_2 e^{-t} \cos 2t \tag{9-130}$$

This expression could be simplified by combining some terms, but it is hardly worth the effort. We now substitute $y = 0$ when $t = 0$ and $y' = 10$ when $t = 0$. Since sin (0) = 0 and cos (0) = 1, the resulting equations become

$$0 = 0 + C_2 \tag{9-131}$$

and

$$10 = 2C_1 - 0 - 0 - C_2 \tag{9-132}$$

This leads quickly to $C_2 = 0$ and $C_1 = 5$. The solution is then

$$y = 5e^{-t} \sin 2t \tag{9-133}$$

In this case, the cosine term vanished. Depending on boundary conditions and the forcing function, a solution may have a sine, a cosine, or both.

Example 9-12

A differential equation is given by

$$\frac{d^2 y}{dt^2} + 2\frac{dy}{dt} + 5y = 20 \tag{9-134}$$

with boundary conditions given by

$$y(0) = 0 \text{ and } y'(0) = 10 \tag{9-135}$$

Determine the solution.

Solution

The left-hand side of the DE is the same as that of Example 9-11, but in this case there is a forcing function. The homogeneous equation and the characteristic equation are the same as the DE of Example 9-11, so the homogeneous solution is

$$y_h = C_1 e^{-t} \sin 2t + C_2 e^{-t} \cos 2t \tag{9-136}$$

We must first determine the particular solution before determining the constants. It is of the form

$$y_p = A = 4 \tag{9-137}$$

where the value 4 was readily determined by substituting A in Equation 9-134. The total solution is

$$y = C_1 e^{-t} \sin 2t + C_2 e^{-t} \cos 2t + 4 \tag{9-138}$$

The derivative is the same as in Example 9-11 and is

$$\frac{dy}{dt} = 2C_1 e^{-t} \cos 2t - C_1 e^{-t} \sin 2t - 2C_2 e^{-t} \sin 2t - C_2 e^{-t} \cos 2t \tag{9-139}$$

Next we substitute $y = 0$ when $t = 0$ and $y' = 10$ when $t = 0$. This leads to

$$0 = 0 + C_2 + 4 \tag{9-140}$$

and

$$10 = 2C_1 - 0 - 0 - C_2 \tag{9-141}$$

Simultaneous solution of the preceding two equations yields

$$C_2 = -4 \tag{9-142}$$

and

$$C_1 = 3 \tag{9-143}$$

The complete solution is then

$$y = 3e^{-t} \sin 2t - 4e^{-t} \cos 2t + 4 \tag{9-144}$$

9-8   APPLICATION PROBLEMS

This section will be devoted to a few applications of differential equations to some practical physical situations.

Example 9-13

This example will deal with *Newton's law of cooling,* which states that the temperature, T, of a relatively small object placed in a medium with temperature $T_{ambient}$ changes at a rate proportional to the difference between the temperature of the object and the ambient temperature. It can be expressed as

$$\frac{dT}{dt} = -k(T - T_{ambient}) \qquad (9\text{-}145)$$

where k is a positive constant measured in kelvins/second (K/s) and t is the time in seconds. Assume that the initial value of the object temperature is T_0. Determine the temperature as a function of time.

Solution

The DE may be rewritten in the standard form of a first-order CCLODE as

$$\frac{dT}{dt} + kT = kT_{ambient} \qquad (9\text{-}146)$$

The homogeneous solution is an exponential function with a negative argument, and the particular solution is a constant. Following the principles discussed in the chapter, the reader is invited to show that the solution of this equation is given by

$$T = T_{ambient} + \left(T_0 - T_{ambient}\right)e^{-kt} \qquad (9\text{-}147)$$

The temperature changes in an exponential fashion. Whether the ambient temperature is greater than or less than the initial surface temperature determines the direction in which the change occurs.

Example 9-14

In this example, we will study the phenomenon of radioactive decay. Most radioactive materials disintegrate at a rate proportional to the amount present at a given time t. Let $Q(t)$ represent the amount at any time t and let Q_0 represent the initial quantity. The rate of change of the material is then governed by the differential equation

$$\frac{dQ}{dt} = -rQ \qquad (9\text{-}148)$$

where r is the positive rate of decay. Determine a solution for the amount of material as a function of time.

Solution

The standard form of this DE can be expressed as

$$\frac{dQ}{dt} + rQ = 0 \qquad (9\text{-}149)$$

Since this DE is of homogeneous form, there is no particular solution and the complete solution can be readily determined as

$$Q = Q_0 e^{-rt} \qquad (9\text{-}150)$$

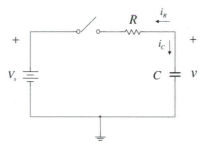

Figure 9-2   Circuit of Example 9-15

Example 9-15

Consider the electrical circuit shown in Figure 9-2, containing a resistance, R, measured in ohms (Ω) and a capacitance, C, measured in farads (F). A capacitor will hold charge, which can be represented by a voltage. Assume that the capacitor is initially charged to V_0 volts (V). Assume that at $t = 0$, the switch connecting the circuit to the external power supply of value V_s is closed. In this example, we will assume for convenience that $V_s > V_0$, although the results will apply for either inequality. Write a differential equation in terms of the unknown voltage $v(t)$ and solve it. Some basic laws required in the process are as follows:

1. Kirchhoff's current law: The **algebraic** sum of the currents leaving a node is zero. (Equivalently, the current entering is equal to the current leaving.)

2. The current i_R in amperes (A) through a resistor is related to the voltage v_R across the resistance by Ohm's law.

$$i_R = \frac{v_R}{R} \tag{9-151}$$

3. The current i_c flowing into a capacitance is related to the voltage v_c across the capacitance by

$$i_C = C\frac{dv_C}{dt} \tag{9-152}$$

Solution

We will assume the reference direction for positive current as leaving the output positive reference node. From Kirchhoff's current law,

$$i_C + i_R = 0 \tag{9-153}$$

The first term is given by Equation 9-152, with the assumption that $v_c = v$. The second term is given by Equation 9-151, but it should be recognized that $v_R = v - V_s$. Substituting these values, we obtain

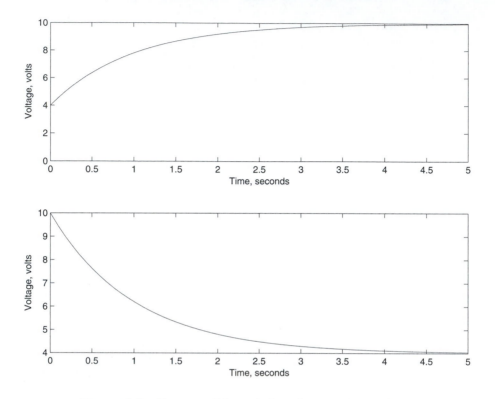

Figure 9-3   Two possible variations in capacitor voltage

$$C\frac{dv}{dt}+\frac{(v-V_s)}{R}=0 \tag{9-154}$$

Rearranging and dividing by C, we have

$$\frac{dv}{dt}+\frac{1}{RC}v=\frac{V_s}{RC} \tag{9-155}$$

This is a first-order CCLODE, and with the initial condition provided, the reader is invited to show that the solution is obtained as

$$v=V_s+\left(V_0-V_s\right)e^{-t/RC}=V_s+\left(V_0-V_s\right)e^{-t/\tau} \tag{9-156}$$

where $\tau = RC$ is known as the *time constant* in seconds. It represents the time required for the change in voltage to be about 63.2% of the total change.

Curves showing two possible situations are illustrated in Figure 9-3. The upper graph displays the capacitor voltage charging from an initial value of 4 V to a final value of 10 V, and the lower graph displays a capacitor voltage discharging from an initial value of 10 V to a final value of 4 V. The time scale provided represents 5 time constants.

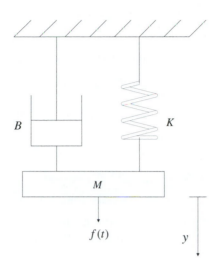

Figure 9-4   Vibration system of Example 9-16

Example 9-16

The model shown in Figure 9-4 can represent certain mechanical vibration systems, such as shock absorbers. The quantity B represents a viscous damping constant, and the quantity K represents a spring constant. The mass is represented by M, and $f(t)$ is some external forcing function. For any displacement y, the viscous damping force f_v opposes a change in motion and is given by

$$f_v = B\frac{dy}{dt} \tag{9-157}$$

Likewise, the spring force f_s resists motion. It is given by

$$f_s = Ky \tag{9-158}$$

Newton's second law states that the algebraic sum of all the forces on a body is equal to the mass times the acceleration. Using the preceding relationships, write a DE that describes the motion of the system.

Solution

The forcing function is acting downward, but the damping and spring forces are in opposition. Hence, Newton's law applied to the mass yields

$$f(t) - f_v - f_s = Ma \tag{9-159}$$

where a is the acceleration given by

$$a = \frac{d^2 y}{dt^2} \tag{9-160}$$

Substitution of the various terms in Equation 9-159 and rearrangement yield

$$M \frac{d^2 y}{dt^2} + B \frac{dy}{dt} + Ky = f(t) \tag{9-161}$$

This result is a second-order CCLODE, which can be solved by the concepts considered in this chapter.

GENERAL PROBLEMS

In Problems 9-1 through 9-18, solve for the variable y in terms of t from the given differential equation based on the methods of this chapter. Initial condition or conditions are provided as applicable.

9-1. $\quad \dfrac{dy}{dt} = 6t^2 \qquad y(0) = 5$

9-2. $\quad \dfrac{dy}{dt} = 20 \cos 5t + 8 \qquad y(0) = 10$

9-3. $\quad \dfrac{dy}{dt} + 4y = 0 \qquad y(0) = 6$

9-4. $\quad \dfrac{dy}{dt} + 50y = 0 \qquad y(0) = 20$

9-5. $\quad \dfrac{dy}{dt} + 4y = 80 \qquad y(0) = 6$

9-6. $\quad \dfrac{dy}{dt} + 50y = 400 \qquad y(0) = 20$

9-7. $\quad \dfrac{dy}{dt} + 4y = 40 \sin 3t \qquad y(0) = 6$

9-8. $\quad \dfrac{dy}{dt} + 50y = 400 \cos 75t \qquad y(0) = 20$

9-9. $\quad \dfrac{d^2 y}{dt^2} + 4 \dfrac{dy}{dt} + 3y = 0 \qquad y(0) = 0 \quad y'(0) = 12$

9-10. $\quad \dfrac{d^2 y}{dt^2} + 7 \dfrac{dy}{dt} + 12y = 0 \qquad y(0) = 20 \quad y'(0) = 0$

9-11. $\quad \dfrac{d^2 y}{dt^2} + 4 \dfrac{dy}{dt} + 3y = 30 \qquad y(0) = 20 \quad y'(0) = 12$

9-12. $\dfrac{d^2y}{dt^2}+7\dfrac{dy}{dt}+12y=96$ $y(0)=20$ $y'(0)=0$

9-13. $\dfrac{d^2y}{dt^2}+3\dfrac{dy}{dt}+2y=24$ $y(0)=0$ $y'(0)=10$

9-14. $\dfrac{d^2y}{dt^2}+3\dfrac{dy}{dt}+2y=24e^{-4t}$ $y(0)=10$ $y'(0)=5$

9-15. $\dfrac{d^2y}{dt^2}+4\dfrac{dy}{dt}+13y=52$ $y(0)=0$ $y'(0)=0$

9-16. $\dfrac{d^2y}{dt^2}+4\dfrac{dy}{dt}+104y=832$ $y(0)=0$ $y'(0)=0$

9-17. $\dfrac{d^2y}{dt^2}+6\dfrac{dy}{dt}+9y=0$ $y(0)=10$ $y'(0)=0$

9-18. $\dfrac{d^2y}{dt^2}+6\dfrac{dy}{dt}+9y=90$ $y(0)=0$ $y'(0)=0$

9-19. Classify the following differential equation as (a) linear or non-linear and (b) ordinary or partial.

$$\sin y\,\dfrac{dy}{dt}+5y=\cos 4t$$

9-20. Classify the following differential equation as (a) linear or non-linear and (b) ordinary or partial.

$$5\dfrac{\partial y}{\partial t}+8\dfrac{\partial y}{\partial x}=12\cos t$$

In Problems 9-21 through 9-28, start with the basic differential equation and solve for velocity and displacement rather than trying to adapt the results developed in the chapter to the solution. Assume that $g=9.80$ m/s^2 or 32.2 ft/s^2. Assume that air resistance is negligible and neglect any curvature of the Earth for horizontal displacement.

9-21. An object is thrown downward from the top of a building with an initial velocity of 30 m/s. Assuming a positive direction of y measured downward from the top, derive an expression for (a) the velocity and (b) the displacement as a function of time. Assume $y(0)=0$.

9-22. An object is thrown downward from the top of a building with an initial velocity of 50 ft/s. Assuming a positive direction of y measured downward from the top, derive an expression for (a) the velocity and (b) the displacement as a function of time. Assume $y(0)=0$.

9-23. An object is propelled directly upward from the ground with an initial velocity of 30 m/s. Assuming a positive direction of y measured *upward*, derive an expression for (a) the velocity and (b) the displacement as a function of time. Assume $y(0) = 0$.

9-24. An object is propelled directly upward from the ground with an initial velocity of 50 ft/s. Assuming a positive direction of y measured *upward*, derive an expression for (a) the velocity and (b) the displacement as a function of time. Assume $y(0) = 0$.

9-25. For the object of Problem 9-23, determine (a) how high the object will go and (b) how long it will be before it returns to the initial position.

9-26. For the object of Problem 9-24, determine (a) how high the object will go and (b) how long it will be before it returns to the initial position.

9-27. Consider the projectile firing model of Example 9-4 and assume that $\theta = 30°$ and $v_0 = 100$ m/s. Determine (a) equations for v_y, v_x, x, and y. (b) Determine the maximum height that the projectile will reach. (c) Determine the time from firing that it takes it to reach the Earth. (d) Determine the maximum horizontal distance.

9-28. Consider the projectile firing model of Example 9-4 and assume that $\theta = 60°$ and $v_0 = 100$ m/s. Determine (a) equations for v_y, v_x, x, and y. (b) Determine the maximum height that the projectile will reach. (c) Determine the time from firing that it takes it to return to the Earth. (d) Determine the maximum horizontal distance.

DERIVATION PROBLEMS

In Problems 9-29 through 9-32, assume that g is a constant and that air resistance is negligible. Neglect any effects of curvature of the Earth.

9-29. For the projectile firing model of Example 9-4, and using equations developed in the text, show that the equation for the trajectory can be expressed as

$$y = \left(\frac{-g}{2v_0^2 \cos^2 \theta} \right) x^2 + (\tan \theta) x$$

9-30. For the projectile firing model of Example 9-4, show that the maximum altitude of the projectile is given by

$$y_{max} = \frac{v_0^2 \sin^2 \theta}{2g}$$

You may use the result of Problem 9-29.

9-31. For the projectile firing model of Example 9-4, determine the angle θ that maximizes the horizontal distance traveled before returning to the earth. You may use the result of Problem 9-29.

9-32. For the projectile firing model of Example 9-4, determine an expression for the maximum horizontal distance x_{max}.

Note: Since the total emphasis in this chapter was analytical in nature, MATLAB problems dealing with differential equations will be delayed to the end of Chapter 11.

Differential Equations: Laplace Transform Methods

10

10-1 OVERVIEW AND OBJECTIVES

In the preceding chapter, we learned how to solve constant coefficient linear ordinary differential equations (CCLODE) by the so-called classical methods. There is one other analytical method that is commonly used to solve such equations, the Laplace transform method. This method is one that falls under the heading of *operational mathematics*. It has been widely used by applied mathematicians, scientists, and engineers. The Laplace transform was developed by the French mathematician Pierre-Simon de Laplace (1749–1827) and was widely adapted to engineering problems in the last century. Its utility lies in the ability to convert differential equations to algebraic forms that are more easily manipulated and solved. Moreover, the notation of Laplace forms has become very common in certain disciplines, so the vocabulary associated with the method is important in understanding many engineering applications.

Objectives

After completing this chapter, the reader should be able to

1. State the definition of the Laplace transform.
2. Discuss the procedure by which a differential equation can be solved by Laplace transforms.
3. Determine the Laplace transforms of common engineering functions using a table.
4. Determine the inverse Laplace transforms of functions that are easily recognizable from a table.
5. Apply partial fraction expansion to simplify a transform function for inverse transformation.
6. Solve a complete constant coefficient linear ordinary differential equation (CCLODE) by Laplace transforms.

10-2 GENERAL APPROACH

The manner in which the Laplace transform is generally applied to solve differential equations is illustrated by the flowchart of Figure 10-1. The differential equation to be solved

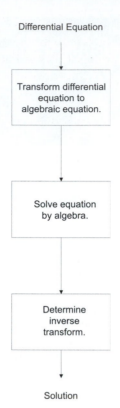

Differential Equation

Transform differential
equation to
algebraic equation.

Solve equation
by algebra.

Determine
inverse
transform.

Solution

Figure 10-1   Solution of a differential equation with the Laplace transform

is first transformed to an algebraic equation using the Laplace transformation. The transform equation can be completely manipulated by basic algebraic methods, and the desired variable is determined. The inverse transform is then determined, and the result is the solution of the differential equation.

In certain disciplines such as electrical engineering and electrical engineering technology, the manner in which the Laplace transform is applied is based on circuit models that actually include Laplace operations in the form of circuit impedances. With such methods, an analyst does not even have to write the differential equation, but instead, the entire solution can be obtained through basic circuit theorems and operations. Since this text is intended for a wider audience, the approach here will be oriented toward the differential equation itself.

Laplace Transformation

In the various developments that follow, we will assume a function of time $f(t)$. This could be an electrical voltage, a mechanical force, a thermal source or any other physical variable that will be part of a differential equation.

The Laplace transform of a time function $f(t)$ is a new function $F(s)$, where s is the Laplace transform variable or operator. Do not worry about what s is at this point. Simply accept it as a variable that must be kept in all equations when it arises. The process of transformation is indicated symbolically as

$$L[f(t)] = F(s) \qquad (10\text{-}1)$$

This equation is read as "the Laplace transform of $f(t)$ is $F(s)$." The process of inverse transformation is indicated symbolically as

$$L^{-1}[F(s)] = f(t) \qquad (10\text{-}2)$$

This latter equation is read as "the inverse Laplace transform of $F(s)$ is $f(t)$."

Definition of Laplace Transform

The mathematical definition of the Laplace transform is

$$F(s) = \int_0^\infty f(t) e^{-st} dt \qquad (10\text{-}3)$$

Based on the definition, the function $f(t)$ is multiplied by e^{-st}, and the product is integrated over positive time to produce the Laplace transform. Although a few transforms will be developed for illustration, most transforms of interest have been derived and are tabulated in numerous references. Most practical problems requiring Laplace transformation can be solved by using tabulated results to determine the transforms. As in the case of derivatives and integrals, transforms may also be determined from MATLAB, which will be illustrated in the next chapter.

Basic Theorems of Linearity

In the derivative and integral tables considered earlier in the text, two basic relationships were provided at the beginning of each table. They were the so-called "obvious" forms, but we have chosen here to address them in more detail. These are the conditions for a system or operation to be **linear**. Assume that K is a constant. In terms of the Laplace transform, the conditions of linearity can be stated by Equations 10-4 and 10-5:

$$L[Kf(t)] = KL[f(t)] = KF(s) \qquad (10\text{-}4)$$

Thus, the Laplace transform of a constant times a function is the constant times the transform of the function.

Next, let $f_1(t)$ and $f_2(t)$ represent any arbitrary functions, and let $F_1(s)$ and $F_2(s)$ represent their respective transforms. We now have

$$L[f_1(t) + f_2(t)] = L[f_1(t)] + L[f_2(t)] = F_1(s) + F_2(s) \qquad (10\text{-}5)$$

Thus, the Laplace transform of the sum of two functions is the sum of the Laplace transforms.

An Incorrect Interpretation

We need to indicate clearly an interpretation sometimes made by beginners, which is how to determine the Laplace transform of a product of two functions. The transform of a product of two time functions is **not** the product of the two transform functions, that is, $L[f_1(t)f_2(t)] \neq F_1(s)F_2(s)$. There is a process called **convolution** that may be used to determine the transform of the product of two functions, but it is far more involved than a simple multiplication and will not be considered in this text. The only way to deal with the product in this text is to consider the product as a new function and determine its transform.

Time Scale

In the definition of the Laplace transform in Equation 10-3, note that $t = 0$ is the starting point for integration. While some functions may be delayed until some later time to begin, the framework for most Laplace transform analysis is based on positive time, and usually with $t = 0$ as the assumed starting point. Keep this point in mind when we begin looking at functions, since they will be assumed to be zero for $t < 0$.

Step Function

In engineering problems in which a forcing function having a constant value is applied to the system, the term *step function* is frequently used. The form of a *unit step function* is shown in Figure 10-2, and it is denoted in many references as $u(t)$. It can be defined as

$$u(t) = 0 \qquad \text{for } t < 0$$
$$= 1 \qquad \text{for } t > 0 \tag{10-6}$$

For example, if a force of 5 N is applied at $t = 0$, it could be described as $5u(t)$. There is nothing mysterious about the $u(t)$ function, but it is a way of describing the switching process. It is treated as a constant in various equations in which its value is used. Any function that starts at $t = 0$ can be considered as being multiplied by the unit step function whether explicitly stated or not. The examples that follow will illustrate the concepts employed in deriving Laplace transforms.

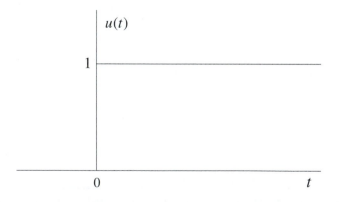

Figure 10-2   Illustration of the unit step function.

Example 10-1

Derive the Laplace transform of the unit step function $u(t)$.

Solution

The basic definition of the Laplace transform was given in Equation 10-3. The unit step function begins at $t = 0$, and since $u(t) = 1$ for positive time, we set up the integral as follows:

$$F(s) = \int_0^\infty (1)e^{-st}\,dt \tag{10-7}$$

Before performing the integration, a few points should be made. First, the Laplace operator s is treated like a constant in the time integration process. Second, the limits of $t = 0$ to $t = \infty$ should be interpreted as telling us to integrate over all positive time rather than necessarily to plug in the two limits. In this case, the limits are in fact the same as in the integral definition. However, in some cases, the function may be 0 over a portion of the positive time interval, and in some cases, the function may be defined differently over different intervals.

The integral is that of an exponential function, and so for the next step, we have

$$F(s) = \left.\frac{e^{-st}}{-s}\right]_0^\infty = 0 - \left(\frac{e^{-0}}{-s}\right) = \frac{1}{s} \tag{10-8}$$

It has been assumed that the exponential function approaches 0 at the upper limit of the preceding equation. This means that s must be interpreted as a positive value for that to be true. A more rigorous basis is to indicate that the integral exists for s to the right of a so-called *abscissa of absolute convergence*. Unless you are a mathematician, you do not need to worry about this fine point.

The conclusion is that the Laplace transform of a constant is the constant divided by s.

Example 10-2

Derive the Laplace transform of the exponential function $e^{-\alpha t}$.

Solution

The transform is derived through the following steps:

$$F(s) = \int_0^\infty e^{-\alpha t}e^{-st}\,dt = \int_0^\infty e^{-(\alpha+s)t}\,dt = \left.\frac{e^{-(s+\alpha)t}}{-(s+\alpha)}\right]_0^\infty = 0 - \frac{e^{-0}}{-(s+\alpha)} \tag{10-9}$$

$$= \frac{1}{s+\alpha}$$

The same comment made in Example 10-1 about the upper limit applies here as well.

Actually, the step function could be considered as a special case of the exponential function for $\alpha = 0$. In that case, $e^{-\alpha t}$ reduces to 1. Simultaneously, the transform $1/(s + \alpha)$

reduces to $1/s$. However, both functions arise so often that it is appropriate to consider them separately.

10-3 TABLE OF LAPLACE TRANSFORM FUNCTIONS

The most common Laplace transforms of interest in applied engineering applications are listed in Table 10-1, at the end of the chapter. The column designated as $f(t)$ in Table 10-1 indicates the basic form of the time function, and the column labeled as $F(s)$ indicates the corresponding form of the Laplace transform. We will provide some interpretations in the discussion that follows.

T-1 and T-2

These were just derived in Examples 10-1 and 10-2, so there is little more to say about them. One common error that students tend to make is to write the transform of a constant incorrectly as a constant, but note that it is a constant divided by s.

T-3 and T-4

These two transforms are those of the sine and cosine functions, respectively. Note that the denominator of both is $s^2 + \omega^2$, but the numerator is different for the two transforms.

T-5 and T-6

These are the transforms for the damped sine function and the damped cosine function, respectively. Many common tables show the exponential factor with a positive argument, but the forms given here are the most common forms encountered. Of course, the signs ultimately take care of the different ways they are stated, but the forms given here are the most useful in engineering applications.

T-7 and T-8

As the degree of t increases, the degree of the s in the denominator of the transform also increases. For the first-degree function t, the transform is $1/s^2$, but for t^2, the transform is $2/s^3$, and so on.

T-9

This transform arises when the characteristic equation, as discussed in Chapter 9, has multiple-order real roots.

T-10

In a number of places throughout the text, we have spoken of impulsive conditions, which has probably left a bit of mystery. Frankly, we are still going to leave it for future consideration, but the so-called impulse function is often denoted as $\delta(t)$. An interesting fact about it is that it has the simplest transform of all functions, namely, the constant 1. Now do not confuse the time function 1, which has a transform of $1/s$, with the transform 1, in which the time function is the impulse $\delta(t)$.

The examples that follow this section are designed solely for practice using Table 10-1.

Example 10-3

A force $f(t)$ in newtons (N) applied to a structural system at $t = 0$ can be described as

$$f(t) = 50u(t) \qquad (10\text{-}10)$$

Determine the Laplace transform.

Solution

With virtually no pause, we use T-1, along with the linearity property, and write

$$F(s) = \frac{50}{s} \qquad (10\text{-}11)$$

Example 10-4

A voltage $v(t)$ in volts (V) beginning at $t = 0$ can be expressed as

$$v(t) = 5e^{-2t} \sin 4t \qquad (10\text{-}12)$$

Determine the Laplace transform.

Solution

From T-5 and the linearity property, we have

$$V(s) = L[v(t)] = 5 \cdot \frac{4}{(s+2)^2 + (4)^2} = \frac{20}{s^2 + 4s + 4 + 16} = \frac{20}{s^2 + 4s + 20} \qquad (10\text{-}13)$$

Example 10-5

A pressure function in pascals (Pa) beginning at $t = 0$ can be expressed as

$$p(t) = 5\cos 2t + 3e^{-4t} \qquad (10\text{-}14)$$

Determine the Laplace transform.

Solution

Along with linearity, the two terms can be transformed according to T-4 and T-2. We have

$$P(s) = L[p(t)] = 5 \cdot \frac{s}{s^2 + (2)^2} + 3 \cdot \frac{1}{s+4} = \frac{5s}{s^2 + 4} + \frac{3}{s+4} \qquad (10\text{-}15)$$

10-4   INVERSE LAPLACE TRANSFORMS BY IDENTIFICATION

While the whole process is still a little puzzling in the minds of many readers, a little practice has shown that the determination of the Laplace transform of a time function of the forms considered has been quite easy. We merely look up the function in Table 10-1, at the end of the chapter, and write down the transform.

Unfortunately, the process of *inverse Laplace transformation*, that is, the process of determining a time function from the transform function, is often more difficult. Once it is mastered, however, the final chore of solving differential equations by this process will quickly fall into place.

Recognizable Functions from the Table

In a few fortunate cases, a Laplace transform will fall exactly into one of the forms in Table 10-1. In that case, the inverse transform can be written down from a fairly simple inspection. The process will be illustrated with several examples that follow.

Example 10-6

Determine the inverse transform of the following function:

$$F(s) = \frac{5}{s} + \frac{12}{s^2} + \frac{8}{s+3} \tag{10-16}$$

Solution

Reviewing Table 10-1, we can deduce that the three terms in respective order are of the forms of transforms T-1, T-7, and T-2, with multiplicative constants. Hence, the inverse transform is

$$f(t) = 5 + 12t + 8e^{-3t} \tag{10-17}$$

Example 10-7

Determine the inverse transform of the velocity function

$$V(s) = \frac{200}{s^2 + 100} \tag{10-18}$$

Solution

The function appears to be close to the transform of T-3, but a little manipulation is in order. First we recognize that the constant term in the denominator is ω^2, and since $\omega^2 = 100$, then $\omega = 10$ rad/s. Note from T-3 that we need ω in the numerator. Therefore, we rearrange the function as

$$V(s) = 20\left(\frac{10}{s^2 + (10)^2}\right) \tag{10-19}$$

The quantity in parentheses is the Laplace transform of sin 10t, so the net inverse transform is

$$v(t) = 20\sin 10t \tag{10-20}$$

Example 10-8

Determine the inverse transform of the voltage function that follows.

$$V(s) = \frac{8s+4}{s^2 + 6s + 13} \tag{10-21}$$

Solution

First it should be emphasized that the roots of a polynomial should be determined before making a decision on the proper method of inversion. The denominator is a quadratic, which would suggest the possibility of T-5 and/or T-6 from Table 10-1. However, as noted by the footnote, those pairs should be used only if the roots are complex (which includes purely imaginary roots). If the roots are real, the partial fraction procedure of the next section should be used.

In the case at hand, the quadratic formula leads to the roots $s_{1,2} = -3 \pm 2i$. Hence, the roots are complex and T-5 and/or T-6 are applicable. In order to apply T-5 and/or T-6, we need to arrange the polynomial in the form $(s + \alpha)^2 + \omega^2$. A process called "completing the square" can do this. To perform this operation, we take half of the coefficient of the s term (3), square it, and both add and subtract it from the polynomial. The procedure follows.

$$s^2 + 6s + 13 = s^2 + 6s + (3)^2 + 13 - (3)^2 = s^2 + 6s + 9 + 4 \tag{10-22}$$

The first three terms constitute a perfect square, and the last term is ω^2. The polynomial is then of the form

$$s^2 + 6s + 13 = (s+3)^2 + (2)^2 \tag{10-23}$$

It is now in the form of the denominator of both T-5 and T-6, and we can write

$$V(s) = \frac{8s+4}{(s+3)^2 + (2)^2} \tag{10-24}$$

The presence of both an s-term and a constant in the numerator suggest that both T-5 and T-6 will be involved. However, further manipulation is required. From T-6, we need to have an $s + 3$ factor in the numerator, and since the multiplier of s is 8, we actually need 8(s + 3). Moreover, we need a 2 in the numerator of the damped sine term. Some manipulation follows.

$$V(s) = \frac{8(s+3)}{(s+3)^2 + (2)^2} + \frac{4-24}{(s+3)^2 + (2)^2} = \frac{8(s+3)}{(s+3)^2 + (2)^2} - \frac{10(2)}{(s+3)^2 + (2)^2} \tag{10-25}$$

Complete the calculation and you will get the original function. We manipulated the terms so that they would fit the forms of T-6 and T-5. The inverse transform is then

$$v(t) = 8e^{-3t} \cos 2t - 10e^{-3t} \sin 2t \qquad (10\text{-}26)$$

10-5　INVERSE TRANSFORMS BY PARTIAL FRACTION EXPANSION

In the last section, we considered inverse Laplace transforms that were immediately recognizable from Table 10-1, at the end of the chapter. In many cases, however, the transforms are more involved. More commonly, the transforms need to be reduced or simplified before the few transforms in the table can be applied.

Forms for CCLODE Forms

In the expressions most frequently encountered in the constant coefficient linear ordinary differential equation (CCLODE) forms, any desired transform function will usually be a ratio of polynomials in the variable s. Assuming a transform variable $F(s)$, the function may be expressed as

$$F(s) = \frac{N(s)}{D(s)} \qquad (10\text{-}27)$$

where $N(s)$ is a numerator polynomial of the form

$$N(s) = a_n s^n + a_{n-1} s^{n-1} + \ldots + a_1 s + a_0 \qquad (10\text{-}28)$$

and $D(s)$ is a denominator polynomial of the form

$$D(s) = b_m s^m + b_{m-1} s^{m-1} + \ldots + b_1 s + b_0 \qquad (10\text{-}29)$$

Poles and Zeros

From our understanding of polynomials, we know that the numerator polynomial contains n roots and the denominator polynomial contains m roots. *The roots of the numerator polynomial are called the **zeros** of F(s), and the roots of the denominator polynomial are called the **poles** of F(s).*

Factored Form

We now consider $D(s)$ to be factored, and thus we can express $F(s)$ in the form

$$F(s) = \frac{N(s)}{b_m(s - p_1)(s - p_2)\ldots.(s - p_m)} \qquad (10\text{-}30)$$

where the p values represent the various poles of $F(s)$.

Relationship to Characteristic Equation

As a general rule, the *poles*, which are the denominator roots in Equation 10-30, correspond to the roots of the characteristic equation of the differential equation when such a solution is the focus of the analysis. As such, the poles of Equation 10-30 may be classified in four groups:

1. real poles of first order
2. complex poles of first order (including purely imaginary poles)
3. real poles of multiple order
4. complex poles of multiple order (including purely imaginary poles)

As was done in the preceding chapter, we will not consider case (4) because of its very rare occurrence, and we will only consider the second-order situation in case (3).

Partial Fraction Expansion

The most common way of dealing with transform functions is that of *partial fraction expansion*. This is the process of "pulling apart" the unwieldy overall function and obtaining a series of simpler functions that can be recognized from Table 10-1.

Real Poles of First Order

The simplest case of all is that of real poles of first order. Assume that the denominator contains r real poles of first order. Although we assumed m total poles, some may belong to other classifications. The partial fraction expansion follows:

$$F(s) = \frac{A_1}{s - p_1} + \frac{A_2}{s - p_2} + ... + \frac{A_r}{s - p_r} + R(s) \tag{10-31}$$

where $R(s)$ is the remaining portion of the expansion due to poles belonging to other classifications. Each of the r terms in Equation 10-31 may be inverse transformed by the use of transform pair T-2 (or by T-1 when $p = 0$).

Consider an arbitrary constant A_k. Let us multiply both sides of Equation 10-31 by $(s - p_k)$ and rearrange some terms. This results in

$$(s - p_k)F(s) = (s - p_k)\left[\frac{A_1}{s - p_1} + \frac{A_2}{s - p_k} + ... + R(s)\right] + A_k \tag{10-32}$$

The $(s - p_k)$ multiplier has canceled the denominator of the A_k term. It has also canceled the $(s - p_k)$ factor in the denominator of $F(s)$. Next, we let $s = p_k$. As a result of that factor in front of the bracketed terms, all terms on the right of Equation 10-32 vanish except the term A_k. Hence, a general expression for A_k is

$$A_k = (s - p_k)F(s)\big]_{s=p_k} \tag{10-33}$$

The notation of this formula means that we multiply $F(s)$ by $(s - p_k)$, which always cancels the $(s - p_k)$ factor in the denominator, and then we let $s = p_k$. The resulting value is the A_k coefficient. This procedure is performed for each of the r real roots of simple order.

After all the A_k terms are determined, the inverse transform of the portion of Equation 10-31 involving the real poles of simple order can be determined from T-2 or T-1. We will designate this portion of the time response as $f_1(t)$ to distinguish it from the total time response $f(t)$. This portion will have the form

$$f_1(t) = A_1 e^{p_1 t} + A_2 e^{p_2 t} + \ldots A_r e^{p_r t} \qquad (10\text{-}34)$$

In most cases of interest, the real poles will be negative in sign, implying that the time function is a sum of decaying exponential terms. Thus, *real negative poles of first order correspond to decaying exponential time functions.*

"Cover-Up" Method

This method is sometimes referred to casually as the "cover-up" method. You will see the logic of this reference when the formula of Equation 10-33 is applied in the example that follows. Multiplication of $F(s)$ by $(s - p_k)$ cancels the denominator factor of the same form, which amounts to "covering it up" or eliminating the factor. This description makes it easy to remember the operation.

Example 10-9

Determine the inverse transform of

$$F(s) = \frac{s+6}{s^2 + 3s + 2} \qquad (10\text{-}35)$$

Solution

One might initially think that this would fit either T-5 or T-6 in Table 10-1. However, the poles, which are the denominator roots of Equation 10-35, are $s_1 = -1$ and $s_2 = -2$. Therefore, T-5 and T-6 are not applicable. However, the partial fraction process of this section can be used.

First, we express the denominator in factored form and the function has the form

$$F(s) = \frac{s+6}{(s+1)(s+2)} \qquad (10\text{-}36)$$

Since the function consists of two real poles of first order, the partial fraction expansion is

$$F(s) = \frac{s+6}{(s+1)(s+2)} = \frac{A_1}{s+1} + \frac{A_2}{s+2} \qquad (10\text{-}37)$$

Using the procedure developed in this section, we can determine A_1 and A_2 as follows:

$$A_1 = (s+1)F(s) \big]_{s=-1} = \frac{s+6}{s+2} \bigg]_{s=-1} = \frac{-1+6}{-1+2} = 5 \qquad (10\text{-}38)$$

Note how it looks like we have "covered up" the (s + 1) factor in the evaluation process.

$$A_2 = (s+2)F(s)\big]_{s=-2} = \frac{s+6}{s+1}\bigg]_{s=-2} = \frac{-2+6}{-2+1} = -4 \tag{10-39}$$

The partial fraction expansion is then

$$F(s) = \frac{5}{s+1} - \frac{4}{s+2} \tag{10-40}$$

The reader is invited to combine the two terms over a common denominator to verify that the original function is obtained. From an application of T-2, the inverse transform is

$$f(t) = 5e^{-t} - 4e^{-2t} \tag{10-41}$$

Example 10-10

Determine the exponential part of the inverse transform of

$$F(s) = \frac{50(s+3)}{(s+1)(s+2)(s^2+2s+5)} \tag{10-42}$$

Solution

The reader can verify that the quadratic factor has complex roots and meets the criteria of classification (2). We will consider that situation in the next section. We will now determine the inverse transform terms corresponding to the real poles of first order: $s = -1$ and $s = -2$. Denote this portion of the function $F_1(s)$ and its inverse transform as $f_1(t)$.

$$F_1(s) = \frac{A_1}{s+1} + \frac{A_2}{s+2} \tag{10-43}$$

The coefficients are determined as follows:

$$A_1 = \frac{50(s+3)}{(s+2)(s^2+2s+5)}\bigg]_{s=-1} = \frac{(50)(2)}{(1)(4)} = 25 \tag{10-44}$$

$$A_2 = \frac{50(s+3)}{(s+1)(s^2+2s+5)}\bigg]_{s=-2} = \frac{(50)(1)}{(-1)(5)} = -10 \tag{10-45}$$

When these coefficients are substituted in Equation 10-43, the inverse transform is determined as

$$f_1(t) = 25e^{-t} - 10e^{-2t} \tag{10-46}$$

10-6   PARTIAL FRACTION EXPANSION FOR FIRST-ORDER COMPLEX POLES

There are several methods for the analytical transform inversion of functions with first-order complex poles. However, with the modern resources of available mathematical software, we do not feel justified in covering more than one analytical method. The method covered is one commonly employed in mathematics texts, and it is part of the partial fraction process.

Assume that the denominator of $F(s)$ contains a factor of the form $(s^2 + bs + c)$ with a pair of complex conjugate poles, $p_{1,2} = -\alpha \pm i\omega$. Note that we are assuming that the real part, when present, will be negative. This is the common situation, and it is a requirement for stability.

Partial Fraction Expansion

The form of the partial fraction expansion is

$$F(s) = \frac{As + B}{s^2 + bs + c} + R(s) \tag{10-47}$$

where $R(s)$ represents all terms in the expansion other than the quadratic factor under consideration.

The concept here is that the denominator of the first term on the right of Equation 10-47 is of second degree, which means that the numerator of the term must contain both a first-degree term and a constant. Therefore, the simple formula for dealing with first-order real poles is not applicable here.

One procedure is as follows: First, determine as much of the total expansion for $R(s)$ as possible by other methods. Next, since Equation 10-47 expresses an equality, select some convenient values of s to substitute into the equation and produce as many simultaneous equations as necessary to solve for the constants. The quadratic term is then inverse transformed by a combination of T-5 and T-6, as was demonstrated earlier. We may select any values of s to substitute into the equation except for the poles of $F(s)$.

Let us reemphasize some points made in Chapter 9 dealing with differential equations: *A pair of complex poles of first order with negative real parts corresponds to a damped sinusoidal time response.* As a special case, *purely imaginary poles correspond to an undamped sinusoidal time response.*

Example 10-11

Complete the partial fraction expansion of the function of Example 10-10.

Solution

The function is repeated here for convenience.

$$F(s) = \frac{50(s+3)}{(s+1)(s+2)(s^2 + 2s + 5)} \tag{10-48}$$

In Example 10-10, we determined the two terms corresponding to the two real poles, but they need to be considered now as part of the total expansion. Factoring of the second-degree polynomial results in the complex poles $p_{1,2} = -1 \pm i2$, so the polynomial will be retained in second-degree form.

Using the coefficients already determined in Example 10-10 for the poles with real values, we have

$$\frac{50(s+3)}{(s+1)(s+2)(s^2+2s+5)} = \frac{25}{s+1} - \frac{10}{s+2} + \frac{As+B}{s^2+2s+5} \tag{10-49}$$

We need two simultaneous equations to determine A and B. We wish to pick simple values, but we cannot choose -1 and -2 since they are poles. One possible choice is $s = 0$, and that will be used first. We have

$$\frac{50(3)}{(1)(2)(5)} = \frac{25}{1} - \frac{10}{2} + \frac{B}{5} \tag{10-50}$$

Solution yields

$$B = -25 \tag{10-51}$$

Another suitable choice is $s = 1$, and this value results in

$$\frac{50(4)}{(2)(3)(8)} = \frac{25}{2} - \frac{10}{3} + \frac{A+B}{8} \tag{10-52}$$

We will not show all the small steps involved in simplifying this equation, but after reducing it and substituting B from Equation 10-51, we obtain

$$A = -15 \tag{10-53}$$

The complete partial expansion then reads

$$F(s) = \frac{25}{s+1} - \frac{10}{s+2} + \frac{-15s-25}{s^2+2s+5} \tag{10-54}$$

We have already determined the inverse transform of the first two terms, which we identified as $f_1(t)$ in Example 10-10. We will denote the last term as $F_2(s)$. It is

$$F_2(s) = \frac{-15s-25}{s^2+2s+5} \tag{10-55}$$

We need to complete the square of the denominator, which yields

$$s^2+2s+5 = s^2+2s+1+5-1 = (s+1)^2+(2)^2 \tag{10-56}$$

We then have

$$F_2(s) = \frac{-15s-25}{s^2+2s+5} = \frac{-15s-25}{(s+1)^2+(2)^2} \tag{10-57}$$

To use T-6, we need an $(s + 1)$ factor in the numerator, and to use T-5, we need ω in the numerator. Some manipulations are required.

$$F_2(s) = \frac{-15s-25}{(s+1)^2+(2)^2} = \frac{-15(s+1)}{(s+1)^2+(2)^2} + \frac{-5(2)}{(s+1)^2+(2)^2} \tag{10-58}$$

The inverse transform is then

$$f_2(t) = -15e^{-t}\cos 2t - 5e^{-t}\sin 2t \tag{10-59}$$

The entire time function is

$$f(t) = f_1(t) + f_2(t)$$
$$= 25e^{-t} - 10e^{-2t} - 15e^{-t}\cos 2t - 5e^{-t}\sin 2t \tag{10-60}$$

10-7 SECOND-ORDER REAL POLES

As has been stated on several occasions, most practical cases of multiple-order poles are those of second-order real values. For anything more elaborate, we will defer to the MAT-LAB operations considered in the next chapter.

Inasmuch as critically damped systems do have second-order real poles, we will consider that one case here. Assume that $F(s)$ contains a denominator factor of the form $(s + \alpha)^2$, corresponding to a second-order pole $p = -\alpha$. The form of the partial fraction expansion is then

$$F(s) = \frac{C_1}{(s+\alpha)^2} + \frac{C_2}{s+\alpha} + R(s) \tag{10-61}$$

where $R(s)$ is the expansion due to all other poles.

Without showing the details, it can be deduced that multiplication by $(s + \alpha)^2$, followed by setting $s = -\alpha$, will lead to the determination of C_1 by the same procedure as for a real pole of first order. Therefore, for C_1, a formula may readily be determined as

$$C_1 = (s+\alpha)^2 F(s)\Big]_{s=-\alpha} \tag{10-62}$$

To determine C_2, we need to determine as much of the remaining part of the partial fraction expansion as possible without any special processes, which would mainly be real poles of first order. We then employ the same procedure as used for complex poles, namely assume as many values of s as necessary to evaluate the function on both sides to obtain one or more equations that can be used to determine C_2 and possible additional terms in $R(s)$. Once C_2 is determined, that portion of the inverse transform can be determined with T-2 and T-9 of Table 10-9. Denoting that portion of the inverse transform momentarily as $F_1(s)$ and its inverse transform as $f_1(t)$, we have

$$F_1(s) = \frac{C_1}{(s+\alpha)^2} + \frac{C_2}{s+\alpha} \tag{10-63}$$

and

$$f_1(t) = C_1 t e^{-\alpha t} + C_2 e^{-\alpha t} = (C_1 t + C_2)e^{-\alpha t} \tag{10-64}$$

Example 10-12

Determine the inverse transform of

$$F(s) = \frac{60}{s(s+2)^2} \tag{10-65}$$

Solution

We have one first-order pole ($p_1 = 0$) and a pair of second-order poles ($p_{2,3} = -2$). The partial fraction expansion will be of the form

$$F(s) = \frac{60}{s(s+2)^2} = \frac{A}{s} + \frac{C_1}{(s+2)^2} + \frac{C_2}{(s+2)} \tag{10-66}$$

The constants A and C_1 are determined as follows:

$$A = sF(s)\big]_{s=0} = \frac{60}{(s+2)^2}\bigg]_{s=0} = \frac{60}{(0+2)^2} = 15 \tag{10-67}$$

$$C_1 = (s+2)^2 F(s)\big]_{s=-2} = \frac{60}{s}\bigg]_{s=-2} = \frac{60}{-2} = -30 \tag{10-68}$$

Based on these values, we can establish the partial fraction expansion as

$$F(s) = \frac{60}{s(s+2)^2} = \frac{15}{s} - \frac{30}{(s+2)^2} + \frac{C_2}{s+2} \tag{10-69}$$

We will now pick a convenient value of s at which to evaluate Equation 10-69. We cannot use $s = 0$ since that is a pole, but $s = 1$ should work fine. We then have

$$\frac{60}{(1)(1+2)^2} = \frac{15}{1} - \frac{30}{(1+2)^2} + \frac{C_2}{(1+2)} \tag{10-70}$$

After a little simplification, the value of C_2 is determined as

$$C_2 = -15 \tag{10-71}$$

The complete partial fraction expansion is then

$$F(s) = \frac{60}{s(s+2)^2} = \frac{15}{s} - \frac{30}{(s+2)^2} - \frac{15}{s+2} \tag{10-72}$$

The time function is

$$f(t) = 15 - 30te^{-2t} - 15e^{-2t} = 15 - 15e^{-2t}(1 + 2t) \tag{10-73}$$

10-8 LAPLACE TRANSFORM OPERATIONS

We have seen how to obtain Laplace transforms from time functions, and we have seen the more tedious process of how to determine the time functions from the transforms. We are almost ready to begin solving differential equations with them, but we need to investigate one more aspect. For each operation applied to the time function, there is a corresponding operation on the Laplace function. These forms are summarized in Table 10-2. Not all of them will be used in this text, but they are included for possible future reference.

O-1

This transform operation states the following:

$$L[f'(t)] = sF(s) - f(0)$$

Momentarily ignoring the initial value term, this operation states that differentiation of the time function is accomplished with the transform function by multiplication by s. Of course, the initial condition $f(0)$ is included in the operation.

O-2

For the second derivative, the transform operation states

$$L[f''(t)] = s^2 F(s) - sf(0) - f'(0)$$

An additional multiplication by s, along with both the initial value of the function and the initial value of the first derivative, yields the Laplace transform of the second derivative. Since we do not plan to go beyond the second derivative without the help of MATLAB, we will leave it at this point, but the pattern is beginning to emerge, so we expect that keen readers might be able to extend the result to higher derivatives.

O-3

For the definite integral of the time function, we have

$$L\left[\int_0^t f(t)dt\right] = \frac{F(s)}{s}$$

In contrast to differentiation, this theorem states that integration of the time function corresponds to division of the transform function by s.

We will not discuss the other theorems at this point since we have enough to begin solving differential equations. However, it should be noted that in some disciplines, particularly electrical and mechanical control systems, the operations considered here are very much like a type of system vocabulary. For example, a processing block with the symbol s is interpreted as a differentiator and a block with the label $1/s$ is interpreted as an integrator. Therefore, the language associated with Laplace transform operations is useful as a means of communication and is widely used in system analysis procedures.

10-9   SOLVING DIFFERENTIAL EQUATIONS WITH LAPLACE TRANSFORMS

After all the preceding preparation, we are finally ready to show how it can be used to solve differential equations. Since we do not intend to go beyond a second-order DE without some help from MATLAB, let us use a second-order form as a means for developing the procedure. Consider the following second-order CCLODE:

$$b_2 \frac{d^2 y}{dt^2} + b_1 \frac{dy}{dt} + b_0 y = f(t) \tag{10-74}$$

Next, apply the Laplace transformation to both sides of the equation.

$$L\left[b_2\frac{d^2y}{dt^2}+b_1\frac{dy}{dt}+b_0y\right]=L[f(t)] \tag{10-75}$$

Based on the linear properties of the Laplace transform, the two sides can be expanded as

$$b_2L\left[\frac{d^2y}{dt^2}\right]+b_1L\left[\frac{dy}{dt}\right]+b_0L[y]=L[f(t)] \tag{10-76}$$

Using results from Table 10-2, we have

$$b_2\left[s^2Y(s)-sy(0)-y'(0)\right]+b_1\left[sY(s)-y(0)\right]+b_0Y(s)=F(s) \tag{10-77}$$

where $Y(s)=L[y(t)]$ and $F(s)=L[f(t)]$.

After a bit of manipulation, Equation 10-77 reduces to

$$Y(s)\left[b_2s^2+b_1s+b_0\right]=F(s)+sb_2y(0)+b_2y'(0)+b_1y(0) \tag{10-78}$$

The desired function in transform form is then

$$Y(s)=\frac{F(s)}{b_2s^2+b_1s+b_0}+\frac{sb_2y(0)+b_2y'(0)+b_1y(0)}{b_2s^2+b_1s+b_0} \tag{10-79}$$

The desired solution is then the inverse Laplace transform of the preceding function.

Do not try to memorize this result, nor is it recommended that you look up the result and use a "plug-in" procedure. Once you have actual values and perform the steps indicated, the process is usually a little smoother. We will illustrate this by several examples that follow.

Example 10-13

Rework Example 9-6 using Laplace transforms.

Solution

The DE is repeated here for convenience. It is

$$\frac{dy}{dt}+2y=12 \tag{10-80}$$

with

$$y(0)=10 \tag{10-81}$$

The procedure is to take the Laplace transforms of both sides of the DE. Skipping a couple of the simple linearity steps (Laplace of the sum is the sum of the Laplaces, etc.), we arrive at the form

$$L\left[\frac{dy}{dt}\right]+2L[y]=L[12] \tag{10-82}$$

The first term is transformed according to operation O-1, and the second term simply involves the definition of the transform. The right-hand side utilizes the transform of a constant, which is transform pair T-1. We have

$$sY(s) - 10 + 2Y(s) = \frac{12}{s} \tag{10-83}$$

The next step involves getting all factors of $Y(s)$ on the left and moving other terms to the right. This yields

$$(s+2)Y(s) = 10 + \frac{12}{s} \tag{10-84}$$

and

$$Y(s) = \frac{10}{s+2} + \frac{12}{s(s+2)} \tag{10-85}$$

The inverse transform of the first term may be determined immediately. However, the second term needs to have a partial fraction expansion. We have

$$\frac{12}{s(s+2)} = \frac{A_1}{s} + \frac{A_2}{s+2} \tag{10-86}$$

The constants are determined as follows:

$$A_1 = s \left[\frac{12}{s(s+2)} \right]_{s=0} = \left[\frac{12}{s+2} \right]_{s=0} = 6 \tag{10-87}$$

$$A_2 = (s+2) \left[\frac{12}{s(s+2)} \right]_{s=-2} = \left[\frac{12}{s} \right]_{s=-2} = -6 \tag{10-88}$$

Combining all the terms, we obtain

$$Y(s) = \frac{10}{s+2} + \frac{6}{s} - \frac{6}{s+2} = \frac{6}{s} + \frac{4}{s+2} \tag{10-89}$$

The inverse transform is then

$$y(t) = 6 + 4e^{-2t} \tag{10-90}$$

Example 10-14

Rework Example 9-7 using Laplace transforms.

Solution

The DE is repeated here for convenience.

$$\frac{dy}{dt} + 2y = 12\sin 4t \tag{10-91}$$

with

$$y(0) = 10 \tag{10-92}$$

With additional practice, we can begin to take larger steps in transforming the DE. We have

$$sY(s) - 10 + 2Y(s) = \frac{12(4)}{s^2 + 16} \tag{10-93}$$

Rearranging and solving for the transform, we obtain

$$Y(s) = \frac{10}{s+2} + \frac{48}{(s+2)(s^2+16)} \tag{10-94}$$

We need to expand the second term in a partial fraction expansion. We have

$$\frac{48}{(s+2)(s^2+9)} = \frac{A}{s+2} + \frac{B_1 s + B_2}{s^2+16} \tag{10-95}$$

The term A is determined as

$$A = \frac{48}{s^2+16}\bigg]_{s=-2} = \frac{48}{20} = 2.4 \tag{10-96}$$

We next write

$$\frac{48}{(s+2)(s^2+16)} = \frac{2.4}{s+2} + \frac{B_1 s + B_2}{s^2+16} \tag{10-97}$$

We now select two convenient values of s to substitute into both sides of Equation 10-97 so that two simultaneous equations can be obtained. Two "nice" values are $s = 0$ and $s = -1$. Substitution of $s = 0$ yields

$$\frac{48}{(2)(16)} = \frac{2.4}{2} + \frac{B_2}{16} \tag{10-98}$$

This leads to

$$B_2 = 4.8 \tag{10-99}$$

Substitution of $s = -1$ yields

$$\frac{48}{(1)(17)} = \frac{2.4}{1} + \frac{-B_1 + B_2}{17} \tag{10-100}$$

Simplification and substitution of the value of B_2 lead to

$$B_1 = -2.4 \tag{10-101}$$

Combining everything, we have

$$Y(s) = \frac{10}{s+2} + \frac{2.4}{s+2} - \frac{2.4s}{s^2+16} + \frac{4.8}{s^2+16} \tag{10-102}$$

The inverse Laplace transform is

$$y(t) = 12.4e^{-2t} - 2.4\cos 4t + 1.2\sin 4t \tag{10-103}$$

Example 10-15

Rework Example 9-9 using Laplace transforms.

Solution

The DE is repeated here for convenience.

$$\frac{d^2y}{dt^2} + 3\frac{dy}{dt} + 2y = 24 \qquad (10\text{-}104)$$

with

$$y(0) = 10 \text{ and } y'(0) = 0 \qquad (10\text{-}105)$$

The Laplace transformation is applied to both sides of Equation 10-104. Since the first term is a second derivative, it is necessary to use O-2, as well as O-1 for the first derivative. We have

$$s^2Y(s) - 10s - 0 + 3[sY(s) - 10] + 2Y(s) = \frac{24}{s} \qquad (10\text{-}106)$$

Regrouping and solving for $Y(s)$, we obtain

$$Y(s) = \frac{24}{s(s^2 + 3s + 2)} + \frac{10s + 30}{s^2 + 3s + 2} = \frac{24}{s(s+1)(s+2)} + \frac{10s + 30}{(s+1)(s+2)} \qquad (10\text{-}107)$$

We could have put everything over one common denominator, but the separation into two terms is probably easier. Each term in Equation 10-107 will now be expanded. All poles are real and of first order, so by now the reader may be able to apply the "cover-up" procedure and determine the constants fairly quickly. The first expansion is

$$\frac{24}{s(s+1)(s+2)} = \frac{A_1}{s} + \frac{A_2}{s+1} + \frac{A_3}{s+2} \qquad (10\text{-}108)$$

Without too much mental strain, we can deduce that $A_1 = 12$, $A_2 = -24$, and $A_3 = 12$. The expansion of Equation 10-108 is then

$$\frac{24}{s(s+1)(s+2)} = \frac{12}{s} - \frac{24}{s+1} + \frac{12}{s+2} \qquad (10\text{-}109)$$

The second term in Equation 10-107 reduces by a similar procedure to

$$\frac{10s + 30}{(s+1)(s+2)} = \frac{20}{s+1} - \frac{10}{s+2} \qquad (10\text{-}110)$$

Combining all terms, we obtain

$$Y(s) = \frac{12}{s} - \frac{4}{s+1} + \frac{2}{s+2} \qquad (10\text{-}111)$$

The inverse time function is

$$y(t) = 12 - 4e^{-t} + 2e^{-2t} \qquad (10\text{-}112)$$

Example 10-16

Rework Example 9-12 using Laplace transforms.

Solution

The DE is repeated here for convenience.

$$\frac{d^2 y}{dt^2} + 2\frac{dy}{dt} + 5y = 20 \tag{10-113}$$

with

$$y(0) = 0 \text{ and } y'(0) = 10 \tag{10-114}$$

Laplace transformation of the DE yields

$$s^2 Y(s) - 0 - 10 + 2\left[sY(s) - 0\right] + 5Y(s) = \frac{20}{s} \tag{10-115}$$

Simplification yields

$$Y(s) = \frac{20}{s(s^2 + 2s + 5)} + \frac{10}{s^2 + 2s + 5} \tag{10-116}$$

The constant over s in the expansion of the first term is readily determined to be 4. The expansion of this term is of the form

$$\frac{20}{s(s^2 + 2s + 5)} = \frac{4}{s} + \frac{As + B}{(s^2 + 2s + 5)} \tag{10-117}$$

We cannot choose s = 0 in this case since it is one of the poles. We will select $s = 1$ and $s = -1$. We have

$$\frac{20}{(1)(1 + 2 + 5)} = \frac{4}{1} + \frac{A + B}{(1 + 2 + 5)} \tag{10-118}$$

and

$$\frac{20}{(-1)(1 - 2 + 5)} = \frac{4}{-1} + \frac{-A + B}{(1 - 2 + 5)} \tag{10-119}$$

Without showing the details, the simultaneous solution yields

$$A = -4 \tag{10-120}$$

$$B = -8 \tag{10-121}$$

We can now combine the terms as follows:

$$Y(s) = \frac{4}{s} + \frac{-4s - 8}{s^2 + 2s + 5} + \frac{10}{s^2 + 2s + 5} = \frac{4}{s} + \frac{-4s + 2}{s^2 + 2s + 5} \tag{10-122}$$

Since the second-degree polynomial has complex roots, we complete the square on it as follows:

$$s^2 + 2s + 5 = s^2 + 2s + 1 + 5 - 1 = (s+1)^2 + (2)^2 \qquad (10\text{-}123)$$

The function is now manipulated so that the proper forms can be realized.

$$Y(s) = \frac{4}{s} + \frac{-4(s+1)}{(s+1)^2 + (2)^2} + \frac{3(2)}{(s+1)^2 + (2)^2} \qquad (10\text{-}124)$$

The time function is then

$$y(t) = 4 - 4e^{-t}\cos 2t + 3e^{-t}\sin 2t \qquad (10\text{-}125)$$

GENERAL PROBLEMS

Problems 10-1 through 10-18 involve differential equations that were to be solved with classical procedures in Chapter 9. For the present chapter, solve each differential equation using Laplace transforms.

10-1. $\dfrac{dy}{dt} = 6t^2 \qquad y(0) = 5$

10-2. $\dfrac{dy}{dt} = 20\cos 5t + 8 \qquad y(0) = 10$

10-3. $\dfrac{dy}{dt} + 4y = 0 \qquad y(0) = 6$

10-4. $\dfrac{dy}{dt} + 50y = 0 \qquad y(0) = 20$

10-5. $\dfrac{dy}{dt} + 4y = 80 \qquad y(0) = 6$

10-6. $\dfrac{dy}{dt} + 50y = 400 \qquad y(0) = 20$

10-7. $\dfrac{dy}{dt} + 4y = 40\sin 3t \qquad y(0) = 6$

10-8. $\dfrac{dy}{dt} + 50y = 400\cos 75t \qquad y(0) = 20$

10-9. $\dfrac{d^2 y}{dt^2} + 4\dfrac{dy}{dt} + 3y = 0 \qquad y(0) = 0 \quad y'(0) = 12$

10-10. $\dfrac{d^2 y}{dt^2} + 7\dfrac{dy}{dt} + 12y = 0 \qquad y(0) = 20 \quad y'(0) = 0$

10-11. $\dfrac{d^2y}{dt^2} + 4\dfrac{dy}{dt} + 3y = 30$ $y(0) = 20$ $y'(0) = 12$

10-12. $\dfrac{d^2y}{dt^2} + 7\dfrac{dy}{dt} + 12y = 96$ $y(0) = 20$ $y'(0) = 0$

10-13. $\dfrac{d^2y}{dt^2} + 3\dfrac{dy}{dt} + 2y = 24$ $y(0) = 0$ $y'(0) = 10$

10-14. $\dfrac{d^2y}{dt^2} + 3\dfrac{dy}{dt} + 2y = 24e^{-4t}$ $y(0) = 10$ $y'(0) = 5$

10-15. $\dfrac{d^2y}{dt^2} + 4\dfrac{dy}{dt} + 13y = 52$ $y(0) = 0$ $y'(0) = 0$

10-16. $\dfrac{d^2y}{dt^2} + 4\dfrac{dy}{dt} + 104y = 832$ $y(0) = 0$ $y'(0) = 0$

10-17. $\dfrac{d^2y}{dt^2} + 6\dfrac{dy}{dt} + 9y = 0$ $y(0) = 10$ $y'(0) = 0$

10-18. $\dfrac{d^2y}{dt^2} + 6\dfrac{dy}{dt} + 9y = 90$ $y(0) = 0$ $y'(0) = 0$

Note: As in the case of Chapter 9, the total emphasis in this chapter was analytical in nature. Hence, MATLAB problems will be delayed to the end of Chapter 11.

Table 10-1   Laplace Transform Pairs Encountered in
Common Engineering Differential Equations

$f(t)$	$F(s) = L[f(t)]$	
1 or $u(t)$	$\dfrac{1}{s}$	T-1
$e^{-\alpha t}$	$\dfrac{1}{s+\alpha}$	T-2
$\sin \omega t$	$\dfrac{\omega}{s^2+\omega^2}$	T-3
$\cos \omega t$	$\dfrac{s}{s^2+\omega^2}$	T-4
$e^{-\alpha t}\sin \omega t$	$\dfrac{\omega}{(s+\alpha)^2+\omega^2}$	T-5*
$e^{-\alpha t}\cos \omega t$	$\dfrac{s+\alpha}{(s+\alpha)^2+\omega^2}$	T-6*
t	$\dfrac{1}{s^2}$	T=7
t^n	$\dfrac{n!}{s^{n+1}}$	T-8
$e^{-\alpha t}t^n$	$\dfrac{n!}{(s+\alpha)^{n+1}}$	T-9
$\delta(t)$	1	T-10

*Use when roots are complex.

Table 10-1   Laplace Transform Operations Encountered with
Common Differential Equations

$f(t)$	$F(s)$	
$f'(t)$	$sF(s) - f(0)$	O-1
$\displaystyle\int_0^t f(t)dt$	$\dfrac{F(s)}{s}$	O-2
$e^{-\alpha t}f(t)$	$F(s+\alpha)$	O-3
$f(t-T)u(t-T)$	$e^{-sT}F(s)$	O-4
$f(0)$	$\lim\limits_{s\to\infty} sF(s)$	O-5
$\lim\limits_{t\to\infty} f(t)$	$\lim\limits_{s\to 0} sF(s)$ *	O-6

*Poles of $sF(s)$ must have negative real parts.

Solution of Differential Equations with MATLAB

11

11-1   OVERVIEW AND OBJECTIVES

MATLAB has some very powerful features for solving differential equations, both linear and non-linear. We will explore some of the features in this chapter as they relate to *constant coefficient linear ordinary differential equation* (CCLODE) forms. The approach that will be used is contained in the **Symbolic Mathematics Toolbox**. The differential equation is formed utilizing symbolic expressions and the initial values are provided through symbolic forms. The result will be the function whose form is desired. If a curve of the solution is desired, a plot may be readily obtained with MATLAB.

Objectives

After completing this chapter, the reader should be able to

1. Encode a CCLODE in symbolic form so that it can be solved with MATLAB.
2. Determine the solution of the DE from a MATLAB analysis using the **dsolve** command.
3. Plot the solution of a DE using **ezplot**.
4. Determine the Laplace transform of a function using the **laplace** command.
5. Determine the inverse Laplace transform of a function using the **ilaplace** command.

11-2   MATLAB SYMBOLIC SOLUTION OF A DIFFERENTIAL EQUATION

MATLAB offers many different operations and commands for solving a variety of differential equations, both linear and non-linear. Many of these capabilities are oriented toward graduate-level work in applied mathematics and engineering and are beyond the scope of this text. However, some of the procedures are quite accessible and simplify greatly the solutions of many common forms of differential equations.

The treatment here will be aimed at the use of differential equation operations that are compatible with Chapters 9 and 10. You will recall that the primary types of DEs considered in those chapters were the constant coefficient linear ordinary differential equation

(CCLODE) forms, which constitute the primary types encountered by most engineers and technologists.

Symbolic Differential Equation Form

The operations to be delineated come from the **Symbolic Mathematics Toolbox**, which is included with the **Student Version** of MATLAB at the time of this writing. If you are using a **Professional Version**, make sure that that toolbox is installed if you wish to employ the commands in this chapter.

In the symbolic representation of derivatives in a MATLAB DE, the following equivalents should be noted:

Mathematical Form	**MATLAB Form**
y	y
$\dfrac{dy}{dt}$	Dy
$\dfrac{d^2 y}{dt^2}$	D2y
$\dfrac{d^3 y}{dt^3}$	D3y
$\dfrac{d^n y}{dt^n}$	Dny

The default independent variable is time as noted, but it may be changed if desired. Likewise, the dependent variable may be changed. However, the Dn forms for the derivatives remain the same.

Representative Form

Let us illustrate the form for encoding a DE by using a particular second-order form. The pattern that emerges should allow the reader to extrapolate to higher-order forms with different forcing functions.

Consider the following second-order CCLODE:

$$b_2 \frac{d^2 y}{dt^2} + b_1 \frac{dy}{dt} + b_0 y = A \sin at \qquad (11\text{-}1)$$

with the boundary conditions

$$y(0) = C_1 \quad \text{and} \quad y'(0) = C_2 \qquad (11\text{-}2)$$

The command for solving the differential equation is **dsolve**. As is true with most MAT-LAB commands, it will be listed simply as **ans** if we do not give it a name. Although we could give it another name if desired, the logical choice in this case is to name it the same as in the DE.

Recognizing that the default form for the MATLAB Command Window will be a different style of type without subscripts, the manner in which Equations 11-1 and 11-2 will be encoded for MATLAB is as follows:

$$>> y = dsolve('b2*D2y + b1*Dy + b0*y = A*sin(a*t)', \qquad (11\text{-}3)$$
$$'y(0) = C1', 'Dy(0) = C2')$$

Assuming that all the constants have values, entering this command will generate a closed-form solution of the DE and list it on the screen. If the reader is experiencing some difficulty with all the symbols, the examples that follow this section will have numbers and that should help clarify the encoding involved.

The result of the preceding form is that one command containing the differential equation and the initial conditions will generate a closed-form solution. After generating the solution, it may be plotted over the domain from t1 to t2 by the command

$$>>ezplot(y, [t1\ t2]) \qquad (11\text{-}4)$$

Example 11-1

Example 9-6 was first solved by the classical approach and then solved by Laplace transforms in Example 10-13. Now use MATLAB **dsolve** to solve the DE and plot it over the domain from 0 to 3 with **ezplot.**

Solution

The original DE and the initial value are repeated for convenience.

$$\frac{dy}{dt} + 2y = 12 \qquad (11\text{-}5)$$

and

$$y(0) = 10 \qquad (11\text{-}6)$$

The manner in which the DE is formulated is as follows:

$$>> y = dsolve('Dy + 2*y = 12', 'y(0)=10') \qquad (11\text{-}7)$$

$$y =$$

$$6+4*exp(-2*t)$$

Note how the DE is first given in the parentheses, followed by the initial value. Do not forget that an asterisk must be used in MATLAB for multiplication, which means that the term $2y$ in the DE must be replaced by 2*y in the command.

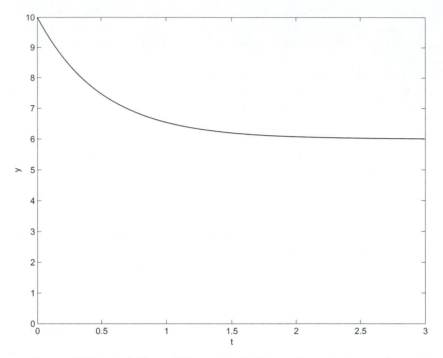

Figure 11-1 Solution of Example 11-1 based on dsolve and ezplot

The result agrees exactly with the solution obtained both by classical methods in Chapter 9 and the Laplace approach in Chapter 10. Next we will use **ezplot** to plot the equation. The command is

$$>> \text{ezplot}(y, [0\ 3]) \tag{11-8}$$

The resulting figure was off the scale at very low values of time, so the axis was modified by the command

$$>> \text{axis}([0\ 3\ 0\ 10]) \tag{11-9}$$

The result is shown in Figure 11-1.

For a simple DE such as this, it is debatable whether using MATLAB is any easier than solving the equation and then plotting it by the processes considered early in the text. However, this was a good "warmup" exercise, and the next few examples will display the real power of MATLAB.

Example 11-2

Example 9-7 was first solved by the classical approach and then solved by Laplace transforms in Example 10-14. Now use MATLAB to solve the DE and plot it.

Solution

The original DE and the initial value are repeated for convenience.

$$\frac{dy}{dt} + 2y = 12\sin 4t \qquad (11\text{-}10)$$

and

$$y(0) = 10 \qquad (11\text{-}11)$$

The encoding for solving the DE and the result follow.

$$\gg \text{y = dsolve('Dy + 2*y = 12*sin(4*t)', 'y(0)=10')} \qquad (11\text{-}12)$$

$$\text{y} =$$

$$-12/5*\cos(4*t)+6/5*\sin(4*t)+62/5*\exp(-2*t)$$

It can be readily verified that the terms and the constants are all equivalent to those determined in Examples 9-7 and 10-14.

The **ezplot** routine is now used to plot the function. After a little trial and error, the domain of time selected was from 0 to 8. The command is then

$$\gg \text{ezplot(y, [0 8])} \qquad (11\text{-}13)$$

The initial vertical axis was not quite satisfactory, so it was modified by the **axis** command

$$\gg \text{axis([0 8 -3 10])} \qquad (11\text{-}14)$$

The result is shown in Figure 11-2.

Example 11-3

Example 9-9 was first solved by the classical approach and then solved by Laplace transforms in Example 10-15. Now use MATLAB to solve the DE and plot it.

Solution

The original DE and the initial value are provided as follows for convenience:

$$\frac{d^2y}{dt^2} + 3\frac{dy}{dt} + 2y = 24 \qquad (11\text{-}15)$$

with $\qquad\qquad y(0) = 10 \quad \text{and} \quad y'(0) = 0 \qquad (11\text{-}16)$

The command and solution follow.

$$\gg \text{y = dsolve('D2y + 3*Dy + 2*y = 24', 'y(0)=10', 'Dy(0)=0')} \qquad (11\text{-}17)$$

$$\text{y} =$$

$$12+2*\exp(-2*t)-4*\exp(-t)$$

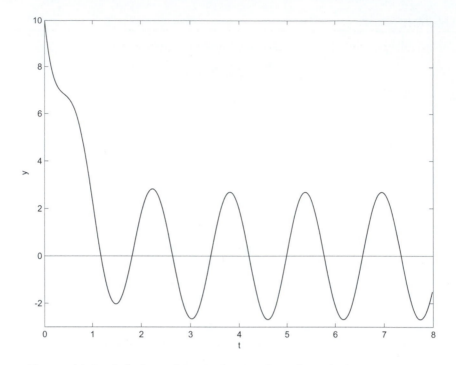

Figure 11-2 Solution of Example 11-2 based on dsolve and ezplot

After some trial and error, the plot of the function was generated by the command

$$>> ezplot(y, [0\ 6]) \qquad (11\text{-}18)$$

The result is shown in Figure 11-3. Since the range lies between the levels of 10 and 12, the resolution would be degraded quite a bit if we scaled it back to a 0 lower level, so we will leave it as is.

Example 11-4

Example 9-12 was first solved by the classical approach and then solved by Laplace transforms in Example 10-16. Now use MATLAB to solve the DE and plot it.

Solution

The original DE and the initial values are

$$\frac{d^2y}{dt^2} + 2\frac{dy}{dt} + 5y = 20 \qquad (11\text{-}19)$$

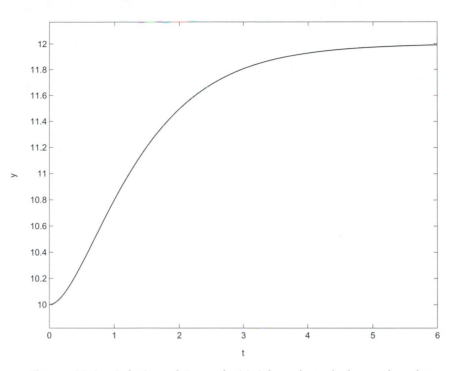

Figure 11-3   Solution of Example 11-3 based on dsolve and ezplot.

with

$$y(0) = 0 \ \text{ and } \ y'(0) = 10 \tag{11-20}$$

Encoding and the solution follow.

$$>> y = \text{dsolve}('D2y + 2*Dy + 5*y = 20', \,'y(0) = 0', \,'Dy(0) = 10') \tag{11-21}$$

$$y =$$

$$4+3*\exp(-t)*\sin(2*t)-4*\exp(-t)*\cos(2*t)$$

The domain selected was from 0 to 5. The plot command follows.

$$>> \text{ezplot}(y, [0 \ 5]) \tag{11-22}$$

The vertical scale required some readjustment. The command used was

$$>> \text{axis}([0 \ 5 \ 0 \ 6]) \tag{11-23}$$

The result is shown in Figure 11-4.

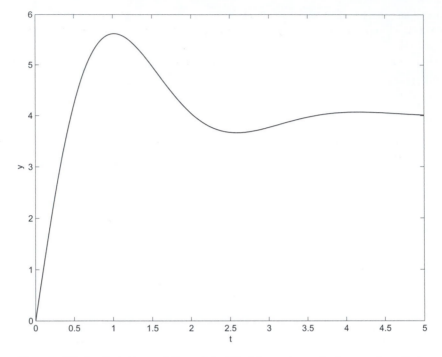

Figure 11-4 Solution of Example 11-4 based on dsolve and ezplot

11-3 MATLAB SYMBOLIC LAPLACE TRANSFORMS

The MATLAB Symbolic Mathematics Toolbox contains a number of Laplace transform and inverse transform operations. They may be used to directly support Laplace transform analysis of differential equations when the framework of analysis is oriented toward that approach.

Inasmuch as derivatives and integrals can be readily determined from MATLAB symbolic operations, so it is with transforms and inverse transforms. In a sense, these operations may be considered as "super look-up tables," but they have the additional feature that the results may be adapted to the specific constants for a given system and plots may be made directly from the functions.

Laplace Transforms of Time Functions

Let us begin with the default variable of t for the time function and the default variable of s for the Laplace transform variable. Since case is important for a MATLAB variable, let $f = f(t)$ for the time function and $F = F(t)$ for the Laplace function. Symbols for the dependent variables are easily changed for different physical variables, but the basic independent variables t and s will be retained throughout this development. It should be noted, however, that even those variables can be changed, and the MATLAB Help file contains the procedures.

The first step required is to declare t and s as symbolic variables. This is accomplished with the simple command

$$>> \text{syms t s} \qquad (11\text{-}24)$$

The time function f is then created according to its form. Standard MATLAB arithmetic and functional operations are used to define the terms involved. The Laplace transform F is then determined by the command **laplace,** as follows:

$$>> \text{F = laplace(f)} \qquad (11\text{-}25)$$

Although not necessary, there are two commands that are sometimes useful in displaying the result. One is the **pretty** operation introduced earlier in the text and is

$$>> \text{pretty(F)} \qquad (11\text{-}26)$$

The other operation that helps in some cases is the **simplify** command, which is

$$>> \text{simplify(F)} \qquad (11\text{-}27)$$

Inverse Transforms

Assume again that the **syms** command of Equations 11-24 is in effect. We are assuming here that the Laplace function F is not in memory and needs to be constructed. F is created according to its form using MATLAB algebraic statements. The inverse Laplace transform is determined by the command **ilaplace** as follows:

$$>> \text{f = ilaplace(F)} \qquad (11\text{-}28)$$

As in the case of the transform operation, the commands **pretty** and **simplify** may be helpful to apply to f in some cases. Moreover, the function f may be readily plotted with **ezplot**.

The examples that follow will illustrate the operations of both transformation and inverse transformation using MATLAB.

Example 11-5

Use MATLAB to determine the Laplace transform of

$$f(t) = 5t \qquad (11\text{-}29)$$

Solution

By now, readers should realize that we usually begin our MATLAB problems with one that could be done much more easily by reverting back to previous methods, in this case either from memory or by looking it up in Table 10-2. However, the intent is to make the entry into a new procedure as simple as possible.

Recall that the first step is the **syms** statement. Strictly speaking, if we are only going from a time function to a Laplace function, we would need only to put **t** in the statement, and if we were going from the Laplace function to a time function we would need only to put **s** in the statement. However, with only one extra key stroke, we establish the possibility of going in both directions, and that will be the procedure used. The statement then is

$$>> \text{syms t s} \tag{11-30}$$

The time function f is then defined in accordance with MATLAB algebraic operations (an asterisk for multiplication in this case) and entered.

$$>> \text{f} = 5*\text{t} \tag{11-31}$$

$$\text{f} =$$

$$5*\text{t}$$

We could, of course, have suppressed the listing of the function by placing a semicolon after the defining relationship, but we chose to have it appear on the screen.

The Laplace transform is then determined by the command

$$>> \text{F} = \text{laplace(f)} \tag{11-32}$$

$$\text{F} =$$

$$5/\text{s}\string^2$$

The function appears in MATLAB form and needs no further simplification.

Example 11-6

A voltage in a certain circuit is given by

$$v(t) = 3e^{-2t}\sin 5t + 4e^{-2t}\cos 5t \tag{11-33}$$

Use MATLAB to determine the Laplace transform $V(s)$.

Solution

The first step is to establish t and s as symbolic variables.

$$>> \text{syms t s} \tag{11-34}$$

Next, the voltage function v is expressed in MATLAB form and entered. Carefully note the form that follows, since it is easy to forget an asterisk for multiplication or the parentheses around the arguments for the exponential and sinusoidal functions. (MATLAB will definitely let you know if you forget either one!)

$$\text{>> v = 3*exp(-2*t)*sin(5*t) + 4*exp(-2*t)*cos(5*t)} \tag{11-35}$$

$$v =$$

$$\text{3*exp(-2*t)*sin(5*t)+4*exp(-2*t)*cos(5*t)}$$

The Laplace transform V is determined from the command.

$$\text{>> V = laplace(v)} \tag{11-36}$$

$$V =$$

$$\text{15/((s+2)\^2+25)+4*(s+2)/((s+2)\^2+25)}$$

This result is a bit unwieldy to untangle, but we have two commands that can help. One is the **pretty** command and the other is the **simplify** command. It is suggested that you try both to see which one, if either, will make the function easier to read. In this case, the **simplify** command worked best. It yields

$$\text{>> V=simplify(V)} \tag{11-37}$$

$$V =$$

$$\text{(23+4*s)/(s\^2+4*s+29)}$$

With this latter form, we can more easily write the equation in standard mathematical terms. It is

$$V(s) = \frac{4s + 23}{s^2 + 4s + 29} \tag{11-38}$$

The reader is invited to apply T-5 and T-6 of Table 10-1 to verify that the result obtained is correct.

Example 11-7

A relatively involved transform function of Chapter 10 was the one used in Examples 10-10 and 10-11 and first given in Equation 10-42. It is repeated here for convenience.

$$F(s) = \frac{50(s + 3)}{(s + 1)(s + 2)(s^2 + 2s + 5)} \tag{11-39}$$

Determine the inverse Laplace transform using MATLAB.

Solution

Recall that there are two real negative poles and a pair of complex conjugate poles with negative real parts. Thus, the solution will contain two decaying exponential functions and exponentially damped sine and cosine terms.

If desired, we could use the **conv** function to multiply the denominator polynomials to create one fourth-degree polynomial, but there is no incentive to do so. Let us allow MAT-LAB to do the work. We begin with

$$\text{>> syms t s} \qquad (11\text{-}40)$$

Next we create the Laplace function F by the statement that follows. Note that in the denominator, we need to use two levels of parentheses, one level for the individual factors, and one level to ensure that all of the appropriate denominator factors are enclosed.

$$\text{>> F=50*(s+3)/((s+1)*(s+2)*(s^2+2*s+5))} \qquad (11\text{-}41)$$

$$F =$$

$$(50*s+150)/(s+1)/(s+2)/(s^2+2*s+5)$$

We now take the inverse Laplace transform with the command

$$\text{>> f = ilaplace(F)} \qquad (11\text{-}42)$$

$$f =$$

$$25*exp(-t)-10*exp(-2*t)-15*exp(-t)*cos(2*t)-5*exp(-t)*sin(2*t)$$

We should be able to interpret this expression, but let us see what the **simplify** command does.

$$\text{>> f = simplify(f)} \qquad (11\text{-}43)$$

$$f =$$

$$25*exp(-t)-10*exp(-2*t)-15*exp(-t)*cos(2*t)-5*exp(-t)*sin(2*t)$$

Well, for this example it did nothing! From the viewpoint of the simplify operation, the function cannot be reduced any further. So, let us see what **pretty** does.

$$\text{>> pretty(f)} \qquad (11\text{-}44)$$

$$25\ exp(-t) - 10\ exp(-2\ t) - 15\ exp(-t)\ cos(2\ t) - 5\ exp(-t)\ sin(2\ t)$$

This form conforms a little more to the standard form and is the same as obtained in Examples 10-10 and 10-11, which in the standard mathematical form is

$$f(t) = 25e^{-t} - 10e^{-2t} - 15e^{-t}\cos 2t - 5e^{-t}\sin 2t \qquad (11\text{-}45)$$

Example 11-8

In the solution of Example 10-14 using Laplace transforms, Equation 10-94 listed the following transform as the solution for *Y(s)*:

$$Y(s) = \frac{10}{s+2} + \frac{48}{(s+2)(s^2+16)} \qquad (11\text{-}46)$$

Use MATLAB to determine the inverse transform $y(t)$.

Solution

As usual we begin with

$$\gg \text{syms t s} \qquad (11\text{-}47)$$

The transform is listed as a sum of two terms with double parentheses in the denominator of the second term.

$$\gg Y = 10/(s+2) + 48/((s+2)*(s^2+16)) \qquad (11\text{-}48)$$

$$Y =$$

$$10/(s+2)+48/(s+2)/(s^2+16)$$

The inverse Laplace transform is determined from the command

$$\gg y = \text{ilaplace}(Y) \qquad (11\text{-}49)$$

$$y =$$

$$62/5*\exp(-2*t)-12/5*\cos(16^\wedge(1/2)*t)+3/10*16^\wedge(1/2)*\sin(16^\wedge(1/2)*t)$$

This is too messy, so let us try the **simplify** command.

$$\gg y=\text{simplify}(y) \qquad (11\text{-}50)$$

$$y =$$

$$62/5*\exp(-2*t)-12/5*\cos(4*t)+6/5*\sin(4*t)$$

This is a lot better and agrees with Equation 10-102. Just for fun, let's try **pretty**.

$$\gg \text{pretty}(y) \qquad (11\text{-}51)$$

$$62/5 \; \exp(-2 \; t) - 12/5 \; \cos(4 \; t) + 6/5 \; \sin(4 \; t)$$

In this example, both **simplify** and **pretty** provided more easily recognizable forms.

MATLAB PROBLEMS

Problems 11-1 through 11-18 involve differential equations that were to be solved with classical procedures in Chapter 9 and Laplace transforms in Chapter 10. For the present chapter, solve each differential equation using MATLAB.

11-1. $\dfrac{dy}{dt} = 6t^2 \qquad y(0) = 5$

11-2. $\dfrac{dy}{dt} = 20\cos 5t + 8 \qquad y(0) = 10$

11-3. $\dfrac{dy}{dt} + 4y = 0 \qquad y(0) = 6$

11-4. $\dfrac{dy}{dt} + 50y = 0 \qquad y(0) = 20$

11-5. $\dfrac{dy}{dt} + 4y = 80 \qquad y(0) = 6$

11-6. $\dfrac{dy}{dt} + 50y = 400 \qquad y(0) = 20$

11-7. $\dfrac{dy}{dt} + 4y = 40\sin 3t \qquad y(0) = 6$

11-8. $\dfrac{dy}{dt} + 50y = 400\cos 75t \qquad y(0) = 20$

11-9. $\dfrac{d^2 y}{dt^2} + 4\dfrac{dy}{dt} + 3y = 0 \qquad y(0) = 0 \quad y'(0) = 12$

11-10. $\dfrac{d^2 y}{dt^2} + 7\dfrac{dy}{dt} + 12y = 0 \qquad y(0) = 20 \quad y'(0) = 0$

11-11. $\dfrac{d^2 y}{dt^2} + 4\dfrac{dy}{dt} + 3y = 30 \qquad y(0) = 20 \quad y'(0) = 12$

11-12. $\dfrac{d^2 y}{dt^2} + 7\dfrac{dy}{dt} + 12y = 96 \qquad y(0) = 20 \quad y'(0) = 0$

11-13. $\dfrac{d^2 y}{dt^2} + 3\dfrac{dy}{dt} + 2y = 24 \qquad y(0) = 0 \quad y'(0) = 10$

11-14. $\dfrac{d^2 y}{dt^2} + 3\dfrac{dy}{dt} + 2y = 24e^{-4t} \qquad y(0) = 10 \quad y'(0) = 5$

11-15. $\dfrac{d^2 y}{dt^2} + 4\dfrac{dy}{dt} + 13y = 52 \qquad y(0) = 0 \quad y'(0) = 0$

11-16. $\dfrac{d^2y}{dt^2} + 4\dfrac{dy}{dt} + 104y = 832$ $y(0) = 0$ $y'(0) = 0$

11-17. $\dfrac{d^2y}{dt^2} + 6\dfrac{dy}{dt} + 9y = 0$ $y(0) = 10$ $y'(0) = 0$

11-18. $\dfrac{d^2y}{dt^2} + 6\dfrac{dy}{dt} + 9y = 90$ $y(0) = 0$ $y'(0) = 0$

11-19. The simplest form of a differential equation is the "directly integrable" form, in which integration of both sides is the method for solution. It follows then that the techniques for numerical integration considered in Chapter 8 are applicable to such cases. Consider the following first-order differential equation:

$$\frac{dy}{dt} = x(t) = te^{-t} + \sin 2\pi t \quad \text{and} \quad y(0) = 0 \quad 0 \le t \le 5$$

Develop an M-file program that will accomplish the tasks that follow.

   a. Define a time vector **t** defined over the domain from 0 to 5 in steps of 0.05.
   b. Evaluate the function **x** over the domain of (a).
   c. Apply appropriate labeling for horizontal and vertical axes.
   d. For **figure(1)**, plot **x** versus **t**. It is suggested that you place a **pause** command at this point.
   e. Obtain an exact mathematical solution using the **dsolve** command and denote it as **y**.

      Note: MATLAB determines the time step for ezplot, and it will be necessary to return to the original time step for the numerical solutions that follow.

   f. Obtain a numerical solution **y0** based on the zero-order integration algorithm **cumsum**.
   g. Obtain a numerical solution **y1** based on the first-order integration algorithm **cumtrapz**.
   h. For **figure (2),** use the **ezplot** command to plot the exact solution **y** over the domain from 0 to 5. Follow the plot with the command **hold on**. This will keep the curve intact while the remaining plots are made on the same figure.
   i. Place the two numerical solutions **y0** and **y1** on the same graph as **y**. Place an **o** at each point for the zero-order solution and a **+** at each point for the first-order solution.
   j. Apply appropriate labeling for the horizontal and vertical axes.

11-20. Write a program that performs a symbolic solution of a second-order CCLODE with a constant forcing function and uses **ezplot** display the response. The DE is of the form

$$b_2\frac{d^2y}{dt^2} + b_1\frac{dy}{dt} + b_0 y = A \quad \text{with } y(0), y'(0), \text{ and } A \text{ specified.}$$

Inputs:
The three constants b_2, b_1, b_0 and A
The constant A
The two initial conditions.
The domain limits t_1 and t_2 for plotting

Outputs:
The mathematical form of $y(t)$
A plot of $y(t)$ versus t

Special Information: When a symbolic solution of a DE is performed with unspecified constants instead of actual numbers, some special procedures are necessary to assign numerical values to the constants. Assume that the DE coefficients and the forcing function are denoted respectively in MATLAB format as **b2, b1, b0,** *and* **A**. However, $y(0)$ and $y'(0)$ are not valid ways to enter these values in an input statement. Instead, denote $y(0)$ as **y0** and $y'(0)$ as **y1** within the input statements.

Once the DE is solved symbolically with constants, it is necessary to employ a **subs** statement to enter the actual numerical values. A format for this purpose within the program is provided by the following command:

y = subs(y, {´b2´, ´b1´, ´b0´, ´A´, ´y0´, ´y1´}, {b2, b1, b0, A, y0, y1})

This command forces the symbolic constants within the first set of brackets to assume the respective values in the second set of brackets, which have been specified in the input statements.

The values of **t1** and **t2** are entered in the **ezplot** statement using the procedure discussed in the chapter. After the plot is obtained, it may be necessary in some cases to employ the **axis** command to observe the full range of the dependent variable.

Introduction to Statistics 12

There are numerous processes in the "real world" whose behavior cannot be predicted with certainty, but in which a prediction can be made concerning the relative likelihood of occurrence. Even the basic laws of physics that are taught as if they are always exact are, in many cases, subject to uncertainties if inspected at a microscopic level.

The term *random variable* describes a variable in which the exact behavior cannot be predicted, but which may be described in terms of probable behavior. A *random process* is any process that involves one or more random variables. The subjects of probability and statistics are based on a mathematical and scientific methodology for dealing with random processes and variables.

An example of a random variable is shown in Figure 12-1. This particular variable $x(t)$ is assumed to be varying with time in some random fashion and is typical of electrical noise that can be heard on a radio receiver when the signal level is very weak. There are complete college curricula devoted to the study of statistics, so this treatment is necessarily limited in scope. However, it should provide a sufficient exposure to acquaint the reader with the concepts and basic definitions associated with the subject.

Objectives

After completing this chapter, the reader should be able to

1. Define probability and discuss its significance.
2. Define the concepts of *mutual exclusiveness* and *statistical independence* and their implications.
3. Define *conditional probability*.
4. Define the *probability density function* (pdf) for either a *discrete* variable or a *continuous* variable and apply it for probability computations.
5. Define the *probability distribution function* for either a *discrete* variable or a *continuous* variable and apply it for probability computations.

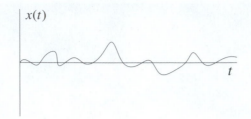

Figure 12-1   A random variable such as electrical noise

6. Using the pdf for either a *discrete* variable or a *continuous* variable, determine the *mean* value, the *mean-square* value, the *root-mean square* value, the *variance,* and the *standard deviation*.
7. State the form of the pdf for a binomial distributed variable and apply it to binary variable situations.
8. Discuss the form of the *gaussian* or *normal* pdf and its significance.
9. Determine probabilities associated with gaussian distributions using curves, tabulated values, and MATLAB.
10. Define the concept of a *sample distribution* and compare it with a *complete population.*
11. Determine various statistical parameters for a sample distribution.
12. Apply MATLAB to determine various statistical parameters.

12-2   PROBABILITY

Probability is associated with the very natural trend of a random event to follow a somewhat regular pattern *if the process is repeated a sufficient number of times.* For example, consider the age-old process of flipping an unbiased coin. If the experiment is continually repeated over a large number of trials, we would expect to get about the same number of heads as tails. Intuitively, we can say that the probability on a given trial of a head is 0.5 and the probability of a tail is 0.5.

Assume that a certain experiment has K possible outcomes and that one and only one outcome will occur as a result of performing the experiment. Let A_1 symbolically represent the condition that the first event occurs and let A_2 symbolically represent the condition that the second event occurs; and so on. We define

$P(A_1)$ = probability that event 1 occurs
$P(A_2)$ = probability that event 2 occurs
.   .   .
$P(A_K)$ = probability that event K occurs

Assume that a large number of trials n of the experiment are performed, which in the limit could be said to approach infinity. Assume that the first event occurs n_1 times, the second event n_2 times, and so on. The various probabilities are then defined as follows:

$$P(A_1) = \lim_{n \to \infty} \frac{n_1}{n}$$

$$P(A_2) = \lim_{n \to \infty} \frac{n_2}{n} \tag{12-1}$$

$$P(A_K) = \lim_{n \to \infty} \frac{n_K}{n}$$

Assume now that the probability values associated with the various outcomes are known. Each value $P(A_k)$ is a number satisfying the condition

$$0 \le P(A_k) \le 1 \tag{12-2}$$

In general, the closer the value of $P(A_k)$ is to 1, the more likely it is that the kth event will occur on a given trial, and the smaller the value of $P(A_k)$, the less likely it is that the event will occur. On an average, the kth event will occur about once every $1/P(A_k)$ trials. For example, if $P(A_k) = 0.1$, the first event in the population will occur about once every ten trials *on an average*. If $P(A_k) = 1$, the kth event is certain; that is, it will always occur when the experiment is performed. Conversely, if $P(A_k) = 0$, the kth event will never occur.

For a population consisting of K possible outcomes, only one of which can occur for each trial of the experiment, the following relationship can be deduced:

$$P(A_1) + P(A_2) + \ldots + P(A_K) = 1 \tag{12-3}$$

Example 12-1

Assume that one card is to be randomly drawn from a standard 52-card deck. Using the relative frequency of occurrence as an intuitive basis, determine the probability that the card is (a) a red card, (b) a heart, (c) an ace, (d) the ace of spades, and (e) the ace of hearts or the ace of diamonds.

Solution

a. Let R represent the condition that a red card is drawn. There are 52 cards, of which 26 are red. Thus,

$$P(R) = \frac{26}{52} = \frac{1}{2} \tag{12-4}$$

b. Let H represent the condition that a heart is drawn. Since there are 13 hearts,

$$P(H) = \frac{13}{52} = \frac{1}{4} \tag{12-5}$$

c. Let A represent the condition that an ace is drawn. Since there are 4 aces,

$$P(A) = \frac{4}{52} = \frac{1}{13} \tag{12-6}$$

d. Let AH represent the condition that the ace of hearts is drawn. Since there is only one ace of hearts,

$$P(AH) = \frac{1}{52} \tag{12-7}$$

e. The desired outcome can either be the ace of hearts or the ace of diamonds, so there are two possible favorable results. Letting E represent the condition that either desirable outcome occurs, we have

$$P(E) = \frac{2}{52} = \frac{1}{26} \tag{12-8}$$

12-3   MUTUAL EXCLUSIVENESS AND STATISTICAL INDEPENDENCE

In this section, the concepts of *mutual exclusiveness* and *statistical independence* will be treated. Although the two concepts are quite different, they are often confused with each other owing to certain similarities in their usage and form. Recall from the last section that when a term such as A_1 is used to represent a certain event occurring, this is done in a symbolic sense. Consequently, we can say that A_1 is "true" if the first event in a population occurs. This inference is the same as encountered in Boolean expressions. Thus, $A_1 = 1$ could represent a Boolean expression indicating that the first event has occurred. Likewise, if the first event does not occur, an appropriate Boolean expression would be $A_1 = 0$. Alternately, $\overline{A_1} = 1$, where $\overline{A_1}$ is the logical complement of A_1 ("not A_1"), and this alternate expression states the same condition.

In the same sense, then, as combinational logic, we define a logical expression such as $A_1 + A_2$ as the condition that either A_1 or A_2 is true. Similarly, an expression such as $A_1 A_2$ represents the condition that both A_1 and A_2 are true. These expressions need not be confused with ordinary algebraic expressions, since it is usually clear in the usage which meaning is implied. From a probability point of view, the expression $P(A_1 + A_2)$ represents the probability that either A_1 *or* A_2 is true, that is, the probability that either the first or the second event has occurred. Similarly, the expression $P(A_1 A_2)$ represents the probability that both A_1 *and* A_2 are true, that is, the probability that both events have occurred.

Mutual Exclusiveness

We now define the concept of *mutual exclusiveness*. Two events are said to be *mutually exclusive* if

$$P(A_1 + A_2) = P(A_1) + P(A_2) \tag{12-9}$$

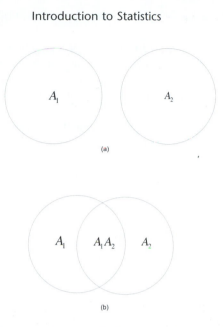

Figure 12-2   Venn diagrams for events that are (a) mutually exclusive and (b) not mutually exclusive

Stated in words, two events are said to be mutually exclusive if the probability of *either* the first event *or* the second event occurring is the sum of the respective probabilities. Note that $A_1 + A_2$ on the left represents an "or" statement, while the sum on the right is a true algebraic sum.

Venn diagrams can illustrate the concept of mutual exclusiveness. In Figure 12-2(a) there is no overlap between the area corresponding to event 1 and the area corresponding to event 2, so the events are mutually exclusive. However, in Figure 12-2(b) there is a common overlapping area between the population corresponding to event 1 and that of event 2, so the two events are not mutually exclusive.

As an example, let A_1 represent the event that a person is in Norfolk, Virginia, at a given time, and let A_2 represent the event that the same person is in New York City at the same time. These two events are mutually exclusive since a person obviously cannot be in both places simultaneously. As a second example, let A_1 represent the event that a person is in Norfolk, and let A_2 represent the event that a person is serving in the U.S. Navy. These two events are not mutually exclusive, since there are many persons in Norfolk who are also in the U.S. Navy. If the two events are not mutually exclusive, some portion of their respective populations overlap, so the probability of one or the other occurring would be weighted too heavily if the respective probabilities are simply added, as in Equation 12-9. From the Venn diagram of Figure 12-2(b), it is seen that the area corresponding to $A_1 A_2$ appears both in the weighting for A_1 and the weighting for A_2 in the respective probabilities. Consequently, if $P(A_1)$ is added to $P(A_2)$, this common area will have been added twice. To correct for this extra weighting, the term $P(A_1\,A_2)$ is subtracted, and the probability of either the first event *or* the second event occurring when the two events are not mutually exclusive is thus

$$P(A_1 + A_2) = P(A_1) + P(A_2) - P(A_1 A_2)$$

$$(12\text{-}10)$$

Observe that this expression reduces to Equation 12-9 when the two events are mutually exclusive since $P(A_1\,A_2) = 0$ for that case.

Statistical Independence

Two events are said to be statistically independent if

$$P(A_1 A_2) = P(A_1) P(A_2)$$

$$(12\text{-}11)$$

Stated in words, two events are said to be statistically independent if the probability that *both* the first event *and* the second event will occur is equal to the product of the respective probabilities. Note that $A_1\,A_2$ on the left represents an "and," while the product on the right is a normal algebraic multiplication. The expression $P(A_1\,A_2)$ is called a *joint probability* of two events and is denoted in some references as $P(A_1, A_2)$.

For events to be statistically independent there must be no cause-effect relationship between them; that is, the occurrence of either event should not in any way influence the occurrence of the other event. As examples of statistical dependence and independence, let A_1 represent the event that the dc voltage in a transmitter power supply suddenly increases to a level exceeding the maximum voltage of a transistor powered by the supply, and let A_2 represent the event that the transistor fails. These two events are not statistically independent since the occurrence of the first event (higher power supply voltage) strongly affects the probable occurrence of the second event (transistor failure). On the other hand, if A_2 represents the event that a similar transistor fails in a completely separate transmitter operating from a different power supply, it is likely that the two events are statistically independent.

Conditional Probability

We next consider the question of evaluating a joint probability when the events are not statistically independent. This requires the use of a concept called *conditional probability*. The expression $P(A_2/A_1)$ is defined to mean "the probability that A_2 is true *given* that A_1 is true." Knowing that A_1 is true may influence the likelihood of A_2 being true, so conditional probability deals with events that may be statistically dependent. For example, the transistor in the example referred to earlier may normally have a low probability of failure under specified operating conditions. However, if we are *given* the fact that its supply voltage has exceeded the maximum rating, the conditional probability of failure may be quite high.

The Venn diagram of Figure 12-2(b) may be used to infer a relationship for conditional probability. Assume that it is known that A_1 is true. This means that any possible outcome must be contained within the area representing A_1. Now what is the probability that A_2 is true if it is known that A_1 is true? The pertinent area is the common area for both A_1 and

A_2, that is, A_1A_2 must be true. The ratio of this area to the total area of A_1 provides a proper formulation for the conditional probability. Thus,

$$P(A_2 / A_1) = \frac{P(A_1A_2)}{P(A_1)} \tag{12-12}$$

The joint probability $P(A_1A_2)$ may be determined from Equation 12-12 as

$$P(A_1A_2) = P(A_1)P(A_2 / A_1) \tag{12-13}$$

The preceding result can be used to determine joint probabilities when events are not statistically independent. In this case, it is necessary to first specify (1) the probability that one of the two events occurs and (2) the conditional probability that the second event occurs given that the first event has occurred. If the events are statistically independent, the fact that A_1 is true will in no way affect the outcome of A_2. In this case, the conditional probability of Equation 12-12 reduces to the simple probability of A_2 being true; that is, $P(A_2/A_1) = P(A_2)$. Thus, Equation 12-13 reduces back to the form of Equation 12-11 for the case of statistical independence.

A few final comments about mutual exclusiveness and statistical independence should help in clarifying the concepts. Most applications of mutual exclusiveness involve probability evaluations in which it is necessary to determine if there are common elements of separate "population" groups representing different outcomes. Of particular concern is the case where one event or the other might occur in the probability evaluation. In contrast, statistical independence is primarily of interest when probabilities involving two or more events all occurring are to be evaluated and when it is desired to determine if the occurrence of any one of the events tends to influence the outcome of other events.

Example 12-2

A single card is to be drawn from a deck of cards. What is the probability that it will be *either* an ace *or* a king?

Solution

Let A represent the outcome of drawing an ace, and let K represent the outcome of drawing a king. The individual probabilities for a single draw are

$$P(A) = \frac{4}{52} = \frac{1}{13} \tag{12-14}$$

$$P(K) = \frac{4}{52} = \frac{1}{13} \tag{12-15}$$

The probability evaluation is $P(A + K)$. Whenever the probability of one event *or* another is desired, it is necessary to determine if the two events are mutually exclusive; that is, can a card simultaneously be an ace and a king? The answer is obviously no, so the two events are mutually exclusive. Thus,

$$P(A+K) = P(A) + P(K) = \frac{1}{13} + \frac{1}{13} = \frac{2}{13} \tag{12-16}$$

Example 12-3

A single card is to be drawn from a deck of cards. What is the probability that it will be *either* an ace *or* a red card?

Solution

As in the previous example, the probability of drawing an ace is

$$P(A) = \frac{1}{13} \tag{12-17}$$

Let R represent the outcome of drawing a red card. This probability is

$$P(R) = \frac{26}{52} = \frac{1}{2} \tag{12-18}$$

Again, we raise the question of mutual exclusiveness, that is, can a card simultaneously be an ace and a red card? The answer is yes in this case, since there are two cards that belong to both groups. Thus, the events are not mutually exclusive. The probability of a card being *both* an ace *and* a red card is

$$P(AR) = \frac{2}{52} = \frac{1}{26} \tag{12-19}$$

From Equation 12-10, we have

$$P(A+R) = P(A) + P(R) - P(AR) = \frac{1}{13} + \frac{1}{2} - \frac{1}{26} = \frac{7}{13} \tag{12-20}$$

Example 12-4

Two cards are drawn in succession from a deck. The first card is replaced and the deck is reshuffled before the second card is drawn. What is the probability that the first card and the second card will both be aces?

Solution

Let A_1 represent the condition that the first card is an ace, and let A_2 represent the condition that the second card is an ace. Since the first card is replaced and the deck is reshuffled before the second card is drawn, there is no dependency between the two favorable outcomes and the individual probabilities are equal. Thus,

$$P(A_1) = \frac{1}{13} \tag{12-21}$$

$$P(A_2) = \frac{1}{13} \tag{12-22}$$

The desired outcome is that an ace is drawn on the first trial *and* an ace is drawn on the second trial, that is, $P(A_1 A_2)$. In view of the statistical independence,

$$P(A_1 A_2) = P(A_1)P(A_2) = \left(\frac{1}{13}\right)^2 = \frac{1}{169} \tag{12-23}$$

Example 12-5

Consider the same experiment and desired outcome as in Example 12-4, but assume that the first card is *not* replaced before drawing the second card.

Solution

The probability of drawing an ace on the first trial is the same as before, that is, $P(A_1)$ = 1/13. However, if an ace is drawn, there is one less ace in the deck and one fewer card. Thus, the occurrence of the first event will influence the occurrence of the second event, and the two events are not statistically independent. Assuming that an ace is obtained on the first draw, there are 51 cards remaining, of which 3 are aces. Thus, the conditional probability that an ace is drawn on the second trial (A_2 is true), given that an ace was drawn on the first trial (A_1 is true), is

$$P(A_2 / A_1) = 3/51 \tag{12-24}$$

Using Equation 12-13, we have

$$P(A_1 A_2) = P(A_1)P(A_2 / A_1) = \left(\frac{1}{13}\right) \times \left(\frac{3}{51}\right) = \frac{1}{221} \tag{12-25}$$

Example 12-6

In this example and the next one, some of the most basic concepts of reliability will be illustrated. Let us assume the existence of some switches that are activated by an external electronic control signal. To illustrate the process with simple values, we will assume *very poor quality* units that only close 90% of the time when the activating signal is applied. First, consider the serial arrangement shown in Figure 12-3 in which *both* switches must close to ensure success. Assume that the actions of closing or not closing for the two switches are statistically independent. Upon applying the control signal, success is achieved if *both* switches close. Determine the probability of success.

Solution

Let $P(S)$ represent the probability of success, and let $P(A)$ and $P(B)$ represent the respective probabilities that the individual switches close. We have

Figure 12-3   Two switches in a serial arrangement for Example 12-5

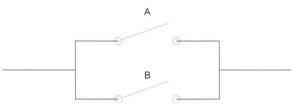

Figure 12-4   Two switches in a parallel arrangement for Example 12-6

$$P(A) = P(B) = 0.9 \qquad (12\text{-}26)$$

Since the actions of the switches are statistically independent, the probability of success (both closing) is

$$P(S) = P(AB) = P(A)P(B) = (0.9)(0.9) = 0.81 \qquad (12\text{-}27)$$

An already bad situation becomes worse with this arrangement. The lesson here is to minimize the number of *and* operations that are required for success if there is one or more that is unreliable.

Example 12-7

Assuming the same switches as in Example 12-6, consider the parallel arrangement depicted in Figure 12-4. In this case, success is achieved if either one *or* the other closes. Determine the probability of success in this case.

Solution

The desired probability in this case is $P(A + B)$, that is, the probability that either *A or B* is true. Without much thought, someone could grab the formula of Equation 12-9 and add 0.9 to 0.9 and get 1.8! Now we have an impossible result. All probability values are bounded between 0 and 1, so something is wrong.

The answer to this dilemma is the fact that even though *A* and *B* are statistically independent, they are *not mutually exclusive*. In the sense of this problem, this means that *A* and *B* will *both* close during some trials, as was determined in Example 12-6. Thus, the correct formulation is that of Equation 12-10, which is placed in the next step.

$$P(S) = P(A + B) = P(A) + P(B) - P(AB) \tag{12-28}$$

The last term in Equation 12-28 involves the same calculation that was made in Example 12-6 and is $(0.9) \times (0.9) = 0.81$, since the two events are statistically independent. Therefore, the probability of success can be determined as

$$P(S) = 0.9 + 0.9 - 0.81 = 0.99 \tag{12-29}$$

While the reliability is still relatively poor, it has improved considerably over the process of the preceding example or the process that would be obtained using just one switch. The process involved here is one of design *redundancy*. If there is a weak link in the design of a complex system, it is prudent to utilize alternate links to accomplish the mission in the event of failure of the weak link.

There is an alternate formulation for this problem that should be noted. Instead of computing the probability of success, suppose we compute first the probability of failure, which we will denote as $P(F)$. For the parallel arrangement, failure requires that *both A and B fail*. Let $P(\overline{A})$ and $P(\overline{B})$ represent these values. We have

$$P(\overline{A}) = P(\overline{B}) = 1 - 0.9 = 0.1 \tag{12-30}$$

Since the two switches are statistically independent, the probability of failure is

$$P(F) = P(\overline{A})P(\overline{B}) = (0.1)(0.1) = 0.01 \tag{12-31}$$

The probability of success is then determined as

$$P(S) = 1 - P(F) = 1 - 0.01 = 0.99 \tag{12-32}$$

This result is the same as determined by the first approach.

12-4   DISCRETE STATISTICAL FUNCTIONS

The first statistical functions to be considered will consist of those involving only *discrete variables*. A *discrete variable* is one that can assume only a finite number of values. For example, a binary digital computer signal is assumed to have only two values: a logical 1 and a logical 0.

To illustrate the process, we will assume a discrete random voltage function that can assume four values. The four values will be assumed to be -5 V, 0 V, 5 V, and 10 V. To maintain notation consistent with the statistical formulation that follows, we will denote the voltage as $x(t)$. A short segment of the random four-level signal is shown in Figure 12-5. Note that the varying transition levels as the pulses are changing are not considered as part of the population. (The transition times are assumed to be very small.) Whatever this signal accomplishes is based solely on the four finite levels. The four finite signal levels will be denoted as x_k where k is an integer that varies from 1 to 4. The values are $x_1 = -5$ V, $x_2 = 0$ V, $x_3 = 5$ V, $x_4 = 10$ V.

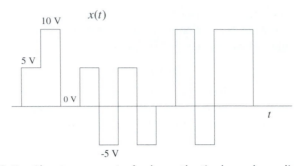

Figure 12-5   Short segment of a hypothetical random discrete voltage
whose statistical behavior is to be determined

The pulses appear in a random order and there is no discernible pattern. However, the question arises from a statistical point of view as to whether certain levels are more likely to appear than others. Assume that over an interval of 100,000 pulses, the four possible levels have the numbers of pulses listed in Table 12-1, at the end of the chapter. The integer k assumes the values 1, 2, 3, and 4, while the corresponding values of x_k are –5 V, 0 V, 5 V, and 10 V. The column labeled "Count" is the number of pulses assuming each value. The last column will be discussed shortly.

The results will now be displayed graphically. Consider first the plot shown in Figure 12-6, in which the number of times each discrete value occurs is shown. The exact number of times each variable occurs could be of interest, but in order to provide a more compact form for specification purposes, it is the *relative* number that is of more interest. Suppose

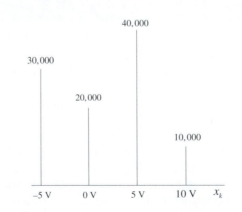

Figure 12-6   Results of the counts for each level of the
hypothetical discrete signal

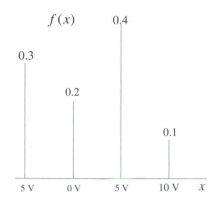

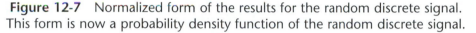

Figure 12-7 Normalized form of the results for the random discrete signal. This form is now a probability density function of the random discrete signal.

each count is normalized or divided by the total number of samples. The results are tabulated in the last column of Table 12-1 and are shown graphically in Figure 12-7. Note that the shape of this graph is the same as those of Figure 12-6, but the values are less than unity. In fact, it can be readily verified that the sum of all these values is exactly unity. Provided that the sampling interval is sufficiently long and the nature of the signal does not change, this function is theoretically independent of the actual number of samples used in establishing the final graph, provided that the statistical pattern remains the same.

To simplify the notation somewhat, we will drop the subscript k from x_k and express the variable simply as x in most formulas that follow. However, it should be clear to the reader when the function is based on a discrete variable. In a few cases, we will revert back to the x_k form for clarity. What has been created through this experiment is a *probability density function*, which will be denoted as a *pdf*. In this case, it is a *discrete probability density function* or *discrete pdf*. It will be denoted as $f(x)$, based on the assumption that there is only one independent variable x. In situations where there is more than one independent variable, it is customary to add a subscript to the function. For example, assume that there are two independent random variables x and y. The two pdfs could be denoted as $f_x(x)$ and $f_y(y)$, respectively.

There is a tendency to get confused about the independent and dependent variables when first encountering the concept of a pdf. The independent (or horizontal) variable is the particular quantity under study (voltage in this case). Thus, while the instantaneous time function had voltage as the dependent (or vertical) variable, when reformulated in terms of a pdf, it becomes the independent (or horizontal) variable. The dependent variable on a pdf is simply a *relative* measure of the likelihood of the corresponding horizontal variable to occur.

Probability Evaluations

The upper-case variable X will be defined as a *random sample* taken from the process from which the pdf $f(x)$ is determined. We know that X can assume only one of four values. The expression $P(X = x)$ is defined to mean *"the probability that a random sample of the process is equal to x."* For a discrete variable, this probability is then related to the pdf by

$$P(X = x) = f(x) \qquad (12\text{-}33)$$

Stated in words, the probability that one random sample of a process is equal to a particular discrete level is the value of the discrete pdf evaluated for that given level. Thus, a discrete pdf may be interpreted as a set of probability values, each of which predicts the probability that a random sample of the process will be the given value.

Remember that the way in which the pdf was established was dividing each number by the total number of samples. Therefore, the net sum of all the pdf values for a discrete variable must satisfy

$$\sum_k f(x) = 1 \qquad (12\text{-}34)$$

where the summation is interpreted to mean over all possible values.

Probability Distribution Function

An alternate formulation is that of the *probability distribution function*, which will be denoted as $F(x)$. It is defined for a discrete variable x by the following equation:

$$F(x) = P(X \le x) \qquad (12\text{-}35)$$

For a discrete pdf, it is determined by summing all the probability density function values up to and including the value associated with x. Mathematically, this process can be represented as

$$F(x_k) = \sum_{n=-\infty}^{k} f(x_n) \qquad (12\text{-}36)$$

where the subscript k has been momentarily added to clarify the summation.

Do not be misled by the presence of $-\infty$ in the summation. It simply means that we begin the summation at the extreme left side of the pdf and continue the sum up to and including $f(x_k)$.

Example 12-8

For the discrete pdf example constructed in this section, determine the following probabilities: (a) $P(X = 5)$, (b) $P[(X = 0) + (X = 5)]$, (c) $P(X > 0)$, (d) $P(X \ge 0)$, (e) $P(-5 \le X \le 10)$, (f) $P(X = 15)$.

Solution

a. The probability that the random sample is a single value is simply the value of the pdf evaluated for that value.

$$P(X = 5) = f(5) = 0.4 \qquad (12\text{-}37)$$

b. This probability is read as "the probability that a random sample has *either* the value 0 *or* the value 5. The two events are *mutually exclusive* since they cannot occur together. Hence,

$$P\big[(X = 0) + (X = 5)\big] = P(X = 0) + P(X = 5) = 0.2 + 0.4 = 0.6 \qquad (12\text{-}38)$$

c. The probability that the random sample is greater than zero amounts to summing all probability values for the independent variable greater than zero. Thus,

$$P(X > 0) = P(X = 5) + P(X = 10) = 0.4 + 0.1 = 0.5 \qquad (12\text{-}39)$$

d. The probability that the random sample is greater than or equal to zero is similar to (c) except that the probability for the variable to be zero is included. Thus,

$$P(X \geq 0) = P(X = 0) + P(X = 5) + P(X = 10) = 0.2 + 0.4 + 0.1 = 0.7 \qquad (12\text{-}40)$$

There is an alternate way to interpret this problem since there is only one point excluded. It is

$$P(X \geq 0) = 1 - P(X < 0) = 1 - P(X = -5) = 1 - 0.3 = 0.7 \qquad (12\text{-}41)$$

Sometimes, this "backwards" point of view is an easier way to deal with a problem.

e. The all-inclusive range of the variable from -5 V to 10 V includes all possible values of the function and is

$$P(-5 \leq X \leq 10) = P(-5) + P(0) + P(5) + P(10) = \sum_{k} f(x) = 1 \qquad (12\text{-}42)$$

f. This last one could be considered as a very simple problem but is inserted as a reminder that the function can be considered to assume only four values. Thus,

$$P(X = 15) = 0 \qquad (12\text{-}43)$$

Example 12-9

Construct the probability distribution function for the pdf of Example 12-8.

Solution

The process utilized is that of Equation 12-36. Refer to Figure 12-8 in the development that follows. For any value of x less than -5 V, the distribution function is $F(x) = 0$. At $x = -5$ V, $F(x)$ "jumps" to a level of 0.3. It remains at that level until we reach $x = 0$, in which case it changes by 0.2 to reach a level of 0.5. At $x = 5$ V, it changes by 0.4 to reach a level of 0.9. Finally, at $x = 10$ V, the final change is 0.1, and $F(x) = 1$ from that point on.

For discrete variables, the probability distribution function is probably less useful than the basic pdf, but it has been considered briefly for completeness. In contrast, for continuous variables, the probability distribution function is more useful and it forms the basis for evaluation of probabilities with certain functions.

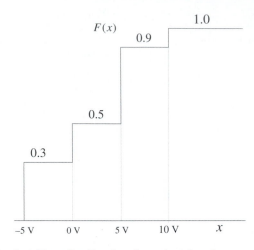

Figure 12-8   Probability distribution function for the random voltage

12-5   STATISTICAL AVERAGES OF DISCRETE VARIABLES

Along with the probability density and distribution functions, there are a number of important statistical parameters that can be determined and specified for a random process. Each parameter may have a particular significance when applied to a random function of a physical process. In this section, we will continue to assume the pdf $f(x)$ of a discrete variable.

Population versus Sample

In dealing with statistical processes, there is a difference between a *complete population* and a *sample* of a population insofar as parameter estimation is concerned. In this section and the next several, we will assume that we are dealing with a complete population.

Expected Value or Expectation

Let $g(x)$ represent some arbitrary function of the variable x. The *expected value* or *expectation* of $g(x)$ is denoted as $E[g(x)]$. It is defined as

$$E\left[g(x)\right] = \sum_{k} g(x) f(x) \tag{12-44}$$

Stated in words, the pdf is multiplied by $g(x)$ evaluated at the various discrete values of x and the product is summed over all possible values of the discrete variable. The reader is probably puzzled at this point, but as we look at specific values of $g(x)$, the concept will be more meaningful.

Mean Value

The *mean value* is determined by setting $g(x) = x$ and is denoted as either $E(x)$ or μ. It is defined as

$$\mu = E(x) = \sum_k xf(x) \tag{12-45}$$

The mean value is thus the expected value of the independent variable x. This is the quantity that is most often referred to as the *average value*. However, in the terminology of statistics, the term "average" is more general and should not be used for a specific parameter. The mean value is also referred to as the *first moment*. Note that for a discrete pdf, it may not correspond to any of the specific independent values.

Mean-Squared Value

The *mean-squared value* is determined by setting $g(x) = x^2$ and is denoted as $E(x^2)$. It is defined as

$$E(x^2) = \sum_k x^2 f(x) \tag{12-46}$$

The mean-squared value is also referred to as the *second moment*. As in the case of the mean value, it may not correspond to any one of the values in the population

Root-Mean-Square Value

The *root-mean-square value* x_{rms} is the positive square-root of the mean-square value, that is,

$$x_{rms} = \sqrt{E(x^2)} \tag{12-47}$$

Variance

The *variance* is determined by setting $g(x) = (x - \mu)^2$ and is denoted as σ^2. It is thus defined as

$$\sigma^2 = E\left[(x-\mu)^2\right] = \sum_k (x-\mu)^2 f(x) \tag{12-48}$$

The variance is also called the *second moment about the mean*.

If the mean value and mean-square value are known, the variance can be determined directly from the following simpler formula:

$$\sigma^2 = E(x^2) - \mu^2 \tag{12-49}$$

Stated in words, *the variance is the mean-square value minus the square of the mean.*

Standard Deviation

The standard deviation σ is simply the square root of the variance, that is,

$$\sigma = \sqrt{\sigma^2} \qquad (12\text{-}50)$$

The variance and the standard deviation are measures of the relative dispersion of the distribution. The larger the values of variance and standard deviation are, the greater will be the spread of the values within the process.

Multiple Variables

It was noted in the last section that when there are two or more random variables involved in an analysis, subscripts may be added to identify them. For example, the pdfs of two variables x and y may be identified as $f_x(x)$ and $f_y(y)$, respectively. The same principle may be applied to the various parameters defined in this section. Thus, the respective means could be identified as μ_x and μ_y, the respective variances could be identified as σ_x^2 and σ_y^2, and so on.

Example 12-10

For the discrete pdf of Example 12-8, determine the (a) mean value, (b) mean-square value, (c) rms value, (d) variance, and (e) standard deviation.

Solution

The various formulas developed in this section will be used freely to determine the statistical parameters.

a. The mean value is

$$\mu = \sum_k xf(x) = \sum_{k=1}^{4} x_k f(x_k) = -5(0.3) + 0(0.2) + 5(0.4) + 10(0.1) = 1.5 \text{ V} \qquad (12\text{-}51)$$

A few comments about the summation in Equation 12-51 should help clarify the procedure. First, note that the first value of the variable (-5) is negative and the product of that value and the corresponding value of the pdf (0.3) produces a negative result. Second, note that one of the possible values of the variable is 0 and the product of that value and the corresponding value of the pdf is also 0. Finally, note that the actual mean value (1.5 V) is different than any of the four voltage values.

As a point of practical interest, the mean value in electrical circuit theory is called the *dc value* and is the value that would be read by an ideal dc voltmeter. (This assumes that the time constant of the meter is sufficiently large to smooth out the instantaneous variations.)

b. The mean-square value is

$$E(x^2) = \sum_k x^2 f(x) = \sum_{k=1}^{4} x_k^2 f(x_k)$$
$$= (-5)^2(0.3) + (0)^2(0.2) + (5)^2(0.4) + (10)^2(0.1) = 27.5 \text{ V}^2 \qquad (12\text{-}52)$$

For the mean-square computation, the negative value is squared, and the result is a positive value. However, zero squared is still zero.

  c. The rms value is the square root of the mean-square value.

$$x_{rms} = \sqrt{E(x^2)} = \sqrt{27.5} = 5.244 \text{ V} \tag{12-53}$$

As another point of practical interest in electrical circuit theory, the rms value is the equivalent power-producing measure or *effective value* of the voltage when used with the appropriate power formula.

  d. Since the mean value and the mean-squared value are known, the easiest way to compute the variance is the following:

$$\sigma^2 = E(x^2) - \mu^2 = 27.5 - (1.5)^2 = 25.25 \text{ V}^2 \tag{12-54}$$

  e. The standard deviation is the square root of the variance.

$$\sigma = \sqrt{\sigma^2} = \sqrt{25.25} = 5.025 \text{ V} \tag{12-55}$$

12-6   BINOMIAL PROBABILITY DENSITY FUNCTION

One of the most common discrete probability density functions is the *binomial* pdf. It is employed in various applications involving two outcomes, and it also serves as a step toward the gaussian continuous pdf, which will be considered later in the chapter.

  Assume that an experiment contains only two possible outcomes, which will be denoted as A and B, respectively. (Flipping a coin would represent an example of this type.) The following definitions are assumed:

  p = probability that A occurs in a given trial
  q = probability that B occurs in a given trial

Since one or the other of the outcomes must occur in a given trial, we have the relationship

$$q = 1 - p \tag{12-56}$$

Finally, it is assumed that each trial is statistically independent from all the others.

Illustration

  Before developing a general formula, let us consider the following situation: What is the probability that in four trials of the experiment, A will occur exactly twice? One way that this could occur would be the following order: AABB. The probability that this occurs is p^2q^2. However, there are other ways in which we can obtain two A's and two B's. Starting with the one already provided, we have the following different orders that could occur:

$$AABB \quad ABAB \quad ABBA \quad BBAA \quad BABA \quad BAAB$$

In other words, there are six different ways in which we could have two A's and two B's. The probability in each case is the same since there are exactly two A's and two B's. Each of these different ways is mutually exclusive of the others, so the net probability of exactly two A's and two B's on four trials is $6p^2q^2$.

Combinations

The various orders discussed above constitute the concept of *combinations* as defined in algebra. In general, there is a formula for determining the number of combinations. The number of combinations of n trials with exactly x occurrences of one of the two outcomes, where x is a non-negative **integer** satisfying $x \leq n$, will be denoted as C_x^n and is determined as

$$C_x^n = \frac{n!}{x!(n-x)!} \tag{12-57}$$

It can be readily shown that the number of combinations of 4 objects taken 2 at a time is 6, which is the case previously discussed.

People often confuse *combinations* with *permutations*. With combinations, it does not matter which A comes before the other A and likewise with B. With permutations, it does matter. The formula for permutations is similar to Equation 12-57, but it does not have the separate factor of $x!$ in the denominator. We will only be utilizing combinations in this work.

Binomial PDF

The binomial pdf is denoted as $f(x)$ and is determined by the formula that follows.

$$f(x) = C_x^n p^x q^{n-x} = \frac{n!}{x!(n-x)!} p^x q^{n-x} \tag{12-58}$$

It can be considered as the probability that the binary event A occurs exactly x times in n trials. Thus, the probability that the random integer sample X is equal to x is

$$P(X = x) = f(x) = \frac{n!}{x!(n-x)!} p^x q^{n-x} \tag{12-59}$$

Remember that x must be an integer bounded by $0 \leq x \leq n$.

Example 12-11

An unbiased coin is flipped three times. What is the probability of getting exactly one head on the three trials?

Solution

In this case $p = q = 0.5$. From Equation 12-59, we have

$$f(1) = C_1^3 (0.5)^1 (0.5)^2 = \frac{3!}{(1!)(2!)} (0.5)^1 (0.5)^2 = 3(0.5)^3 = 0.375 \tag{12-60}$$

Example 12-12

For the coin-flipping experiment of Example 12-11, what is the probability of getting *at least* one head in the three trials?

Solution

The slight difference in wording makes a big difference in the outcome. Success in this case could mean one head, two heads, or three heads. Since all of these outcomes are mutually exclusive, we have

$$P(H) = f(1) + f(2) + f(3)$$

$$= \frac{3!}{(1!)(2!)}(0.5)^1(0.5)^2 + \frac{3!}{(2!)(1!)}(0.5)^2(0.5)^1 + \frac{3!}{(3!)(0!)}(0.5)^1(0.5)^2 \qquad (12\text{-}61)$$

$$= 3(0.5)^3 + 3(0.5)^3 + (0.5)^3 = 0.375 + 0.375 + 0.125 = 0.875$$

An alternate approach that is slightly easier in this case is to determine first the probability that there will be no heads. We have

$$P(\bar{H}) = P(X = 0) = f(0) = \frac{3!}{(0!)(3!)}(0.5)^0(0.5)^3 = 0.125 \qquad (12\text{-}62)$$

We can then say that the probability that we obtain at least one head is one minus the probability that we obtain no heads. Hence,

$$P(H) = 1 - P(\bar{H}) = 1 - 0.125 = 0.875 \qquad (12\text{-}63)$$

This is another example of taking the "back-door" approach.

12-7 CONTINUOUS STATISTICAL FUNCTIONS

The second type of statistical functions to be considered will consist of those involving *continuous variables*. A *continuous variable* is one that assumes an infinite number of possible values within a specified range. The range itself need not be infinite, but within the range specified, it can assume any value. In the context of electrical signals, a *continuous signal* usually is the same thing as an *analog* signal. The probability density function of a continuous signal is a continuous function of the independent variable. A typical example of the pdf of a continuous variable is shown in Figure 12-9. The pdf is indicated as $f(x)$, and x is the independent continuous variable. As in the case of the discrete pdf, the continuous pdf is a relative measure of how likely a value will occur. In the example of Figure 12-9, voltage values close to 6 V are much more likely to occur than values near 0 V and 8 V.

To simplify the presentation, we are using $f(x)$ for a continuous variable. This is the same notation as was employed for a discrete variable, but the manner in which the continuous case is handled is quite different. Therefore, be sure to note in a given case which form applies.

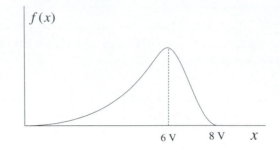

Figure 12-9   Typical example of the probability density function
of a continuous variable

A fundamental difference between a discrete pdf and a continuous pdf is that the probability of a given *exact* value occurring is 0 for a continuous pdf. Thus, the probability that a random value is some *exact* value x_0 is given by

$$P(X = x_0) = 0 \qquad\qquad (12\text{-}64)$$

Instead, a range must always be provided for a continuous variable. Thus, a statement such as $P(x_1 \leq X \leq x_2)$, which is interpreted to mean "the probability that a sample lies within the range from x_1 to x_2," will produce a non-zero value provided that this range covers some portion of the pdf.

Probability Formulation

For a continuous pdf, the probability that a sample lies between x_1 and x_2 is the *area under the pdf curve* as illustrated in Figure 12-10. This result can be stated as

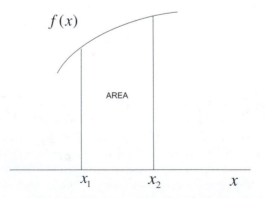

Figure 12-10   Probability of a random sample of a continuous variable being
between two points is the area under the pdf curve between the points

$$P(x_1 \leq X \leq x_2) = \int_{x_1}^{x_2} f(x)dx \qquad (12\text{-}65)$$

For a continuous function, it really does not matter if the "less than or equal" symbols are replaced by "less than" symbols since the results will be the same. Recall that for a discrete pdf, however, there can be a significant difference.

In general, the pdf integrated over all possible values of x must satisfy

$$\int_{-\infty}^{\infty} f(x)dx = 1 \qquad (12\text{-}66)$$

Stated in words, the net area under the curve of the pdf must be unity.

Probability Distribution Function of a Continuous Variable

The probability distribution of a continuous pdf will be denoted by $F(x)$. It is defined as

$$F(x) = \int_{-\infty}^{x} f(u)du \qquad (12\text{-}67)$$

Note that x is momentarily replaced by the "dummy variable" u for integration purposes. As in the case of the discrete pdf, the lower limit $-\infty$ is symbolic in that it indicates that integration begins at the far left side of the pdf and continues as a running integral up to an arbitrary value x.

If the probability distribution function is known, a probability computation of the form of Equation 12-65 can be formulated directly from the properties of definite integrals. We have

$$P(x_1 \leq X \leq x_2) = F(x_2) - F(x_1) \qquad (12\text{-}68)$$

This form will prove to be the most practical way to deal with the gaussian function in a later section.

Example 12-13

The velocity $v(t)$ associated with a random wind is *uniformly distributed* between 0 m/s and 10 m/s. This means that all velocity values between the limits are equally likely to occur. Plot the pdf and label all pertinent values associated with it.

Solution

The form of the pdf is shown in Figure 12-11. The horizontal variable is the velocity v, and since all values between 0 m/s and 10 m/s are equally likely, the pdf is simply a constant value over that range. To show the process fully, the level has been labeled as K, which is initially unknown.

The value of K is determined from the requirement that the net area under the pdf curve must be unity. For this simple function, the area under the curve is readily determined as the base times the height, which yields $10K$, and we require that $10K = 1$. This leads to

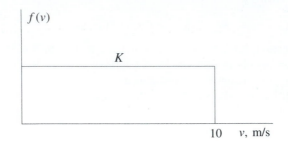

Figure 12-11   Uniform probability density function of Examples 12-13 and 12-14

$$K = \frac{1}{10} = 0.1 \qquad (12\text{-}69)$$

We can then express the pdf as

$$f(v) = 0.1 \text{ for } 0 \leq v \leq 10 \text{ m/s}$$
$$= 0 \text{ elsewhere} \qquad (12\text{-}70)$$

Example 12-14

For the pdf of Example 12-13, assume that a random sample V of the wind velocity is to be measured. Determine the following probabilities: (a) $P(V = 5 \text{ m/s})$, (b) $P(4.9 \text{ m/s} \leq V \leq 5.1 \text{ m/s})$, (c) $P(V \leq 2 \text{ m/s})$, (d) $P(V \geq 7 \text{ m/s})$, and (e) $P(0 \leq V \leq 10 \text{ m/s})$.

Solution

Refer to Figure 12-12 for support in some of the calculations that follow.

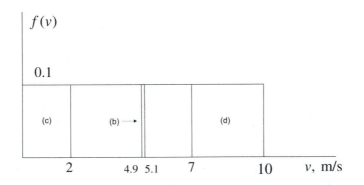

Figure 12-12   Some areas pertaining to Example 12-14

(a) $P(V = 5 \text{ m/s})$

As explained earlier, for any continuous pdf, the probability of a *single exact value* occurring is zero. Thus,

$$P(V = 5 \text{ m/s}) = 0 \tag{12-71}$$

(b) $P(4.9 \text{ m/s} \leq V \leq 5.1 \text{ m/s})$

This probability is the area under the curve from 4.9 to 5.1, which is easily evaluated by arithmetic.

$$P(4.9 \leq V \leq 5.1) = 0.1 \times (5.1 - 4.9) = 0.02 \tag{12-72}$$

This area is shown as the narrow strip (b) on Figure 12-12.

(c) $P(V \leq 2 \text{ m/s})$

This probability is the area under the curve from 0 to 2. It is labeled (c).

$$P(V \leq 2) = 0.1 \times 2 = 0.2 \tag{12-73}$$

(d) $P(V \geq 7 \text{ m/s})$

This probability is the area under the curve from 7 to 10. It is labeled (d).

$$P(V \geq 7) = 0.1 \times (10 - 7) = 0.3 \tag{12-74}$$

(e) $P(0 \leq V \leq 10 \text{ m/s})$

The range involved includes all possible values of force. Hence,

$$P(0 \leq V \leq 10) = 1 \tag{12-75}$$

Example 12-15

Determine the probability distribution function for the uniform pdf of Example 12-14.

Solution

The basic process is that of Equation 12-67. The beginning point indicated as "$-\infty$" is really 0 in this case. Hence the integral becomes

$$F(v) = \int_{-\infty}^{v} f(u)\,du = \int_{0}^{v}(0.1)\,du = 0.1u\Big]_{0}^{v} = 0.1v \quad \text{for} \quad 0 \leq v \leq 10 \tag{12-76}$$

This function is shown in Figure 12-13. The reader may find it instructive to use this function to repeat some of the computations of Example 12-14 utilizing Equation 12-68.

12-8   STATISTICAL AVERAGES OF CONTINUOUS VARIABLES

The same statistical averages defined for discrete probability density functions can be calculated for continuous functions. The main difference is that summations are replaced by integration. The various averages are defined in the steps that follow. As in the case of discrete variables, we will assume a complete *population* at this point as opposed to a sample of the population.

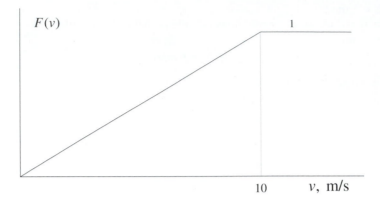

Figure 12-13   Probability distribution function of Example 12-15

Expected Value or Expectation

Let $g(x)$ represent some arbitrary function of the continuous variable x. The expected value or expectation of $g(x)$ is denoted as $E[g(x)]$. It is defined as

$$E\big[g(x)\big] = \int_{-\infty}^{\infty} g(x)f(x)dx \qquad (12\text{-}77)$$

The infinite limits on the integral tell us that it must be performed over the entire range over which the pdf is defined. In the actual integration, the values used correspond to those at the ends of the range.

Mean Value

The *mean value* is determined by letting $g(x) = x$ and is denoted as either $E(x)$ or μ. It is defined as

$$\mu = E(x) = \int_{-\infty}^{\infty} xf(x)dx \qquad (12\text{-}78)$$

Mean-Squared Value

The *mean-squared value* is determined by letting $g(x) = x^2$ and denoted as $E(x^2)$. It is defined as

$$E(x^2) = \int_{-\infty}^{\infty} x^2 f(x)dx \qquad (12\text{-}79)$$

Root-Mean-Square Value

The *root-mean-square value* x_{rms} is the positive square root of the mean-square value, that is,

$$x_{rms} = \sqrt{E(x^2)} \tag{12-80}$$

Variance

The *variance* is determined by letting $g(x) = (x - \mu)^2$ and is denoted as σ^2. It is thus defined as

$$\sigma^2 = E\left[(x-\mu)^2\right] = \int_{-\infty}^{\infty} (x-\mu)^2 f(x)dx \tag{12-81}$$

As in the case of the discrete pdf, the variance can be expressed in terms of the mean-square and mean values as

$$\sigma^2 = E(x^2) - \mu^2 \tag{12-82}$$

Standard Deviation

The standard deviation σ is simply the square root of the variance, that is,

$$\sigma = \sqrt{\sigma^2} \tag{12-83}$$

Example 12-16

Consider the random velocity $v(t)$ of Examples 12-13 and 12-14. Determine the following quantities for the force: (a) mean value, (b) mean-square value, (c) total rms value, (d) variance, and (e) standard deviation.

Solution

For convenience, the pdf is repeated here.
$$\begin{aligned} f(v) &= 0.1 \quad \text{for } 0 \le v \le 10 \text{ m/s} \\ &= 0 \quad \text{elsewhere} \end{aligned} \tag{12-84}$$

a. The mean value is

$$\mu = \int_{-\infty}^{\infty} vf(v)dv = \int_{0}^{10} 0.1vdv = \left. \frac{0.1v^2}{2} \right|_{0}^{10} = 5 \text{ m/s} \tag{12-85}$$

b. The mean-square value is

$$E(v^2) = \int_{-\infty}^{\infty} v^2 f(v)dv = \int_{0}^{10} 0.1v^2 dv = \left. \frac{0.1v^3}{3} \right|_{0}^{10} = 33.33 \text{ m}^2/\text{s}^2 \tag{12-86}$$

c. The total rms value is the square root of the mean-square value, that is,

$$v_{rms} = \sqrt{33.33} = 5.774 \text{ m/s} \tag{12-87}$$

d. The variance is most easily determined as follows:

$$\sigma^2 = E(v^2) - (\mu)^2 = 33.33 - (5)^2 = 8.333 \text{ m}^2/\text{s}^2 \qquad (12\text{-}88)$$

e. The standard deviation is

$$\sigma = \sqrt{8.333} = 2.887 \text{ m/s} \qquad (12\text{-}89)$$

12-9   GAUSSIAN DISTRIBUTION

The *gaussian* probability density function is probably the most important function used in statistical analysis. This function, which is also referred to as a *normal* distribution, seems to arise as a natural phenomenon throughout the universe. A very basic result from statistical theory is the *central limit theorem*, which states that the distribution of the sum of a number of statistically independent variables tends to approach a gaussian distribution in the limit irrespective of the nature of the individual distributions.

Gaussian PDF

The general shape of a gaussian pdf is shown in Figure 12-14. The gaussian curve is sometimes referred to a "bell-shaped" curve, for obvious reasons. The mathematical form of the gaussian pdf is

$$f(x) = \frac{1}{\sqrt{2\pi\sigma^2}} e^{-(x-\mu)^2/2\sigma^2} \qquad \text{for } -\infty < x < \infty \qquad (12\text{-}90)$$

where μ and σ^2 are immediately identified as the mean and variance, respectively.

As noted in Equation 12-90, the theoretical limits cover an infinite range on either side of the mean. However, the actual area under the curve at a reasonable distance from the mean approaches zero rapidly. In many actual applications of the gaussian function, infinite values never occur, so the theoretical model must be tempered with real-world limitations.

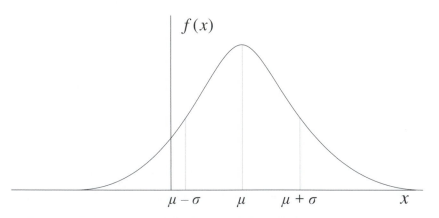

Figure 12-14   Form of the gaussian probability density function.

Probability Formulation

The probability that a random sample X of a gaussian pdf lies between x_1 and x_2 is given by the integral

$$P(x_1 \leq X \leq x_2) = \int_{x_1}^{x_2} \frac{1}{\sqrt{2\pi\sigma^2}} e^{-(x-\mu)^2/2\sigma^2} dx \tag{12-91}$$

Unfortunately, this integral cannot be expressed easily in closed form. If one were approaching this problem for the first time, it would be necessary to employ a numerical integration approach. Fortunately, this has already been done, and it is only necessary to employ widely tabulated results.

There are several different approaches that have been used in assembling and representing the results of the numerical integration. Most utilize a normalized gaussian function and we will begin with that approach.

Normalized Gaussian pdf

The normalized gaussian pdf is a gaussian function having a mean value of zero and a variance of one. Letting z represent the normalized variable, the normalized function is

$$f(z) = \frac{1}{\sqrt{2\pi}} e^{-z^2/2} \quad \text{for} -\infty < z < \infty \tag{12-92}$$

where it is readily noted that $\mu = 0$ and $\sigma^2 = 1$ for the normalized function. The form of the normalized gaussian pdf is shown in Figure 12-15.

It can be shown that z is related to x of Equation 12-91 by

$$z = \frac{x-\mu}{\sigma} \tag{12-93}$$

or conversely

$$x = \sigma z + \mu \tag{12-94}$$

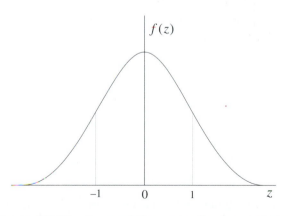

Figure 12-15   Form of the normalized gaussian pdf

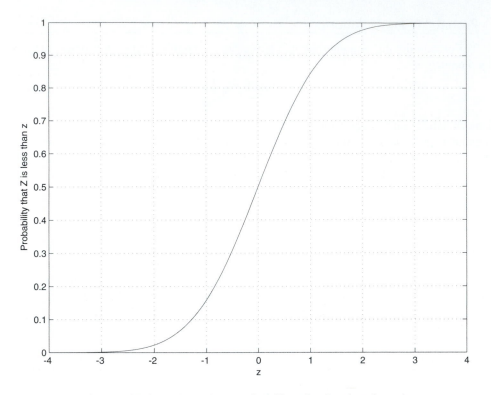

Figure 12-16   Gaussian probability distribution function

By tabulating the area under the normalized curve, probabilities involving any values of x can be evaluated by translating the problem into corresponding computations involving z. This process will be illustrated with some of the examples that follow this section.

Probability Distribution Function

The probability distribution function $F(z)$ can be expressed as

$$F(z) = \int_{-\infty}^{z} f(u)\,du \tag{12-95}$$

This function is shown in Figure 12-16 and can be obtained by numerical integration of the pdf. Some tabulated values of $F(z)$ over the most important range encountered in many practical applications are provided in Table 12-2, at the end of the chapter. The probability that a random sample Z lies between z_1 and z_2 may be determined as

$$P(z_1 \leq Z \leq z_2) = F(z_2) - F(z_1) \tag{12-96}$$

Although the theoretical infinite limits cannot be seen, several limiting values should be noted.

$$F(-\infty) = 0 \tag{12-97}$$

$$F(0) = 0.5 \tag{12-98}$$

$$F(\infty) = 1 \tag{12-99}$$

Evaluating Probabilities

For rough estimates of probabilities associated with the gaussian function, the curve of Figure 12-16 and Equation 12-96 can be employed. More accurate values for certain cases can be evaluated with the data of Table 12-2, at the end of the chapter. Finally, MATLAB may be employed for any possible values of the limits, and this process will be illustrated in Section 12-12. All examples at the end of this section have been "rigged" so that the data of Table 12-2 may be employed.

Example 12-17

As a "warmup" exercise, this first example will utilize the normalized function $f(z)$ throughout. Therefore, assume a gaussian distributed variable z with a mean value of 0 and a standard deviation of 1. Determine (a) $P(Z \geq 0)$, (b) $P(1 \leq Z \leq 3)$, (c) $P(-1 \leq Z \leq 2)$, (d) $P(-2 \leq Z \leq 2)$, (e) $P(|Z| \leq 2)$, (f) $P(Z \geq 2)$, and (g) $P(|Z| \geq 2)$.

Solution

The results of this section and Table 12-2 will be used extensively, with comments added as appropriate.

a. $P(Z \geq 0)$

This somewhat trivial case is considered to ensure that the reader is clear on the different forms of notation that could be employed here. One could immediately deduce that the probability is 0.5, but it is instructive to go through some formal steps involved.

$$P(Z \geq 0) = P(0 \leq Z \leq \infty) = P(\infty) - P(0) = 1 - 0.5 = 0.5 \tag{12-100}$$

b. $P(1 \leq Z \leq 3)$

$$P(1 \leq Z \leq 3) = F(3) - F(1) = 0.998650 - 0.841345 = 0.1573 \tag{12-101}$$

c. $P(-1 \leq Z \leq 2)$

$$P(-1 \leq Z \leq 2) = F(2) - F(-1) = 0.977250 - 0.158655 = 0.8186 \tag{12-102}$$

d. $P(-2 \leq Z \leq 2)$

$$P(-2 \leq Z \leq 2) = P(2) - P(-2) = 0.977250 - 0.022750 = 0.9545 \tag{12-103}$$

There is a special point that should be made here. Due to the even nature of the normalized function, any probability evaluation that is symmetrical with respect to the origin can be made by evaluating only the positive side and doubling it. This process is slightly simpler and usually eliminates one lookup. Thus, an alternate approach is

$$(12\text{-}104)$$

We will employ this approach in subsequent cases where it applies.

e. $P(|Z| \leq 2)$

A statement implying that the magnitude is less than or equal to a certain value is equivalent to

$$P(|Z| \leq 2) = P(-2 \leq Z \leq 2) = 0.9545 \qquad (12\text{-}105)$$

Although stated differently, this case is basically the same as the preceding one so the result is the same.

f. $P(Z \geq 2)$

$$P(Z \geq 2) = P(2 \leq Z \leq \infty) = F(\infty) - F(2) = 1 - 0.97725 = 0.02275 \qquad (12\text{-}106)$$

g. $P(|Z| \geq 2)$

This is equivalent to

$$P(|Z| \geq 2) = P\big[(Z \geq 2) + (Z \leq -2)\big] = 2P(Z \geq 2) = 2(0.02275) = 0.0455 \quad (12\text{-}107)$$

This could also be expressed as

$$P(|Z| \geq 2) = 1 - P(|Z| \leq 2) = 1 - 0.9545 = 0.0455 \qquad (12\text{-}108)$$

Example 12-18

A certain random voltage $x(t)$ has a gaussian distribution with a dc value of zero and an rms value of 5 V. A single random sample X of the voltage is taken. Determine the following probabilities:

(a) $P(X \geq 10 \text{ V})$, (b) $P(|X| \geq 10 \text{ V})$ and (c) $P(|X| \leq 15 \text{ V})$.

Solution

Since the mean (DC) value is zero, the standard deviation is equal to the rms value. Thus, $\mu = 0$ and $\sigma = 5$ V. The actual sample space must be converted to the normalized space by the transformation

$$z = \frac{x - \mu}{\sigma} = \frac{x - 0}{\sigma} = \frac{x}{5} \qquad (12\text{-}109)$$

Pertinent values of x can then be transformed to corresponding values of z. The key x values are 10 and 15, and the corresponding z values are 2 and 3, respectively. The calculations follow:

a. $P(X \geq 10 \text{ V})$

This can be expressed as

$$P(X \geq 10 \text{ V}) = P(Z \geq 2) = F(\infty) - F(2) = 1 - 0.977250 = 0.02275 \quad (12\text{-}110)$$

b. $P(|X| \geq 10 \text{ V})$

This value is twice the preceding value.

$$P(|X| \geq 10) = 2P(|Z| \geq 2) = 2 \times 0.02275 = 0.0455 \tag{12-111}$$

c. $P(|X| \leq 15 \text{ V})$

This can be expressed as

$$P(|X| \leq 15 \text{ V}) = P(|Z| \leq 3) = 2P(0 \leq Z \leq 3) = 2 \times [0.998650 - 0.5] = 0.9973 \tag{12-112}$$

Example 12-19

A certain random force, $x(t)$, has a gaussian distribution with a mean value of 10 N and a standard deviation of 5 N. Determine the probability that a single sample X exceeds 20 N.

Solution

We are given that $\mu = 10$ N and $\sigma = 5$ N. The transformation is

$$z = \frac{x - \mu}{\sigma} = \frac{x - 10}{5} \tag{12-113}$$

The value $x = 20$ is transformed to

$$z = \frac{20 - 10}{5} = 2 \tag{12-114}$$

$$P(X \geq 20) = P(Z \geq 2) = F(\infty) - F(2) = 1 - 0.97725 = 0.02275 \tag{12-115}$$

12-10　SAMPLING STATISTICS

The emphasis thus far in the chapter has been directed toward processes in which the statistics could be determined directly from a probability density function or a probability distribution function. The assumption is that the properties of the complete *population* are known. In contrast, there are numerous situations in which only a *sample* of the complete population can be obtained. This is the most common situation encountered in many of the most important applications of applied statistics. For example, in industrial quality control, it is often not practical to perform a test on each individual product. Rather, a representative sample of products is taken, and the measurements performed on the sample are used to infer properties of all of the products. Likewise, in such everyday applications as opinion surveys, only a sample of the total population can be observed.

Sampling Statistical Parameters

Since the statistical parameters obtained from a sample may or may not equal the true statistical parameters being sought, it is customary to use different symbols for the parameters obtained from a finite sample. The symbols that will be used follow, and the "true" parameters follow in parentheses.

$\overline{x}=$ sample mean (true mean = μ)

$\overline{x^2}=$ sample mean-square value [true mean-square value = $E(x^2)$]

$s^2=$ sample variance (true variance = σ^2)

$s=$ sample standard deviation (true standard deviation = σ)

Note in each case that the particular term is preceded by the adjective *sample*. In this manner, we can distinguish between the true parameter and the estimate based on sampling. Ideally, the goal is to determine the true parameters of the total population from the sample, but there is always some uncertainty based on using a sample, and this is one of the major subjects in applied statistics.

Assume that n independent discrete samples of a process are taken and are designated as x_k, with k assuming the integer values from 1 to n. The various estimates are defined in the paragraphs that follow.

Sample Mean

The sample *mean* is designated as $\overline{x}$ and defined as follows:

$$\overline{x}=\frac{1}{n}\sum_{k=1}^{n}x_k \qquad (12\text{-}116)$$

This formula is likely familiar to the reader from past averaging processes. To apply it, the various samples are added together and divided by the number of samples. This is the parameter that is most often referred to as the *average*, but remember that in statistical analysis, the term *average* is more general. While this operation may or may not produce the true mean of the process, it is the best estimate in a statistical sense.

Sample Mean-Square Value

The sample *mean-square* value is determined from the formula

$$\overline{x^2}=\frac{1}{n}\sum_{k=1}^{n}x_k^2 \qquad (12\text{-}117)$$

Sample Variance

The *sample variance* is defined as

$$s^2=\frac{1}{n-1}\sum_{k=1}^{n}(x_k-\overline{x})^2 \qquad (12\text{-}118)$$

Now the perceptive reader probably has an important question. The question has to do with why we divide by $n-1$ in Equation 12-118 instead of by n. To be honest, this question still bothers the author, who is an engineer rather than a statistician or a mathematician.

However, using mathematical analysis beyond the scope of this text, it turns out that division by $n-1$ provides a slightly better estimate for the variance when the data involved are a sample of the population rather than the complete population. However, it should be noted that some people use n rather than $n-1$ because it seems to be more compatible with respect to the definition of sample mean. Provided that the sample size is reasonably large, it really does not make much difference with respect to the outcome, and since it is an estimate anyway, there is no need to make an overwhelming case for either form. The reader should be aware of the difference, however, when using calculators having statistical functions and when using software such as MATLAB. It is the author's experience that most use the form of Equation 12-118, which is said to be an *unbiased estimate* of the variance.

Alternate Formula for Sample Variance

It can be shown that the formula that follows is equivalent to Equation 12-118.

$$s^2 = \frac{1}{n-1}\left[\sum_{k=1}^{n}x_k^2 - \frac{\left(\sum_{k=1}^{n}x_k\right)^2}{n}\right]$$

(12-119)

While this formula may appear more foreboding than the first form, it is often easier to apply. Note that the first term within brackets is very similar to Equation 12-117 and involves summing the square values. The second term in brackets involves first summing all the sample values, squaring that result, and then dividing by the number of terms.

Sample Standard Deviation

The *sample standard deviation* is defined as

$$s = \sqrt{s^2}$$

(12-120)

Other Statistical Parameters

There are a few other parameters that are useful in characterizing a set of data values. The most common of these will be explored in the paragraphs that follow.

Median

The *median* of a set of data is a value such that the number of data points above the value is equal to the number below the value. For an odd number of data points, it is the middle value. For an even number of data points, it is a hypothetical point located halfway between the two adjacent middle values. For example, consider the five data points: 3, 3, 5, 8, 9. The median in this case is 5 since it is exactly in the middle, with two points above and two points below. Now consider the six data points: 3, 3, 5, 8, 9, 10. In this case, the median is defined as 6.5 since it is halfway between the third point, 5, and the fourth point, 8.

The median value is useful as an alternative to the mean in situations where the latter would result in a misrepresentation of the data. For example, suppose that in a small company of 10 employees, 9 are paid at close to the same salary, but 1 is paid at a much higher pay than the others. The mean value would be much greater than the representative pay of most employees, so the median would be a better indicator.

Mode

The mode is the most common value in a collection of data. It may or may not be a significant parameter depending on the type of data and the application. For example, if all data points are different except for two points, which are the same, there may be little significance in calling that one value the mode. However, out of 50 data points, if 15 are the same and no other data values occur that often, it would be quite appropriate to define the value that occurs 15 times as the mode.

Theoretically, for large quantities of data approaching a normal distribution, the mean value, the median, and the mode all tend to converge to the same value.

Range

Let x_{max} represent the maximum value of any data point and let x_{min} represent the minimum value. The range is defined as

$$\text{range} = x_{max} - x_{min} \tag{12-121}$$

Example 12-20

In this example and others involving data, we will consider a modest size data set in order to keep the analysis to a manageable level for purposes of explanation. Based on a test given to a large class in which the maximum score is 100 points, assume that 25 grades taken at random from the entire set produced the results that follow.

50 52 56 60 63 68 70 73 76 76 76 77 79 81 82 82 83 85 88 89 89 90 92 94 96

Determine the (a) sample mean, (b) sample mean-square value, (c) sample rms value, (d) sample variance, (e) sample standard deviation, (f) median, (g) mode, and (h) range.

Solution

Since the set of numbers is a sample of a larger population, we will employ the sample statistical parameters. To minimize long expressions, we will omit the details of some of the actual sums that follow, but the reader may wish to verify the values provided.

a. The sample mean is

$$\bar{x} = \frac{1}{n} \sum_{k=1}^{n} x_k = \frac{1}{25} \sum_{k=1}^{25} x_k = \frac{1927}{25} = 77.08 \tag{12-122}$$

where the sum of all the grades is 1927. It should be noted that as far as accuracy is concerned, the sample mean could be rounded to 77 since the individual values were known only to two significant digits. Nevertheless, it is desirable, at least initially, to carry out the values obtained to more places in order to minimize round-off when combined with other parameters. Thus, we will list the values as they appear with a calculator, but be aware that the values listed may appear to be more accurate than they really are.

b. The sample mean-square value is

$$\overline{x^2} = \frac{1}{n}\sum_{k=1}^{n} x_k^2 = \frac{1}{25}\sum_{k=1}^{25} x_k^2 = \frac{152,585}{25} = 6103.40 \tag{12-123}$$

c. The sample variance is

$$s^2 = \frac{1}{n-1}\sum_{k=1}^{n}(x_k - \overline{x})^2 = \frac{1}{24}\sum_{n=1}^{24}(x_k - 77.08)^2 = \frac{4051.84}{24} = 168.83 \tag{12-124}$$

To minimize round-off, the initial value of the mean in the preceding parentheses was carried out to four significant digits.

An alternate way to compute the sample variance is

$$s^2 = \frac{1}{n-1}\left[\sum_{k=1}^{n} x_k^2 - \frac{\left(\sum_{k=1}^{n} x_k\right)^2}{n}\right] = \frac{1}{24}\left[\sum_{k=1}^{25} x_k^2 - \frac{\left(\sum_{k=1}^{25} x_k\right)^2}{25}\right]$$

$$= \frac{1}{24}\left[152,585 - \frac{(1927)^2}{25}\right] = \frac{1}{24}[4051.84] = 168.83 \tag{12-125}$$

d. The sample standard deviation is

$$s = \sqrt{s^2} = \sqrt{168.83} = 12.99 \tag{12-126}$$

e. Since the number of data points is odd (25), the median value is the 13th value from either end. Hence,

$$\text{median} = 79 \tag{12-127}$$

This result means that half of the class had scores above 79 and about half had scores below 79. This value can be compared with the mean, which is about 77. Since the mean value is less than the median, one can conclude that the lower grades tend to weight more heavily on the data than the upper grades.

f. In searching for the mode we note that there are three values that appear more than once. There are two values of 82, two values of 89, and three values of 76. Therefore, it is technically correct to say that the mode is 76. However, this could be considered as a relatively "weak" mode since it does not stand out among the other values. A mode is more meaningful when it appears with a relatively high frequency compared with other values.

g. The range is simply the maximum value minus the minimum value. The former is 96 and the latter is 50. Thus,

$$range = 96 - 50 = 46 \qquad (12\text{-}128)$$

12-11   MATLAB STATISTICAL OPERATIONS

We will now turn our attention to some of the statistical operations that can be performed with MATLAB. It should be stated at the outset that The MathWorks offers a complete **Statistics Toolbox** as a MATLAB add-on that is capable of many advanced statistical analysis functions. However, as has been the practice throughout the text, we will limit our consideration here to some of the functions that are contained in the **Student Version**.

Assume that a set of n data points x_k, with $k = 1$ to n, are available and that some basic statistical parameters are desired. First, the data are entered as a vector using the default MATLAB terminology.

$$>> x = [x1 \ x2 \ x3 \ . \ . \ . \ . \ xn] \qquad (12\text{-}129)$$

Various commands for determining some statistical parameters, along with some other data processing operations, are summarized in the table that follows:

Function	Format
Sample Mean	mean(x)
Sample Median	median(x)
Sample Standard Deviation	std(x)
Maximum Value	max(x)
Minimum Value	min(x)
Sum of All Values	sum(x)
Product of All Values	product(x)
Sort in Ascending Order	sort(x)

The application of some of these operations is illustrated with the example that follows.

MATLAB Example 12-1

Use MATLAB to determine some of the statistical parameters of the data of Example 12-20.

Solution

The most tedious part of the analysis is entering the 25 data points. While we could enter it as one vector, it is easier to keep track by first separating the data into smaller vectors.

We will arbitrarily divide the process into five vectors, with each containing five values. The operations follow.

$$>> xa = [50\ 52\ 56\ 60\ 63]; \tag{12-130}$$

$$>> xb = [68\ 70\ 73\ 76\ 76];$$

$$>> xc = [76\ 77\ 79\ 81\ 82];$$

$$>> xd = [82\ 83\ 85\ 88\ 89];$$

$$>> xe = [89\ 90\ 92\ 94\ 96];$$

The complete vector x containing 25 points is now formed by the command

$$>> x = [xa\ xb\ xc\ xd\ xe]; \tag{12-131}$$

Some of the parameters will be determined in a different order than in Example 12-20 in order to emphasize the more common and easier to apply MATLAB operations. In each case, the corresponding symbol for the result will be denoted by x followed by some letters suggesting the operation.

The mean value xmean is

$$>> xmean = mean(x) \tag{12-132}$$

xmean =

77.0800

The sample median xmedian is

$$>> xmedian = median(x) \tag{12-133}$$

xmedian =

79

The sample standard deviation xstd is

$$>> xstd = std(x) \tag{12-134}$$

xstd =

12.9933

The sample variance could be easily determined by forming the square of the sample standard deviation, but we will omit that step.

In order to determine the mean-square value, the **mean** operation is combined with the array product **x.^2** as the argument. Denoting this parameter as xmean_square, we have

$$>> \text{xmean\_square} = \text{mean(x.\textasciicircum 2)} \qquad (12\text{-}135)$$

xmean_square =

6.1034e+003

The other operations in the table are either unnecessary for the purpose of the example or trivial. The latter description would be true for the **max** and **min** operations, since they can be readily observed from the data.

The values obtained with MATLAB can be compared with those obtained in Example 12-20, and the results are the same in all cases in which computations were made.

12-12 GAUSSIAN PROBABILITIES WITH MATLAB

Probability evaluations for gaussian functions may be performed to a high degree of accuracy with MATLAB, but it takes a slight amount of manipulation with the basic mathematical package. The reason is that the gaussian function is not part of the basic mathematical package, but some usable alternate functions are available. These functions will be discussed in this section and their application will be demonstrated.

Error Function

A mathematical function known as the *error function* **erf(x)** is defined as

$$\text{erf}(x) = \sqrt{\frac{2}{\pi}} \int_0^x e^{-u^2} \, du \qquad (12\text{-}136)$$

This function is tabulated in various mathematical tables and is available in the MATLAB basic mathematical package. While it is not exactly the same as the gaussian probability distribution function, it may be modified to accommodate an evaluation for that function.

Probability Evaluation with Error Function

The details will not be shown here, but the mathematically oriented reader may wish to study how an appropriate change of variables in the gaussian definition results in the following formula for the *gaussian probability distribution function*:

$$F(z) = P(Z \le z) = \frac{1}{2} \left[1 + \text{erf} \left(\frac{z}{\sqrt{2}} \right) \right] \qquad (12\text{-}137)$$

The MATLAB command for the error function for any argument x is simply **erf(x)** for either positive or negative values of **x**. Therefore, the result of Equation 12-137 for MATLAB works for either positive or negative values of the argument.

The probability that a sample lies between two values, z_1 and z_2, with $z_2 > z_1$, is given by

$$P(z_1 \le Z \le z_2) = \frac{1}{2}\left[\text{erf}\left(\frac{z_2}{\sqrt{2}}\right) - \text{erf}\left(\frac{z_1}{\sqrt{2}}\right) \right] \qquad (12\text{-}138)$$

Complementary Error Function

Many applications of the gaussian pdf involve determining the probability that a random sample lies **above** a certain value. The *complementary error function* erfc (x) is more convenient to use in this case. It is defined as

$$\text{erfc}(x) = 1 - \text{erf}(x) \qquad (12\text{-}139)$$

The MATLAB command for the complementary error function is **erfc(x)**. The following relationship can be developed using the complementary error function:

$$P(Z \ge z) = \frac{1}{2}\text{erfc}\left(\frac{z}{\sqrt{2}}\right) \qquad (12\text{-}140)$$

The relationships of Equations 12-137, 12-138, and 12-140 may be applied with MATLAB to determine various probabilities for gaussian variables with MATLAB. If any computation involves infinity in an argument, a value of 10 or greater for the argument should suffice.

Inverse Error Functions

Suppose we are given a probability value and we wish to determine a statistical parameter that satisfies the given value. This involves the inverse process depicted by the probability distribution function. Both the error function and the complementary error function have inverse functions available in MATLAB. The inverse of the error function is denoted as **erfinv(x)**, and the inverse of the complementary error function is denoted as **erfcinv(x)**.

Suppose you desire to find the value of z such that $P(Z \le z) = p$, where p is a specified probability. It can be shown that in MATLAB terms, the value of z is given by

$$>>z = \text{sqrt}(2)*\text{erfinv}(2*p - 1) \qquad (12\text{-}141)$$

Next, suppose you desire to find the value of z such that $P(Z \ge z) = p$. Again, using MATLAB notation, the value of z is given by

$$>>z = \text{sqrt}(2)*\text{erfcinv}(2*p) \qquad (12\text{-}142)$$

MATLAB Example 12-2

Rework the first four parts of Example 12-17 using MATLAB.

Solution

We will not repeat a statement of the complete problem here, but all the computations involve determining probabilities associated with a normalized gaussian variable z. In this case, however, the MATLAB results using the error function will be utilized.

Since the value of $\sqrt{2}$ appears so often with the use of the error function, it will be defined as a separate constant k as follows:

$$>> k = sqrt(2);\hspace{3cm}(12\text{-}143)$$

a. $P(Z \geq 0)$

We know of course that this is a trivial case with the result being 0.5, but how would we determine the value with Equation 12-138? The answer is that any value of z_1 above about 10 or so would appear as "infinity" insofar as any practical results are concerned. Therefore, let us try 10. Denoting the result as pa, we have

$$>> pa=.5*(erf(10/k) - erf(0))\hspace{2cm}(12\text{-}144)$$

$$pa =$$

$$0.5000$$

b. $P(1 \leq Z \leq 3)$

$$>> pb = 0.5*(erf(3/k) - erf(1/k))\hspace{2cm}(12\text{-}145)$$

$$pb =$$

$$0.1573$$

c. $P(-1 \leq Z \leq 2)$

$$>> pc = 0.5*(erf(2/k) - erf(-1/k))\hspace{2cm}(12\text{-}146)$$

$$pc =$$

$$0.8186$$

d. $P(-2 \leq Z \leq 2)$

$$>> pd = 0.5*(erf(2/k) - erf(-2/k))\hspace{2cm}(12\text{-}147)$$

$$pd =$$

$$0.9545$$

All of these values agree with those of Example 12-17.

MATLAB Example 12-3

In a certain Internet digital communication system, the noise has a gaussian distribution with a mean value of 0 and a standard deviation of 2 V. A "false alarm" or incorrect interpretation will occur if a noise spike exceeds a certain threshold level. Determine the minimum voltage of a digital threshold level such that the probability of error is no greater than 10^{-5}.

Solution

We first determine the value of z that satisfies the relationship that $P(Z \geq z) = 10^{-5}$. This is best achieved with the computational form of Equation 12-142. Assuming $p = 1 \times 10^{-5}$, we have

$$\text{>> z = sqrt(2)*erfcinv(2*1e-5)} \tag{12-148}$$

$$z =$$

$$4.2649$$

The variable z is the normalized variable and is related to the actual variable, x, by

$$z = \frac{x - \mu}{\sigma} = \frac{x - 0}{2} = \frac{x}{2} \tag{12-149}$$

The minimum value of the voltage threshold level x is then required to be

$$x = 2z = 2 \times 4.2649 = 8.53 \text{ V} \tag{12-150}$$

where realistic rounding has been performed.

GENERAL PROBLEMS

12-1. A single card is drawn from a deck of cards. What is the probability that it will be (a) an ace of hearts, (b) a heart, (c) either a heart or a diamond, (d) either a red king or any queen, and (e) a face card?

12-2. A single card is drawn from a deck of cards. What is the probability that it will be (a) either a king or a face card, (b) either a king or a spade, and (c) either a heart or a face card?

12-3. Two cards are drawn in succession from a deck of cards. The first card is replaced and the deck is reshuffled before the second card is drawn. What is the probability that the two cards are (a) both kings, (b) an ace on the first draw and a king on the second draw, and (c) an ace and a king in either order?

12-4. Consider the same experiments and desired outcomes as in Problem 12-3, but assume that the first card is not replaced before drawing the second card. Repeat the probability evaluations.

12-5. Consider the same unreliable switches of Example 12-6, but assume that *three* are connected in a *serial* arrangement, that is, all three have to close to make the system work. Determine the probability of success.

12-6. Consider the serial switching arrangement depicted in Figure 12-3 but assume that the probability of A closing is 0.95 and the probability of B closing is 0.85. Determine the probability of success.

12-7. Consider the same unreliable switches of Example 12-7, but assume that *three* are connected in a *parallel* arrangement. Determine the probability of success. *Hint:* With three, it is easier to use the "backwards" approach, that is, first determine the probability of failure.

12-8. Consider the parallel switching arrangement depicted in Figure 12-4, but assume the two separate probability values of Problem 12-6. Determine the probability of success.

12-9. A two-level voltage x assumes only the values of -5 V and 5 V. Experimental data taken over a period of time result in 8000 samples of -5 V and 12000 samples of 5 V. Define a pdf and sketch it.

12-10. A four-level voltage x assumes only the following levels: 0 V, 1 V, 2 V, and 3 V. Experimental data taken over a period of time result in the following data:

Voltage, V	0	1	2	3
Number of Samples	4000	6000	8000	2000

Define a pdf and sketch it.

12-11. Construct a probability distribution function for the pdf of Problem 12-9 and sketch it.

12-12. Construct a probability distribution function for the pdf of Problem 12-10 and sketch it.

12-13. For the pdf of Problem 12-9, determine the (a) mean value, (b) mean-square value, (c) rms value, (d) variance, and (e) standard deviation.

12-14. For the pdf of Problem 12-10, determine the (a) mean value, (b) mean-square value, (c) rms value, (d) variance, and (e) standard deviation.

12-15. A random *unipolar* binary voltage signal produces zeros and ones with equal probability. A "0" appears as 0 V and a "1" appears as E volts. Determine the pdf and sketch it.

12-16. A random *bipolar* binary voltage signal produces zeros and ones with equal probability. A "0" appears as $-E$ volts and a "1" appears as E volts. Determine the pdf and sketch it.

12-17. For the signal of Problem 12-15, determine the (a) mean value, (b) mean-square value, (c) total rms value, (d) variance, and (e) standard deviation.

12-18. For the signal of Problem 12-16, determine the (a) mean value, (b) mean-square value, (c) total rms value, (d) variance, and (e) standard deviation.

12-19. Repeat Problem 12-15 if there are three times as many ones as zeros in a long data stream.

12-20. Repeat Problem 12-16 if there are three times as many ones as zeros in a long data stream.

12-21. Repeat the analysis of Problem 12-17 for the data stream of Problem 12-19.

12-22. Repeat the analysis of Problem 12-18 for the data stream of Problem 12-20.

12-23. An unbiased coin is flipped five times. What is the probability of getting exactly one head in the five trials?

12-24. In the experiment of Problem 12-23, what is the probability of getting at least one head in the five trials?

12-25. The pressure associated with a certain fluid on a structure is a random variable $x(t)$ and is *uniformly distributed* between the levels of 0 Pa and 20 Pa. Plot the pdf and determine its level.

12-26. The force $x(t)$ associated with a certain random process is *uniformly distributed* between the levels of –2 N and 6 N. Plot the pdf and determine its level.

12-27. For the random pressure of Problem 12-25, assume that a random sample X is to be taken. Determine the following probabilities: (a) $P(X = 4\ \text{Pa})$, (b) $P(8\ \text{Pa} \le X \le 10\ \text{Pa})$, (c) $P(X \le 4\ \text{Pa})$, (d) $P(X \ge 9\ \text{Pa})$.

12-28. For the random force of Problem 12-26, assume that a random sample X is to be taken. Determine the following probabilities: (a) $P(X = 0\ \text{N})$, (b) $P(1.95\ \text{N} \le X \le 2.05\ \text{N}$, (c) $P(X \le 1\ \text{N})$, (d) $P(X \ge 1\ \text{N})$.

12-29. For the random pressure of Problems 12-25 and 12-27, determine the following quantities: (a) mean value, (b) mean-square value, (c) total rms value, (d) variance, and (e) standard deviation.

12-30. For the random force of Problems 12-26 and 12-28, determine the following quantities: (a) mean value, (b) mean-square value, (c) total rms value, (d) variance, and (e) standard deviation.

12-31. A random voltage $x(t)$ has a gaussian distribution with a mean value of 0 V and a standard deviation of 10 V. A single random sample X is taken. Determine the following probabilities: (a) $P(X \ge 0\ \text{V})$, (b) $P(X \ge 10\ \text{V})$, (c) $P(X \le -10\ \text{V})$, (d) $P(-10\ \text{V} \le X \le 10\ \text{V})$, (e) $P(|X| \ge 10\ \text{V})$, (f)$P(X \ge 20\ \text{V})$, and (g) $P(|X| \ge 20\ \text{V})$.

12-32. A random voltage $x(t)$ has a gaussian distribution with a mean value of 12 V and a standard deviation of 3 V. A single random sample, X, is taken. Determine the following probabilities: (a) $P(X \geq 15$ V$)$, (b) $P(X \leq 9$ V$)$, (c) $P(X \geq 21$ V$)$.

12-33. Assume that the voltage of Problem 12-31 represents an undesirable noise signal and that it will trigger a "false alarm" in a circuit if the level exceeds 10 V. Suppose that the bandwidth of the circuit is such that two independent samples of the voltage have the opportunity to trigger the circuit in a certain time interval. Determine the probability of a false alarm in the time period.

12-34. Assume that the voltage of Problem 12-32 represents an undesirable noise signal superimposed on the mean or dc level and that it will trigger a "false alarm" in a circuit if the net level exceeds 21 V. Suppose that the bandwidth of the circuit is such that three independent samples of the voltage have the opportunity to trigger the circuit in a certain time interval. Determine the probability of a false alarm in the time period.

12-35. The weekly earnings (in dollars) of a random sample of 10 persons are as follows:

100 200 300 300 300 400 400 500 600 2000

Determine the (a) sample mean, (b) sample mean-square, (c) sample rms value, (d) sample variance, (e) sample standard deviation, (f) median, (g) mode, and (h) range.

12-36. Based on an examination given to a large class in which the maximum score is 100 points, assume that 20 grades taken at random from the entire set produced the results that follow.

43 49 56 61 65 68 70 70 74 74 74 76 79 83 84 87 88 92 96 100

Determine the (a) sample mean, (b) sample mean-square, (c) sample rms value, (d) sample variance, (e) sample standard deviation, (f) median, (g) mode, and (h) range.

12-37. Assume that a given binary signal is disturbed by such a large level of noise that errors in received bits occur about half the time. For three consecutive bits, calculate the following probabilities concerning the number of errors occurring in the bit stream: (a) no errors, (b) exactly one error, (c) exactly two errors, (d) exactly three errors.

12-38. Repeat Problem 12-37 if the noise level is reduced such that errors in received bits occur about one-tenth the time.

12-39. Each of the possible conditions in parts (a) through (d) of Problem 12-37 may be considered as mutually exclusive since a specific number of errors is implied in each case. Based on the data of that problem, determine the probability that there will be no more than one error in three digits.

12-40. Repeat Problem 12-39 for the lower noise level of Problem 12-38.

MATLAB PROBLEMS

12-41. Rework Problem 12-31 using MATLAB.

12-42. Rework Problem 12-32 using MATLAB.

12-43. Rework Problem 12-35 using MATLAB. It is suggested that you use the bank format.

12-44. Rework Problem 12-36 using MATLAB.

12-45. Write an M-file program that evaluates a probability for a binomial probability density function based on a specific number of occurrences x, in which the probability of a single occurrence in a single trial is p.

> *Inputs:* The probability p
> The number of trials n
> The value x, where x is an integer bounded by $0 \leq x \leq n$
>
> *Output:* The probability of a single occurrence x

12-46. Write an M-file program that evaluates the net probability for a binomial probability density function based on all possible numbers of occurrences from 0 up to and including x, in which the probability of a single occurrence in a single trial is p.

> *Inputs:* The probability p
> The number of trials n
> The value x, where x is an integer bounded by $0 \leq x \leq n$
>
> *Output:* The probability of any number of occurrences up to and including x

12-47. Write an M-file program that evaluates for a gaussian distributed variable the following probability:

$$P(x_1 \leq X \leq x_2)$$

> *Inputs:* The mean value μ
> The standard deviation σ
> The value x_1
> The value x_2
> *Output:* The probability value

12-48. Write an M-file program that evaluates for a gaussian distributed variable the following probability:

$$P(X \geq x)$$

> *Inputs:* The mean value μ
> The standard deviation σ
> The value x
> *Output:* The probability value

Table 12-1   Tabular data for the hypothetical random discrete signal of Figure 12-3

x_k, volts	Count	$f(x_k)$
-5	30,000	0.3
0	20,000	0.2
5	40,000	0.4
10	10,000	0.1

Table 12-2   Gaussian probability distribution function values

z	$F(z)$	z	$F(z)$
-3.1	0.000968	0.0	0.500000
-3.0	0.001350	0.1	0.539828
-2.9	0.001866	0.2	0.579260
-2.8	0.002555	0.3	0.617911
-2.7	0.003467	0.4	0.655422
-2.6	0.004661	0.5	0.691462
-2.5	0.006210	0.6	0.725747
-2.4	0.008198	0.7	0.758036
-2.3	0.010724	0.8	0.788145
-2.2	0.013903	0.9	0.815940
-2.1	0.017864	1.0	0.841345
-2.0	0.022750	1.1	0.864334
-1.9	0.028717	1.2	0.884930
-1.8	0.035930	1.3	0.903200
-1.7	0.044565	1.4	0.919243
-1.6	0.054799	1.5	0.933193
-1.5	0.066807	1.6	0.945201
-1.4	0.080757	1.7	0.955435
-1.3	0.096800	1.8	0.964070
-1.2	0.115070	1.9	0.9712-83

Table 12-2　Gaussian probability distribution function values *(continued)*

z	$F(z)$	z	$F(z)$
-1.1	0.135666	2.0	0.977250
-1.0	0.158655	2.1	0.982136
-0.9	0.184060	2.2	0.986097
-0.8	0.211855	2.3	0.989276
-0.7	0.241964	2.4	0.991802
-0.6	0.274253	2.5	0.993790
-0.5	0.308538	2.6	0.995339
-0.4	0.344578	2.7	0.996533
-0.3	0.382089	2.8	0.997445
-0.2	0.420740	2.9	0.998134
-0.1	0.460172	3.0	0.998650

Table 12-3. Summary of Statistical Parameters.

Parameter	Continuous PDF	Discrete PDF	Sample
Mean	$\mu = \int_{-\infty}^{\infty} xf(x)dx$	$\mu = \sum_k x_k f(x_k)$	$\bar{x} = \frac{1}{n}\sum_{k=1}^{n} x_k$
Mean-Square	$E(x^2) = \int_{-\infty}^{\infty} x^2 f(x)dx$	$E(x^2) = \sum_k x_k^2 f(x_k)$	$\overline{x^2} = \frac{1}{n}\sum_{k=1}^{n} x_k^2$
Root-Mean-Square	$x_{rms} = \sqrt{E(x^2)}$	$x_{rms} = \sqrt{E(x^2)}$	$x_{rms} = \sqrt{\overline{x^2}}$
Variance	$\sigma^2 = \int_{-\infty}^{\infty} (x-\mu)^2 f(x)dx$ $= E(x^2) - \mu^2$	$\sigma^2 = \sum_k (x_k - \mu)^2 f(x_k)$ $= E(x^2) - \mu^2$	$s^2 = \frac{1}{n-1}\sum_{k=1}^{n} (x_k - \bar{x})^2$
Standard Deviation	$\sigma = \sqrt{\sigma^2}$	$\sigma = \sqrt{\sigma^2}$	$s = \sqrt{s^2}$

Curve Fitting and Correlation 13

All phases of scientific, engineering, and human service practices involve the acquisition, processing, and interpretation of data. We have seen in the preceding chapter how the basic statistical properties of such data may be determined and utilized. However, the processes introduced may be further refined and extended to produce important measures that can be used to assist in the determination of relationships between variables and to assess how different variables might affect each other.

This chapter will be concerned primarily with two separate but closely interrelated processes: (1) the fitting of experimental data to mathematical forms that describe their behavior and (2) the correlation between different experimental data to assess how closely different variables are interdependent.

The fitting of experimental data to a mathematical equation is called *regression*. Regression may be characterized by different adjectives according to the mathematical form being used for the fit and the number of variables. For example, *linear regression* involves using a straight-line or linear equation for the fit. As another example, *multiple regression* involves a function of more than one independent variable.

Objectives

After completing this chapter, the reader should be able to
1. Discuss the concept of regression and how it can be used to determine best-fit equations.
2. Explain the minimum mean-square error process utilized in regression analysis.
3. Determine the best linear equation fit using the formula provided.
4. Apply the **polyfit** command of MATLAB to determine the best-fitting polynomial of an arbitrary degree.
5. Apply MATLAB to determine the best-fit multiple regression linear formula.
6. Define *cross-correlation* and discuss its significance.
7. Define *covariance*, discuss its significance, and show how it relates to *cross-correlation*.
8. Define the *correlation coefficient* and discuss its significance.

9. Apply MATLAB to determine the *correlation coefficient matrix* relating an arbitrary number of variables.

13-2   LINEAR REGRESSION

The first case that will be considered is that of a two-dimensional set of data in which the "best" straight-line or linear equation fit will be determined. This may or may not make sense for the data involved, depending on its behavior. If it is obvious from a simple inspection that the variation is drastically different from that of a linear equation, the procedure may yield results that make very little sense. However, if the general trend of the data appears to approximate a straight line, the procedure may yield meaningful results.

Assume that there are n points in which each point is characterized by values of both the independent variable x and the dependent variable y. Thus, let $x_1, x_2, x_3,..., .x_n$ represent the values of x and let $y_1, y_2, y_3,....,y_n$ represent the corresponding values of y. Earlier in the text, we used m to represent the slope and b to represent the vertical intercept. However, to conform to notation that will be used in subsequent sections for higher-degree functions, the desired representation will be modified somewhat and the linear equation will be assumed in the form

$$y = a_1x + a_0 \tag{13-1}$$

where a_1 is the slope and a_0 is the vertical intercept. These two values are to be determined to represent an "optimum" fit for the given set of points. Understand that the equation obtained may produce few if any **exact** matches at the given points, but the objective is to represent the "best" overall estimate in some sense.

Mean-Square Error Criterion

The determination of an appropriate equation is somewhat subjective since there are various criteria that could be used. For example, one might determine the linear equation that produces the smallest maximum error or one that produces zero error at a specific point. However, the most widely employed criterion is that the net mean-square error between points produced by the approximation and the actual points is to be minimized. This process is referred to as a *mean-square error criterion*.

The reason that the square is most often used is that all values are non-negative and minimization tends to be optimum. If the simple algebraic differences were used, relatively large positive and negative differences could tend to cancel, producing a possible appearance of a small error when, in fact, the difference values could be large in magnitude. By squaring the differences, this possibility is eliminated.

Let $y_1^a, y_2^a, y_3^a,....,y_n^a$ represent the approximate values of y as computed from the linear approximation of Equation 13-1. Since the mean value of a quantity is determined simply by dividing the sum by the number of samples, minimization of the sum of the squared errors will also minimize the mean square, so we can work directly with the sum. The sum of the squared-error terms E is then determined as

$$E = (y_1^a - y_1)^2 + (y_2^a - y_2)^2 + \ldots + (y_n^a - y_n)^2$$
$$= \sum_{k=1}^{n} (y_k^a - y_k)^2 \tag{13-2}$$

The minimization process involves the substitution of the linear approximation and an application of differential calculus. While this process involves material that was covered earlier in the text, the presence of the summation makes the derivation a bit messy and a little tricky. We will state the results here without proof.

Some Preliminary Computations

The approach that we will use is to first calculate four parameters, three of which are based on definitions introduced in the preceding chapter. This approach differs slightly from the treatment provided in some texts, but the end results are exactly the same. Moreover, the use of these intermediate steps ties in very closely with the work of the preceding chapter and provides additional measures of possible interest. Therefore, the four parameters are calculated as follows:

$$\bar{x} = \text{sample mean of the } x \text{ values} = \frac{1}{n} \sum_{k=0}^{n} x_k \tag{13-3}$$

$$\bar{y} = \text{sample mean of the } y \text{ values} = \frac{1}{n} \sum_{k=0}^{n} y_k \tag{13-4}$$

$$\overline{x^2} = \text{sample mean-square of the } x \text{ values} = \frac{1}{n} \sum_{k=1}^{n} x_k^2 \tag{13-5}$$

$$\overline{xy} = \text{sample mean of the product } xy = \frac{1}{n} \sum_{k=1}^{n} x_k y_k \tag{13-6}$$

The last parameter is the only form that was not introduced in the last chapter. As will be seen later in the chapter, this term can also be called the *cross-correlation* between the variables x and y.

Determination of the Constants

Based on the preceding definitions and the mean-square error criterion, the best-fit values for the slope a_1 and the vertical intercept a_0 are

$$a_1 = \frac{(\overline{xy}) - (\bar{x})(\bar{y})}{(\overline{x^2}) - (\bar{x})^2} \tag{13-7}$$

$$a_0 = \frac{\left(\overline{x^2}\right)(\overline{y}) - (\overline{x})(\overline{xy})}{\left(\overline{x^2}\right) - (\overline{x})^2} \qquad (13\text{-}8)$$

An alternate more convenient formula for a_0 utilizing the calculated value of a_1 is

$$a_0 = \overline{y} - a_1\overline{x} \qquad (13\text{-}9)$$

Example 13-1

Some data concerning an independent variable, x, and a dependent variable, y, are given below.

x	0	1	2	3	4	5	6	7	8	9
y	4.00	6.10	8.30	9.90	12.40	14.30	15.70	17.40	19.80	22.30

(a) Determine the best-fitting straight line for the data. (b) Plot the actual data and the best-fitting straight line on the same scale.

Solution

---

a. Strictly speaking, computations involving data should be inspected to determine the accuracy in terms of the number of significant digits, but we will not get sidetracked on that type of analysis here. Instead, we will assume that the values of x are "exact" and that the values of y are known to two decimal places, as shown.

As will be seen later, the process can be implemented most easily with MATLAB. Even if we did it manually, we could still use MATLAB to perform some of the steps involved. However, to illustrate the process fully, we will carry out the operations in a step-by-step fashion. First, the four parameters initially required will be computed.

$$\overline{x} = \frac{1}{10}\sum_{k=1}^{10} x_k = \frac{0+1+2+3+4+5+6+7+8+9}{10} = \frac{45}{10} = 4.50 \qquad (13\text{-}10)$$

$$\overline{y} = \frac{1}{10}\sum_{k=1}^{10} y_k = \frac{4+6.1+8.3+9.9+12.4+14.3+15.7+17.4+19.8+22.3}{10}$$

$$= \frac{130.2}{10} = 13.02 \qquad (13\text{-}11)$$

$$\overline{x^2} = \frac{1}{10}\sum_{k=1}^{10} x_k^2$$

$$= \frac{(0)^2 + (1)^2 + (2)^2 + (3)^2 + (4)^2 + (5)^2 + (6)^2 + (7)^2 + (8)^2 + (9)^2}{10} \qquad (13\text{-}12)$$

$$= \frac{285}{10} = 28.50$$

$$\overline{xy} = \frac{1}{10}\sum_{k=1}^{10} x_k y_k$$

$$= \frac{0+6.1+16.6+29.7+49.6+71.5+94.2+121.8+158.4+200.7}{10} \qquad (13\text{-}13)$$

$$= \frac{748.6}{10} = 74.86$$

The step involving the display of the quantities being multiplied in the last equation was omitted due to its length, but the reader may readily check each term in the numerator of Equation 13-13.

   We are now ready to determine the best values of a_1 and a_0; Equations 13-7 and 13-9 will be used.

$$a_1 = \frac{(\overline{xy})-(\overline{x})(\overline{y})}{(\overline{x^2})-(\overline{x})^2} = \frac{74.86-(4.50)(13.02)}{28.50-(4.50)^2} = \frac{16.27}{8.250} = 1.9721 \qquad (13\text{-}14)$$

$$a_0 = \overline{y} - a_1\overline{x} = 13.02 - 1.972 \times 4.50 = 4.1455 \qquad (13\text{-}15)$$

The equation for the best-fitting straight-line equation is thus

$$y = 1.9721x + 4.1455 \qquad (13\text{-}16)$$

b. Since the x-values are equally spaced in this example, the corresponding vector may be easily created by the simple command

$$>> x = 0:9; \qquad (13\text{-}17)$$

We will denote the vector of calculated approximate values from Equation 13-16 as yapp. It is readily generated by

$$>> yapp = 1.9721*x + 4.1455; \qquad (13\text{-}18)$$

The vector representing the actual dependent variables will be denoted simply as y. It is determined by forming all the values as a row vector.

$$>> y = [4\ 6.1\ 8.3\ 9.9\ 12.4\ 14.3\ 15.7\ 17.4\ 19.8\ 22.3]; \qquad (13\text{-}19)$$

Plots are now generated by the command

$$>> plot(x, yapp, x, y, \text{'o'}) \qquad (13\text{-}20)$$

Note that the **'o'** following y will result in that function being displayed by a series of o's rather than as a continuous curve. After additional labeling and some slight scaling, the results are shown in Figure 13-1. It is clear that the approximate straight-line equation fits very well with the experimental data.

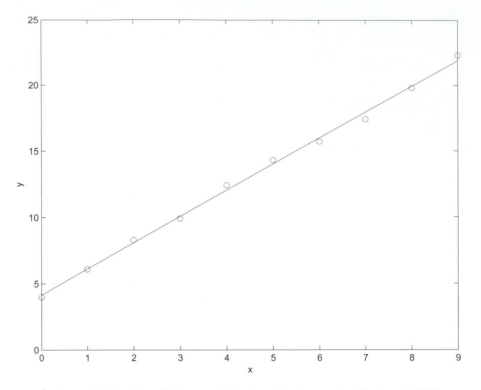

Figure 13-1   Best-fitting straight line for Examples 13-1 and 13-2

13-3   MATLAB POLYNOMIAL CURVE FITTING

We will now turn our attention to the process of utilizing MATLAB to determine a best-fitting curve. MATLAB allows one to choose a polynomial of arbitrary degree m. Thus, the linear straight-line fit of the previous section may be readily determined or some higher-degree polynomial may be used.

Choice of Polynomial Degree

The choice of a polynomial degree is somewhat arbitrary and may require some trial and error for an optimum fit. One can usually determine visually if the data seem to fit a straight line, and that was the situation in Example 13-1. However, for functions with greater curvature, several degrees could be tried and the results carefully evaluated. This is clearly an area that requires experience and careful judgment to fully exploit its benefits. For the particular application involved, it may be necessary to concentrate only on a limited portion of the domain, in which case data outside this domain may be ignored.

The presence of significant maxima and minima in the function can affect the choice of the degree. As a general rule, the degree chosen should be at least one degree higher than the total number of maxima and minima over the domain in question. Thus, if there is one significant maximum and one significant minimum in the domain involved, one might try at least a cubic equation at the start. By "significant," it is implied that there is a clear pattern rather than just simple slight variations about the observed trend.

The Polyfit Command

As in the last section, assume that there are *n* points in which each point is characterized by both a value of the independent variable *x* and the dependent variable *y*. Let x_1, x_2, x_3,....x_n represent the values of *x* and let y_1, y_2, y_3,....y_n represent the corresponding values of *y*. The function is to be represented by a polynomial $y = p(x)$ of degree *m* in the form

$$y = p(x) = a_m x^m + a_{m-1} x^{m-1} + ... + a_1 x + a_0 \qquad (13\text{-}21)$$

Define a row vector x with n values as

$$>> x = [x1 \ x2 \ x3........xn]; \qquad (13\text{-}22)$$

If the x values are equally spaced, as was the case with Example 13-1, the x vector may be generated by one of the simple constant step commands. However, one may have a situation in which the independent variable values are somewhat random, and they may need to be entered manually.

Define a row vector y with n values as

$$>> y = [y1 \ y2 \ y3......yn]; \qquad (13\text{-}23)$$

The polynomial p is then determined by the command

$$>> p = polyfit(x, y, m) \qquad (13\text{-}24)$$

The quantity p is a row vector with m + 1 coefficients representing the coefficients of the polynomial arranged in descending degree from left to right and the constant term Thus, p(1) is the coefficient of the highest-degree term in Equation 13-21; that is, a_m, and so on.

Evaluating the Polynomial

An evaluation of the approximating polynomial at all of the values of x may be achieved most easily with the **polyval** command, which was introduced back in Chapter 5. Let yapp denote the vector representing the values calculated with the approximation. We have

$$>> yapp = polyval(p, x) \qquad (13\text{-}25)$$

The two functions may then be plotted on the same scale by a **plot** command. For the experimental data, it may be more meaningful to apply specific symbols at the points, as was demonstrated in Example 13-1.

If the same values of x are used for both the approximation and the actual values, a suitable plot command would be

$$>> \text{plot}(x, \text{yapp}, x, y, \text{'o'}) \hspace{3cm} (13\text{-}26)$$

This was the form used in Example 13-1. It worked very well since the approximating function was a straight line. On the other hand, suppose that the approximating function has a great deal of curvature and that only a limited number of points were used in determining the function. In that case, a continuous plot of the approximating function using a limited number of points may display sudden changes in the slope that degrade the presentation. In that case, it may be desirable to create a new independent variable, say x1, having closely spaced points for the purpose of plotting yapp. Since yapp can now be considered as a continuous function, the number of points employed is arbitrary. With this approach, the plot command could be of the form

$$>> \text{plot}(x1, \text{yapp}, x, y, \text{'o'}) \hspace{3cm} (13\text{-}27)$$

Example 13-2

Rework Example 13-1 using the **polyfit** command and a straight-line equation.

Solution

For convenience and completeness, some of the steps will be repeated here. Since the x-values have a linear spacing, the 10 values will be generated by the simple command

$$>> x = 0:9; \hspace{4cm} (13\text{-}28)$$

The y values are now entered as a 10-element row vector.

$$>> y = [4\ 6.1\ 8.3\ 9.9\ 12.4\ 14.3\ 15.7\ 17.4\ 19.8\ 22.3]; \hspace{1cm} (13\text{-}29)$$

The polynomial p of degree 1 is determined as

$$>> p = \text{polyfit}(x, y, 1) \hspace{3cm} (13\text{-}30)$$

$$p =$$

$$1.9721 \quad 4.1455$$

The first value is the coefficient of x and the second value is the constant. These are exactly the same values that were determined manually in Example 13-1, so what else can be said? We could plot the results, but they will be the same as in Example 13-1, so there is no need to repeat the process.

Example 13-3

As a matter of curiosity, investigate the effect of approximating the data of the previous two examples by a polynomial of degree 2.

Solution

Assuming that the vectors x and y of the previous example are still in memory, the command in this case is

$$\text{>> p = polyfit(x, y, 2)} \tag{13-31}$$

$$p =$$

$$0.0011 \quad 1.9619 \quad 4.1591$$

Note that the linear term (1.9619) and the constant term (4.1591) differ very little from the corresponding values in the previous example. However, there is now a second-degree term (0.0011), albeit it is very small. Let us see if these results make much difference. Let yapp2 represent the second-degree approximation. It is easily generated by the command

$$\text{>> yapp2 = polyval(p, x);} \tag{13-32}$$

We will now plot the second-degree data and the actual data by the command

$$\text{>> plot(x, yapp2, x, y, 'o')} \tag{13-33}$$

The results are shown in Figure 13-2. A comparison with Figure 13-1 reveals that the second-degree approximation did not change the results enough to note any discernible differ-

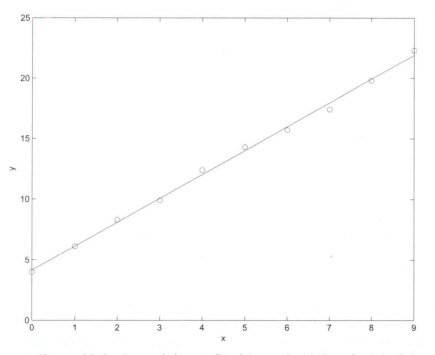

Figure 13-2 Second-degree fit of Example 13-3 and original data

ence. Clearly, the data provided are close enough to a straight line that a linear fit should be adequate for most purposes.

Example 13-4

Instead of experimental data, this example will be used to investigate how well the **polyfit** function can be used to represent a sinusoidal function over a full cycle by several different polynomials. Consider the sinusoidal function y with a period of 2 s defined over the domain from -1 to 1 by the equation

$$y = \sin \pi t \quad \text{for} \quad -1 \leq t \leq 1 \tag{13-34}$$

Based on a step size of 0.05 between points, investigate curve fits of the following degrees: (a) 1, (b) 2, (c) 3, (d) 4, (e) 5.

Solution

Prior to developing the approximations, an actual plot of the function will be made to review the form of the function being approximated. We begin by defining the vector representing the domain, which is

$$\text{>> t = -1:0.05:1;} \tag{13-35}$$

The function y is generated by

$$\text{>> y = sin(pi*t);} \tag{13-36}$$

The plot command is

$$\text{>> plot(t, y)} \tag{13-37}$$

The result is shown in Figure 13-3, with some additional labeling provided by standard procedures.

On the curves that follow, we will show the true values of y with o's. The various approximations will be shown as continuous curves.

a. $m = 1$

The attempt to approximate a sinusoidal function over a cycle with a straight line could be considered as a ridiculous exercise in futility, but it is being done for educational purposes and to make an interesting point for later reference. The polynomial coefficient vector will be denoted as p1. It is given by

$$\text{>> p1 = polyfit(t, y, 1)} \tag{13-38}$$

p1 =

0.8854   0.0000

Note that the constant is 0. There is a reason for this that will be discussed in part (b). The corresponding first-order approximation will be denoted as yapp1 and is evaluated as

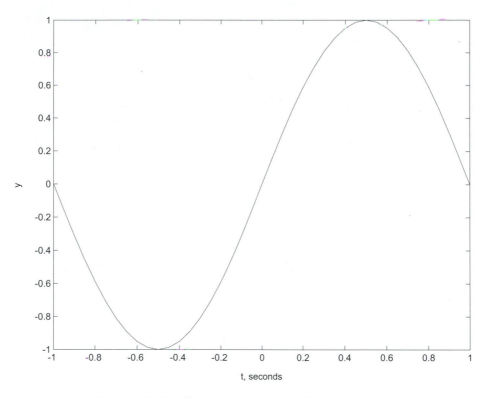

Figure 13-3 Sinusoidal function of Example 13-4

$$>> \text{yapp1} = \text{polyval(p1, t)}; \hspace{3cm} (13\text{-}39)$$

Plots of y and yapp1 on the same scale are achieved by the command

$$>> \text{plot(t, yapp1, t, y, }'o') \hspace{3cm} (13\text{-}40)$$

The two curves are shown in Figure 13-4. There is obviously a great disparity between the two functions, but the straight-line function is doing the best it can considering its limitations.

b. $m = 2$

The polynomial coefficient vector will be denoted as p2. It is generated by the command

$$>> \text{p2} = \text{polyfit(t, y, 2)} \hspace{3cm} (13\text{-}41)$$

$$\text{p2} =$$

0.0000   0.8854   -0.0000

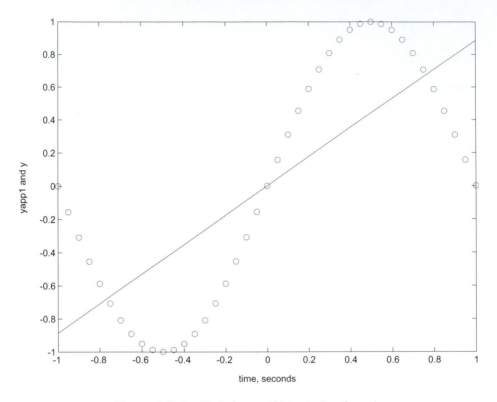

Figure 13-4 First-degree fit to a sine function

Whoa! Considering that the second-degree term is 0, the result is the same as obtained in part (a). What is going on? The answer to this question lies in the fact that the function being approximated is an odd function, that is, $y(-t) = -y(t)$. Since a second-degree term is an even function, its optimum value for the given function is 0. Likewise, any other even-degree terms and a constant term will all be zero. Therefore, for the function being considered, the second-degree approximation is the same as the first-degree approximation and there is no need to make a plot.

c. $m = 3$

This polynomial coefficient vector will be denoted as p3 and is generated by

$$\gg \text{p3} = \text{polyfit(t, y, 3)} \tag{13-42}$$

$$\text{p3} =$$

$$-2.8139 \quad -0.0000 \quad 2.6568 \quad 0.0000$$

The third-degree approximation is denoted as yapp3.

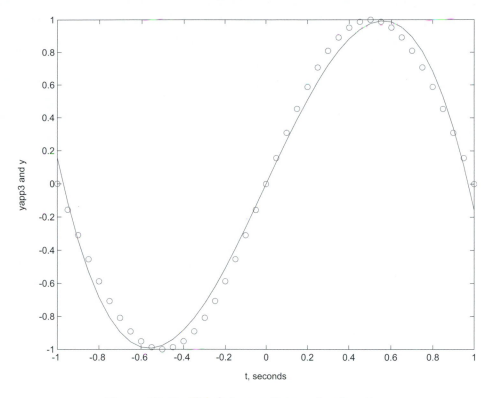

Figure 13-5 Third-degree fit to a sine function

$$\text{>> yapp3 = polyval(p3, t);} \qquad (13\text{-}43)$$

A plot of this function and the original is obtained by the command

$$\text{>> plot(t, yapp3, t, y, 'o')} \qquad (13\text{-}44)$$

After some additional labeling, the results are shown in Figure 13-5. While the two curves clearly differ, the third-order approximation has taken a giant step toward the ultimate goal.

    d. $m = 4$

In a similar fashion to the second-degree approximation, the fourth-degree approximation provides 0 for the highest-degree term and the other coefficients are the same as for the third-degree case, so we will bypass this step.

    e. $m = 5$

This polynomial coefficient vector will be denoted as p5 and results from

$$\text{>> p5 = polyfit(t, y, 5)} \qquad (13\text{-}45)$$

$$p5 =$$

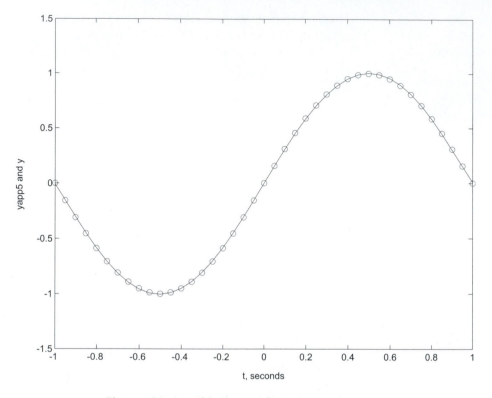

Figure 13-6 Fifth-degree fit to a sine function

$$1.6982\ 0.0000\ -4.7880\ -0.0000\ 3.0990\ 0.0000$$

The fifth-degree approximation is denoted as yapp5 and is evaluated by

$$>> \text{yapp5} = \text{polyval(p5, t)};\qquad\qquad\text{(13-46)}$$

Plots of this function and the original are generated by

$$>> \text{plot(t, yapp5, t, y, 'o')}\qquad\qquad\text{(13-47)}$$

The results are shown in Figure 13-6. Now we are getting somewhere! The difference between the two curves is not discernible on the scale provided.

Most applications of curve-fitting involve experimental data rather than fitting a polynomial to a well-known given function, so in most cases the results will be neither even nor odd and, in general, will contain all degree terms up to the highest degree. Therefore, this exercise could be considered as a bit artificial since we were dealing with a well-known function. However, it illustrates how different degrees of representation can influence the ultimate outcome.

Before leaving this example, one point should be made very clear. The fifth-degree polynomial obtained in part (e) provides a good approximation to the sine function *only over the interval used in determining the coefficients.* In fact, as the magnitude of time increases without bound, the polynomial magnitude likewise increases without bound. However, the sine function continues to oscillate between −1 and 1. Therefore, in general, one must be very careful about using a polynomial approximation obtained over one domain interval to extrapolate the behavior over a different interval. If one could, it would be easy to get rich in the stock market by fitting the data over some interval and using it to predict the behavior in future intervals!

Finally, some readers may be familiar with the power series approximations of functions and may realize that a sine function may be represented over any arbitrary domain *if a sufficient number of terms is used.* Indeed, the infinite series for sin *x* has the form

$$\sin x = x - \frac{x^3}{3!} + \frac{x^5}{5!} - \frac{x^7}{7!} + \dots = \sum_{n=0}^{\infty} (-1)^n \frac{x^{2n+1}}{(2n+1)!} \tag{13-48}$$

This result does not contradict the statement of the last paragraph. Our last polynomial used in the approximation was a fifth-degree form, and it was optimized only over the domain used. The infinite series of Equation 13-48 is applicable for all values of *x*, but as the magnitude of *x* increases, more and more terms in that equation must be used to allow the function to converge to the correct value.

Example 13-5

The steady-state operating temperature T in a certain chemical process is a non-linear but fairly stable function of the distance x within the chemical mass. It is believed to follow a second-degree variation. A relationship is desired that will predict the temperature at any point within the system. Data obtained from some measurements of the distance x in feet and the temperature in degrees Fahrenheit are as follows:

x (ft)	0	1	2	3	4	5
T (deg F)	71	76	86	100	118	140

Determine the best-fitting second-degree equation fit and plot the curve on the same scale with the actual values.

Solution

Since the step size for the distance variable is 1 ft, the vector x is generated by the simple command

$$>> x = 0:5; \tag{13-49}$$

The temperature values are placed in a row vector T.

$$>> T = [71\ 76\ 86\ 100\ 118\ 140]; \tag{13-50}$$

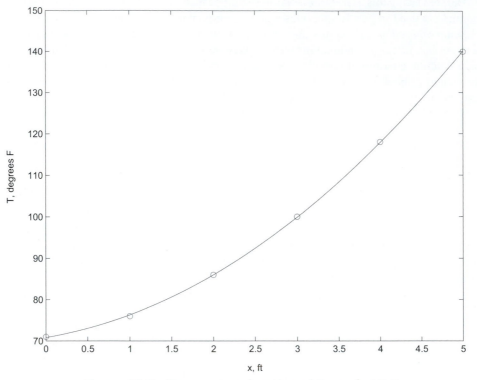

Figure 13-7 Temperature function of Example 13-5

A second-degree polynomial p is determined from the command

$$>> p = polyfit(x,T,2) \qquad (13\text{-}51)$$

$$p =$$

$$2.0893 \ 3.4107 \ 70.8214$$

This means that the best-fitting temperature equation in second-degree form is

$$T = 2.0893x^2 + 3.4107x + 70.8214 \qquad (13\text{-}52)$$

If this equation is evaluated only at the six points, the resulting curve will appear a bit "jagged." As discussed earlier, to smooth the presentation, a new independent variable x1 will be introduced that covers the same domain, but with much smaller steps. We will arbitrarily choose steps of 0.1 ft. This vector is generated by the command

$$>> x1 = 0:0.1:5; \qquad (13\text{-}53)$$

Let T1 represent the polynomial approximation evaluated at all points of x1. It is generated by

$$\gg \text{T1} = \text{polyval(p, x1);} \tag{13-54}$$

A plot command will now be used and the actual points will be identified with o's.

$$\gg \text{plot(x1, T1, x, T, 'o')} \tag{13-55}$$

The results with additional labeling are shown in Figure 13-7. The result involves a second-degree term, a first-degree term, and a constant, but the fit appears to be quite good.

13-4 MULTIPLE REGRESSION

Thus far, we have considered curve-fitting strategies only for the case of one dependent variable and one independent variable. However, we have considered both first-order and higher-order polynomial fits.

In the most general case, there may be several dependent variables and several independent variables. In that situation, each dependent variable may be considered as a function of all the independent variables. In general, the dependency could involve any arbitrary degrees for the approximating polynomials or more arbitrary functions such as exponential forms, logarithmic forms, and so on. As one might expect, the situation can become quite complex as the number of variables increases and the types of functions become more general. Therefore, most realistic applications of the process have to consider the complexity and ultimate limitations of the situation.

The most common application of multiple regression is based on a linear dependency of the variables involved, and we will limit our consideration to that case. To that end, assume that there are m independent variables indicated initially as $x_1, x_2, \ldots., x_m$. Assuming that each dependent variable may be expressed as a linear function of the m independent variables, we will consider only one dependent variable at this point and denote it as y. Thus, it is to be represented as

$$y = a_0 + a_1 x_1 + a_2 x_2 + \ldots + a_m x_m \tag{13-56}$$

where the various a coefficients are to be determined. Assume that there are k values of each of the independent variables. For x_1, the values will be denoted as $x_{11}, x_{12}, x_{13}, \ldots.,$ x_{1k}, with similar notation for the samples of the other variables with the subscript 1 replaced by appropriate values.

The process involves the use of the backslash operator (\ ) and turns out to involve the *pseudo-inverse matrix*, a term that was mentioned in passing in Chapter 2. The procedure using MATLAB will be provided in the steps that follow.

1. Form **m** column vectors, each of length **k**, representing the values for the independent variables **x**. Assuming the samples of x1 are represented in the MATLAB format as x11, x12, and so on, we have

$$\text{>> x1 = [x11 x12 x13......x1k]';} \qquad (13\text{-}57)$$

$$\text{>> x2 = [x21 x22 x23......x2k]';} \qquad (13\text{-}58)$$

$$\text{>> xm = [xm1 xm2 xm3.....xmk]';} \qquad (13\text{-}59)$$

Note that each of the vectors could have been entered directly in column form by placing semicolons between entries, but the approach used with spaces is simpler. However, the transpose of each row vector must be calculated in order to produce the required column vectors.

2. Form a column vector of length **k** representing the values for the dependent variable y. It is generated by

$$\text{>> y = [y1 y2 y3.....yk]';} \qquad (13\text{-}60)$$

3. Form a rectangular matrix X of size **k** by **m+1** with the following command:

$$\text{>> X= [ones(size(x1)) x1 x2 ......xm];} \qquad (13\text{-}61)$$

4. The least square column matrix will be denoted as **a** and will have a length **m + 1**. It is determined by the backslash operator using the following command:

$$\text{>> a = X\backslash y} \qquad (13\text{-}62)$$

5. Let Y represent the column vector of length **k** displaying the values generated by the approximation. It can be determined by the command

$$\text{>> Y = X*a;} \qquad (13\text{-}63)$$

6. A relative measure of the error can be determined by evaluating the absolute value of the maximum deviation. Let Error_Maximum represent this value. It can be determined by the command

$$\text{>> Error\_Maximum = max(abs(Y-y))} \qquad (13\text{-}64)$$

Example 13-6

A small college department in a hypothetical university has eight faculty members. The Dean of the college decides to run a multiple regression analysis to determine how well the faculty salaries tend to vary with certain parameters. The eight faculty are identified as A through H. Five criteria are used in the evaluation and they are identified by the symbols x1 through x5 that follow. The tabulated results are based on the long-term accumulated data for the faculty.

1. Number of years of full-time employment in the department: x1
2. Average teaching evaluations based on a scale of 1 to 6:   x2
3. Number of publications: x3
4. Volume of external research funding in thousands of dollars: x4
5. Number of committees and special service assignments: x5

The presumed dependent variable is the salary in thousands of dollars per year and it is indicated as y.

The data are tabulated as follows:

Faculty	x1	x2	x3	x4	x5	y
A	1	3.8	2	0	1	45
B	3	4.7	5	22	4	50
C	4	4.1	7	60	6	60
D	5	3.7	3	25	6	52
E	8	4.9	12	120	4	70
F	12	5.5	5	50	8	58
G	15	4.5	4	0	12	52
H	18	5.0	5	20	15	56

Determine a multiple regression linear formula that best fits the data involved.

Solution

The form of the linear relationship required will be

$$y = a_0 + a_1 x_1 + a_2 x_2 + a_3 x_3 + a_4 x_4 + a_5 x_5 \qquad (13\text{-}65)$$

Switching to the default MATLAB notation, we define the following column vectors:

$$>> \text{x1} = [1\ 3\ 4\ 5\ 8\ 12\ 15\ 18]'; \qquad (13\text{-}66)$$

$$>> \text{x2} = [3.8\ 4.7\ 4.1\ 3.7\ 4.9\ 5.5\ 4.5\ 5.0]'; \qquad (13\text{-}67)$$

$$>> \text{x3} = [2\ 5\ 7\ 3\ 12\ 5\ 4\ 5]'; \qquad (13\text{-}68)$$

$$>> \text{x4} = [0\ 22\ 60\ 25\ 120\ 50\ 0\ 20]'; \qquad (13\text{-}69)$$

$$>> \text{x5} = [1\ 4\ 6\ 6\ 4\ 8\ 12\ 15]'; \qquad (13\text{-}70)$$

$$>> \text{y} = [45\ 50\ 60\ 52\ 70\ 58\ 52\ 56]'; \qquad (13\text{-}71)$$

We next form the matrix X as

$$>> \text{X} = [\text{ones(size(x1))} \ \text{x1 x2 x3 x4 x5}]; \qquad (13\text{-}72)$$

The matrix a is then determined by

$$>> \text{a} = \text{X} \backslash \text{y} \qquad (13\text{-}73)$$

a =

47.3887

0.1305

-0.9262

0.2492

0.1800

0.4400

Although we do not actually need to write it out, the form of the equation in normal mathematical notation is

$$y = 47.3887 + 0.1305x_1 - 0.9262x_2 + 0.2492x_3 + 0.1800x_4 + 0.4400x_5 \qquad (13\text{-}74)$$

Before making any evaluations, we need to indicate one potential misinterpretation. Do not be initially misled by the relative magnitudes of these numbers. For example, suppose the funding level x_4 had been expressed in actual dollars rather than thousands of dollars. For the same data, the constant would have been 10^{-3} times the value obtained in order to make the results come out the same. Therefore, the actual sizes of the constants may not necessarily initially tell us much until the data have been analyzed further. This issue will be dealt with through correlation properties in the next section.

Although the sign of a coefficient can begin to make a definite impression upon us, it does not tell the whole story since all the variables are working together to determine the outcome. If we looked only at the negative sign for x_2, it might suggest that the best teachers will make the lowest salaries, which could be a possibility in the "real world." However, it is too premature to make a full judgment until the correlation properties are determined in the next section.

During the meantime, let us see how well the formula obtained predicts the salaries by computing the results obtained by it. Let Y represent the values computed with the regression formula. Then, the vector is

$$>> Y = X*a \qquad (13\text{-}75)$$

Y =

44.9381

50.3923

59.2955

52.5011

70.2401

57.6250

51.4555

56.5525

The maximum difference between predicted and actual values is

$$\text{>> Error\_Maximum} = \max(\text{abs}(Y\text{-}y)) \qquad (13\text{-}76)$$

$$\text{Error\_Maximum} =$$

$$0.7045$$

Since the salaries are expressed in thousands of dollars, the maximum difference is about $705.

13–5   CORRELATION

We are now ready to investigate a means for determining the relative correlation between different variables. The major consideration is how the variation in one or more variables affects the remaining variables. To illustrate the process, let us begin with the situation of two variables, x and y. Once the concept is understood with two variables, it will be extended to the case of an arbitrary number of variables.

Cross-Correlation

The **cross-correlation** between two variables will be denoted as corr(x, y) and can be defined as

$$\text{corr}(x, y) = E(xy) \qquad (13\text{-}77)$$

Stated in words, the cross-correlation between two variables is simply the expected value of their product. In terms of a set of samples, it could be indicated as $\overline{xy}$. Indeed, this was one of the terms used in the determination of the linear regression formula, so we have already been introduced to the concept.

Covariance

While the cross-correlation function is quite useful in many applications, the presence of relatively large mean values for the two processes may tend to obscure the study of variations in the variables. For studying the interrelationships of changes in variables, it is desirable to strip the variables of their mean values so that only deviations from the means are of interest. For that purpose, the **covariance** is of primary interest. The covariance function will be denoted as cov(x, y)and is defined as

$$\text{cov}(x, y) = E\left[(x - \overline{x})(y - \overline{y})\right] \qquad (13\text{-}78)$$

where $\overline{x}$ and $\overline{y}$ are the sample means of the two variables. By subtracting these means from the two variables, only the variations affect the outcomes.

It can be shown that an alternate form for Equation 13-78 is

$$\text{cov}(x, y) = E(xy) - (\overline{x})(\overline{y}) = \text{corr}(x, y) - (\overline{x})(\overline{y}) \qquad (13\text{-}79)$$

Thus, the covariance and correlation functions for two variables differ only by the products of the means.

Normalized Covariance or Correlation Coefficient

Finally, we need to carry this process one more step. A covariance value of 10 for one set of variables may show a greater relative correlation than a value of 100 for another set of variables. Said differently, the correlation and covariance functions may vary considerable in magnitudes depending on the levels of the variables involved. In the final analysis, it is convenient to normalize the variables in a form that can be interpreted the same way irrespective of the actual levels involved

For this purpose, let us define the **normalized covariance** or **correlation coefficient** of two variables x and y as $C(x, y)$. It is defined as

$$C(x, y) = \frac{E\left[(x - \overline{x})(y - \overline{y})\right]}{\sigma_x \sigma_y} = \frac{\text{cov}(x, y)}{\sqrt{\text{cov}(x, x)\,\text{cov}(y, y)}} \tag{13-80}$$

The value is a number bounded between -1 and 1 with the following interpretations:

1. If $C(x, y) = 1$, the two variables are totally correlated in a **positive** sense. This means that y always has a positive direct relationship to x. It does not necessarily mean that y is a linear function of x, but it means that if either variable is specified, the other will have a unique value.

2. If $C(x, y) = -1$, the two variables are totally correlated in a **negative** sense. This means that y always has some type of inverse relationship to x, meaning that an increase of x results in a decrease in y in some sense. Again, the exact relationship may assume different forms, but a positive change in one will result in a unique negative change in the other.

3. If $C(x, y) = 0$, the two variables are said to be ***uncorrelated***. This means that there is no measurable relationship between the two variables and neither has an effect on the other.

Based on the extremes listed, the relative relationship may be inferred by the value. Thus, a correlation coefficient near 1 would imply a very strong positive correlation and a correlation coefficient near -1 would suggest a very strong negative correlation. Finally a relationship near 0 would imply a very low degree of correlation.

Let us illustrate the preceding concepts with some examples. The number of shovels sold in a city would probably be highly correlated in a positive sense with the number of inches of snowfall in a given time interval. The number of heavy jackets sold would probably be highly correlated in a negative sense with high temperature increases. Finally, the number of dishes sold would probably show virtually zero correlation with the number of inches of rainfall.

MATLAB Covariance for Two Variables

In the case of two scalar variables, x and y, the correlation coefficient matrix C may be determined by the command

$$>> C = \text{corrcoef}(x,y); \qquad\qquad (13\text{-}81)$$

The result is a 2x2 matrix with the following format:

$$C = \begin{bmatrix} c_{11} & c_{12} \\ c_{21} & c_{22} \end{bmatrix} \qquad\qquad (13\text{-}82)$$

The coefficient c_{11} is the expected value of the product of x with itself divided by the product of σ_x and itself. Therefore, this value is necessarily $c_{11} = 1$, and similarly $c_{22} = 1$. These two values constitute the main diagonal of the matrix. There is symmetry about the main diagonal in that $c_{12} = c_{21}$ and this value constitutes the normalized covariance or correlation coefficient between the variables x and y.

Normalized Covariance Matrix for More than Two Variables

When there are more than two variables, the covariance matrix will contain values representing both normalized self and cross-correlation terms. Let M represent a $k \times m$ matrix in which each k element column represents the sample values of a different variable. The correlation coefficient matrix is generated in MATLAB by the following command:

$$>> C = \text{corrcoef}(M) \qquad\qquad (13\text{-}83)$$

The result will be a square matrix of size $m \times m$. The pattern is basically the same as for the two-element correlation coefficient. Along the main diagonal, the values are

$$c_{11} = c_{22} = ... = c_{mm} = 1 \qquad\qquad (13\text{-}84)$$

All other values are the cross-correlation coefficients and there is symmetry about the main diagonal. For example, $c_{12} = c_{21}$ is the correlation coefficient linking the variable of the first column with the variable of the second column and so on. The concept will be illustrated in the examples that follow.

Correlation versus Cause and Effect

Does a high degree of positive correlation between an assumed independent variable x and a dependent variable y mean that x tends to cause y? For example, suppose there is a positive correlation between the number of hours of television watched per day by a married couple and the divorce rate. Does that mean that if an individual couple watches more television each day that they have a higher probability of divorce? The answer is "not necessarily." Correlation does not necessarily imply causation. It simply shows some sort of mathematical link based on the data employed but it absolutely does not identify that the independent variable causes or contributes to increased likelihood of the occurrence of the dependent variable in all cases. There may be many other variables that enter into the interrelationship, which is where human judgment and experience are required. Many physical and social scientists earn their living by dealing with such complex issues. All we can do here is to introduce the mathematical tools, but they must be tempered with many other factors in a real-life situation.

Example 13-7

This example will continue studying the results of the eight-person university department of Example 13-6. The relative correlation between the salary y and the five parameters x1, x2, x3, x4, and x5 will be studied. Form the five correlation coefficient matrices based on the correlation of y with each of the five parameters.

Solution

We will denote the correlation coefficient matrix involving y and x1 as Cyx1, the correlation coefficient involving y and x2 as Cyx2, etc. The five matrices will first be calculated and then comments will be made.

$$>> Cyx1 = corrcoef(y,x1) \tag{13-85}$$

Cyx1 =

1.0000 0.2589

0.2589 1.0000

$$>> Cyx2 = corrcoef(y,x2) \tag{13-86}$$

Cyx2 =

1.0000 0.4773

0.4773 1.0000

$$>> Cyx3 = corrcoef(y,x3) \tag{13-87}$$

Cyx3 =

1.0000 0.9396

0.9396 1.0000

$$>> Cyx4 = corrcoef(y,x4) \tag{13-88}$$

Cyx4 =

1.0000 0.9394

0.9394 1.0000

$$>> Cyx5 = corrcoef(y,x5) \qquad (13\text{-}89)$$

$$Cyx5 =$$

1.0000 0.0991

0.0991 1.0000

As expected, each of the matrices has two rows and two columns, the value along the main diagonal is one, and there is symmetry about the main diagonal.

Now for some interesting comments: In spite of the presence of a negative sign in the linear regression formula, all five of the cross-correlation coefficients are positive. This means that a "better" performance for that particular parameter correlates with a higher salary. It might be a startling revelation if either of the constants turned out to be negative, since this might suggest that a "better" performance could lead to a lower salary. (Understand that we are using increasing years of service as a sort of "better" parameter, which could be debated.)

Yes, all the cross-correlation coefficients are positive, but look at their relative magnitudes. For publications and the generation of external funding, the correlation coefficients are very high, at 0.9396 and 0.9394, respectively. The number of years of service has only a moderately positive value of 0.2589, and the number of service activities has the weakest positive effect with a value of only 0.0991.

How about teaching, the indicator that many feel is the most important activity of a university? Well, the value is 0.4773, which certainly is a reasonably good positive correlation parameter. Yet, it remains well below the values for publications and external funding. Yes, the data presented here are hypothetical, but the author's extensive experience at the university level might suggest that they are not too far from reality!

au: is .4773 above correct?

Example 13-8

Combine the results of the preceding two examples into a single correlation coefficient matrix that provides all the results (and more).

Solution

For this purpose, we will form a single matrix M, in which the first column is y and the next five columns are the variable x1 through x5. Since there are eight faculty, the resulting matrix will be an 8×6 matrix. The correlation coefficient matrix will then be a 6×6 matrix. The steps follow.

$$>> M = [y \ x1 \ x2 \ x3 \ x4 \ x5] \qquad (13\text{-}90)$$

$$M =$$

45.0000	1.0000	3.8000	2.0000	0	1.0000
50.0000	3.0000	4.7000	5.0000	22.0000	4.0000
60.0000	4.0000	4.1000	7.0000	60.0000	6.0000
52.0000	5.0000	3.7000	3.0000	25.0000	6.0000
70.0000	8.0000	4.9000	12.0000	120.0000	4.0000
58.0000	12.0000	5.5000	5.0000	50.0000	8.0000
52.0000	15.0000	4.5000	4.0000	0	12.0000
56.0000	18.0000	5.0000	5.0000	20.0000	15.0000

The correlation coefficient matrix is denoted as C. It is

$$>> C = \text{corrcoef(M)} \tag{13-91}$$

$$C =$$

1.0000	0.2589	0.4773	0.9396	0.9394	0.0991
0.2589	1.0000	0.6277	0.0854	-0.0583	0.9330
0.4773	0.6277	1.0000	0.4191	0.3518	0.4304
0.9396	0.0854	0.4191	1.0000	0.9389	-0.0816
0.9394	-0.0583	0.3518	0.9389	1.0000	-0.2377
0.0991	0.9330	0.4304	-0.0816	-0.2377	1.0000

The five coefficients determined in the preceding example are identified as the second through the sixth elements along either the first row or the first column. Other than the main diagonal, all other elements are cross-correlation terms between various independent variables, and some of these values may make little or no sense. For example, the value $C(6,5) = C(5,6) = -0.2377$ is the relative correlation between the volume of external funding and the number of service activities. Well, it might make some sense since those who are heavily involved in external funding may have little time for service activities. At any rate, there is probably more "information" here than any normal person would want to try to absorb.

MATLAB PROBLEMS

For illustration, Example 13-1 was evaluated without the use of MATLAB. However, due to the tedious nature of the procedures of this chapter, the problems provided here can all be evaluated with MATLAB.

13-1. Data concerning an independent variable, x, and a dependent variable, y, are shown in the table that follows.

x	0	0.2	0.4	0.6	0.8	1.0	1.2	1.4	1.6	1.8	2.0	2.2
y	6	5.45	4.93	4.52	4.03	3.53	3.02	2.61	2.03	1.58	1.03	0.62

 a. Use MATLAB to determine the best fitting straight-line equation for y versus x.

 b. Plot the best fitting straight line and the actual data on the same scale, with the latter represented by o's.

13-2. The current i in amperes (A) versus the voltage v in volts (V) for a somewhat non-linear resistor are provided in the table that follows.

v(V)	0	3	6	9	12	15	18	21
i(A)	0	0.5	1.0	1.48	1.90	2.32	2.64	2.98

 a. Use MATLAB to determine the best-fitting straight line equation for i versus v.

 b. Plot the best-fitting straight line and the actual data on the same scale, with the latter represented by o's.

13-3. Apply a second-degree fit to the data of Problem 13-1 and plot it along with the actual data.

13-4. Apply a second-degree fit to the data of Problem 13-2 and plot it along with the actual data.

13-5. Consider the function $y = \cos \pi t$ for $-1 \le t \le 1$. Based on a step size of 0.05 between points, determine curve fits for the following degrees: (a) 1, (b) 2, (c) 3, (d) 4. For each case, plot the approximation and the actual points, with the latter represented by o's. What can you conclude about the terms obtained using a polynomial of *odd degree* in determining a fit for an *even* function?

13-6. Consider the function $y = e^{-x^2/2}$ for $-3 \le x \le 3$. Based on a step size of 0.1 between points, determine curve fits of the following degrees: (a) 2, (b) 4, (c) 6. For each case, plot the approximation and the actual points, with the latter represented by o's.

13-7. The force f acting on a certain mechanical part is a non-linear function of the distance x from the end. Data concerning some measurements of the force in newtons (N) versus the distance in meters (m) are provided in the table that follows.

x, m	0	1	2.5	3.3	5.7	8
f, N	3.10	4.28	6.91	8.62	15.3	24.6

Determine the best-fitting second-degree equation for f versus x and plot the curve on the same scale with the actual values, with the latter represented by o's. For the approximation, employ a new independent variable, xa, defined over the domain from 0 to 8, but with a much finer resolution than x.

13-8. The input voltage v_1 and the output voltage v_2 in a certain electronic device is a non-linear function. Some measurements of both variables in volts (V) are provided in the table that follows.

v_1, V	0	1	2.3	5.7	6.8	8.2
v_2, V	0	3.84	8.52	18.9	21.4	25.0

Determine the best-fitting second-degree equation for v_2 versus v_1 and plot the curve on the same scale with the actual values, with the latter represented by o's. For the approximation, employ a new independent variable v1a defined over the domain from 0 to 8.2, but with a much finer resolution than v1.

Note: MATLAB contains a file with the title **census.mat**, which contains the population of the United States in the years 1790 through 1990 at 10-year intervals. The data may be loaded into MATLAB by the command **load census**. Two 21×1 column matrices are loaded into the Command Window. One vector is denoted as **cdate** and contains the values 1790, 1800, 1810, etc. The other vector is denoted as **pop** and contains the corresponding population in *millions of people* for each of the years in the previous vector.

13-9. Plot the population as a function of the actual years from 1790 through 1990. While the data are actually discrete, you may utilize a continuous curve for convenience.

13-10. Theoretically, it would be possible to fit the population of Problem 13-9 to a mathematical equation, with the independent variable assuming the actual value of the year and the dependent variable assuming the value of the population. In practice, this process is somewhat difficult due to the large values for the years and the limited spread, and a "poorly conditioned" warning may result. Instead, the independent variable can be represented by a value x representing the year measured from 1790. Thus x = 0 is actually 1790, and each step of x represents a 10-year increment. The last value, x = 200, corresponds to 1990. Since the data are arranged in columns, x should be defined as a column vector. It is recommended that the display format be changed by the command **format e**. After completing this problem, the default format may be restored by the command **format**.

    a. Determine a first-degree equation fit based on the shifted independent variable x.

    b. Plot a continuous curve obtained from the first-degree fit along with the original data as specific points represented by o's. The actual horizontal variable can now be the original **cdate**.

    c. Determine the maximum value of the difference between the two curves.

13-11. Repeat Problem 13-10 with a second-degree fit.

13-12. Repeat Problem 13-10 with a third-degree fit.

Note: MATLAB contains a file with the name **count.dat** that contains the results of some traffic observations at three locations over a 24-hour period. The data are organized in a 24 × 3 matrix in which each column represents the number of cars passing that particular intersection, and each row represents the number of cars observed over a particular hour starting at midnight and continuing for one full day. For example, count (8,2) represents the number of vehicles observed in the 8th hour at Location 2. This would correspond to the time from 7 A.M. to 8 A.M.

Load the data into the Command Window by the command **load count.dat**. These data will be used in Problems 13-13 and 13-14. Define the independent variable t as the hour varying from 1 to 24 in steps of 1.

13-13. Plot the three curves of traffic volume as a function of x. Label them simply as #1, #2, and #3.

13-14. Determine the nine correlation coefficients for the data. What can you conclude about the relative correlation of traffic in this city?

Introduction to Spatial Vector Analysis

14

14.1 OVERVIEW AND OBJECTIVES

The term **vector** has slightly different meanings in different areas of mathematics, engineering, and science. Throughout the text thus far, the term has been used to refer to row or column matrices according to the standard conventions of matrix algebra, and these conventions are in turn employed by MATLAB.

There is another widely used definition for **vector**, which is associated with spatial quantities that have specific directions in terms of the three-dimensional coordinate system in which we live. Examples of such quantities are forces, velocities, displacements, electric fields, magnetic fields, and many other physical variables. These variables have appeared throughout the text, but we have always defined the directions in scalar terms and have used the term *vector* only as defined for matrix and MATLAB purposes.

The use of the term *vector* in regard to spatial quantities is not mutually exclusive from the use with matrix theory and MATLAB. As we will see, a three-dimensional spatial vector can be represented in terms of a row or column matrix with three elements. However, there are certain mathematical processes and conventions that are peculiar to each type. Therefore, the term *spatial vector* will be used in the context of those vectors that have an orientation in space when it is necessary to make a distinction.

For the purposes of this chapter, a *spatial vector* will be defined as a quantity that has both a magnitude and a direction. Since the focus throughout the chapter will be on spatial vectors, the adjective *spatial* will often be omitted. At any point at which a MATLAB vector is created, the terms *row vector* and *column vector* will be used as appropriate.

Objectives

After completing this chapter, the reader should be able to

1. Describe the *rectangular coordinate system.*
2. Express a three-dimensional vector in terms of the rectangular coordinate system.
3. Determine the *magnitude* or *absolute value* of a vector in terms of the three components.
4. Determine the *direction angles* in terms of the projections on the three axes.

5. Determine the projections on the three axes in terms of the magnitude and direction angles.
6. Define and calculate the *scalar* or *dot product* of two vectors and discuss its physical interpretation.
7. Define and calculate the *vector* or *cross product* of two vectors and discuss its physical interpretation.
8. Explain the right-hand rule.
9. Define and calculate the *triple scalar product* of three vectors and discuss its physical interpretation.
10. Discuss some of the applications of spatial vector analysis.
11. Use MATLAB to perform some of the basic vector operations.

14-2   RECTANGULAR COORDINATES

Our sense of the universe is based on the awareness of *space* as it exists in three mutually perpendicular dimensions. Therefore, most spatial vectors are based on defining a coordinate system in which a vector quantity may be uniquely specified.

Rectangular Coordinate System

The most logical, and easiest to comprehend, coordinate system is the *rectangular coordinate system*. All of this chapter will be devoted to that system. However, there are other useful coordinate systems that are covered in more advanced books, two of which are the *cylindrical coordinate system* and the *spherical coordinate system.*

The *rectangular coordinate* system employs the three basic dimensions of x, y, and z, as illustrated in Figure 14-1. The three axes are mutually perpendicular to each other. The system is called a *right-hand* coordinate system for the following reason: Take your right hand with the thumb pointed outward and the fingers pointed in the x direction. Then rotate the fingers from x to y, and the thumb will point in the direction of z.

Some applications of spatial vector analysis are easier to visualize by assuming only two dimensions for illustration purposes. When this is done, the choice will usually be to use the x and y coordinates, and the resulting two-dimensional system will be referred to as the *x-y plane*. One could also use a two-dimensional system based on either the *y-z plane* or the *x-z plane*.

Vector Notation

Vectors will be identified with nonitalicized, boldface type. For example, a force vector could be denoted as **F**.

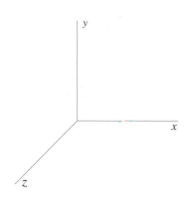

Figure 14-1   Rectangular coordinate system in three dimensions

Unit Vectors

It is convenient in spatial vector analysis to utilize *unit vectors* along the three axes. These vectors are assumed to have a length of 1 in whatever units are being considered, but they "tag" the elements of the vector. One of the most common schemes is defined by the following references:

Axis	Unit Vector
x	**i**
y	**j**
z	**k**

Note that the unit vectors are also expressed in non-italicized bold print. Some books use notation such as $\mathbf{u}_x$, $\mathbf{u}_y$, and $\mathbf{u}_z$ for the three unit vectors, but the choices used here are simpler in form and probably more commonly used.

A three-dimensional representation displaying the three unit vectors **i, j,** and **k** is shown in Figure 14-2. Clearly, the format provides a unique and meaningful way of describing the vector quantity.

Association with Complex Number Theory

Don't confuse the unit spatial vectors **i** and **j** with the imaginary numbers i or j in complex number theory. There are certain similarities in that i (or j) is a vector-like quantity and some references even define complex numbers as vectors in two dimensions. However, there are some significant differences in their meanings and the corresponding mathematical operations. In fact, electric and magnetic fields are often represented in terms of complex numbers and they are simultaneously spatial vectors, meaning that both spatial vector analysis and complex variable theory are used in their analysis.

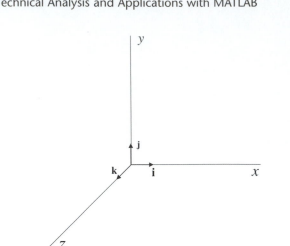

Figure 14-2   Unit vectors in rectangular coordinate system

The context of subsequent developments should clarify whether spatial vectors or complex numbers are being considered. Of course, the styles of the symbols will be different and that should insure that they aren't confused with each other.

Vector Representation

Consider an arbitrary vector **A** as illustrated in Figure 14-3. Assume that the three projections along the three axes are A_x, A_y, and A_z, respectively. This vector can then be represented in rectangular form as

$$\mathbf{A} = A_x\mathbf{i} + A_y\mathbf{j} + A_z\mathbf{k} \tag{14-1}$$

The orientation (but not the length) of a vector can also be described in terms of the *direction angles*. The three angles in Figure 14-3 are denoted as ϕ_x, ϕ_y, and ϕ_z, respectively. Each angle is measured between the vector and the corresponding axis denoted by the subscript.

Magnitude or Absolute Value of a Vector

The magnitude of **A** will be denoted as A. Since the three axes are mutually perpendicular, the magnitude is

$$A = |\mathbf{A}| = \sqrt{A_x^2 + A_y^2 + A_z^2} \tag{14-2}$$

The vertical bars surrounding the vector indicate "absolute value of," which is equivalent to magnitude.

Generally, the magnitude of a spatial vector represents the net effect in the direction of the vector. For example, if the vector is a force, the result would be the net force in the

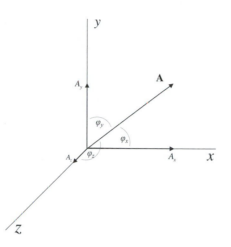

Figure 14-3 Vector showing direction angles and the three components

direction of the vector. If the vector represents a distance quantity, the magnitude is the net length of the distance.

Direction Angles

The cosine of each direction angle may be determined by forming the ratio of the projection on the given axis to the net length. Hence, expressions for the three cosines associated with **A** are as follows:

$$\cos \phi_x = \frac{A_x}{A} = \frac{A_x}{\sqrt{A_x^2 + A_y^2 + A_z^2}} \tag{14-3}$$

$$\cos \phi_y = \frac{A_y}{A} = \frac{A_y}{\sqrt{A_x^2 + A_y^2 + A_z^2}} \tag{14-4}$$

$$\cos \phi_z = \frac{A_z}{A} = \frac{A_z}{\sqrt{A_x^2 + A_y^2 + A_z^2}} \tag{14-5}$$

The corresponding angles may then be determined from the inverse cosines of the three values.

Projections on the Axes

If the length A and the direction angles are known, the three projections may be readily determined by changing the forms of the previous three equations. Thus,

$$A_x = A\cos\phi_x \qquad (14\text{-}6)$$

$$A_y = A\cos\phi_y \qquad (14\text{-}7)$$

$$A_z = A\cos\phi_z \qquad (14\text{-}8)$$

Example 14-1

A force has components in the x, y, and z directions of 3, 4, and –12 N, respectively. Express the force as a vector, **F**, in rectangular coordinates.

Solution

The vector is given by
$$\mathbf{F} = 3\mathbf{i} + 4\mathbf{j} - 12\mathbf{k}$$

$$(14\text{-}9)$$

The negative sign for the last term indicates that the z component of force is in the negative direction.

Example 14-2

Determine the magnitude of the force in Example 14-1.

Solution

The magnitude F is given by
$$F = \sqrt{(3)^2 + (4)^2 + (-12)^2} = 13 \text{ N}$$

$$(14\text{-}10)$$

As far as the magnitude is concerned, it would not make any difference whether –12 or 12 were entered in the formula since the squares of both are the same. However, it makes a significant difference in the direction.

Example 14-3

Determine the three direction angles for the force of Examples 14-1 and 14-2.

Solution

The pertinent relationships are Equations 14-3, 14-4, and 14-5. First, each ratio is formed, and then the inverse cosine is determined.

$$\cos\phi_x = \frac{A_x}{A} = \frac{3}{13} = 0.2308 \qquad (14\text{-}11)$$

$$\phi_x = \cos^{-1} 0.2308 = 76.66° = 1.338 \text{ rad} \qquad (14\text{-}12)$$

$$\cos\phi_y = \frac{A_y}{A} = \frac{4}{13} = 0.3077 \tag{14-13}$$

$$\phi_y = \cos^{-1}0.3077 = 72.08^\circ = 1.258 \text{ rad} \tag{14-14}$$

$$\cos\phi_z = \frac{A_z}{A} = \frac{-12}{13} = -0.9231 \tag{14-15}$$

$$\phi_z = \cos^{-1}(-0.9231) = 157.4^\circ = 2.747 \text{ rad} \tag{14-16}$$

Note that since the z-component is negative, the angle of the vector with respect to the positive z axis is greater than 90°. One could, of course, measure the angle with respect to the negative z axis, in which case the angle would be $180.0° - 157.4° = 22.6°$.

14-3   SCALAR OR DOT PRODUCT

We will now consider the scalar product or dot product of two vectors. Along with vector **A** as defined in the previous section, assume a vector **B** defined by

$$\mathbf{B} = B_x\mathbf{i} + B_y\mathbf{j} + B_z\mathbf{k} \tag{14-17}$$

Assume that there is an angle θ between the vectors, as illustrated in Figure 14-4.

The scalar or dot product of two vectors is defined as

$$\mathbf{A} \cdot \mathbf{B} = AB\cos\theta \tag{14-18}$$

Physical Interpretation

There are two equally valid interpretations of the dot product, which are illustrated in Figure 14-5. The projection of vector **A** on vector **B** is given by $A\cos\theta$, which is illustrated in part (a). The product of this projection times the length B is the dot product. Alternately, the projection of vector **B** on vector **A** is given by $B\cos\theta$,

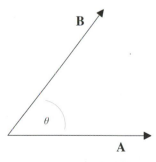

Figure 14-4   Two spatial vectors oriented at different angles

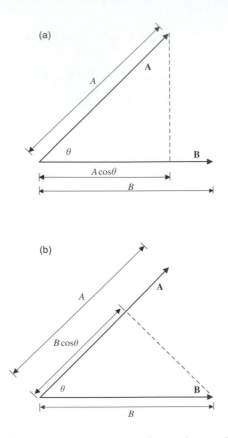

Figure 14-5   Two ways to interpret the scalar or dot product.

which is illustrated in part (b). The product of this projection times the length A produces the same result as in (a).

Two limiting cases are as follows:

1. $\theta = 0°$. In this case, the two vectors are parallel to each other and $\cos\theta = 1$. Therefore,

$$\mathbf{A} \cdot \mathbf{B} = AB \tag{14-19}$$

2. $\theta = 90°$. In this case, the vectors are perpendicular to each other and $\cos\theta = 0$. Therefore,

$$\mathbf{A} \cdot \mathbf{B} = 0 \tag{14-20}$$

The converses of the preceding two conditions are also true.

1. If $\mathbf{A} \cdot \mathbf{B} = AB$, the two vectors are parallel to each other.
2. If $\mathbf{A} \cdot \mathbf{B} = 0$ and neither A nor $B = 0$, the two vectors are perpendicular to each other.

Formula for Determining Scalar or Dot Product

The formula for determining the scalar product is the following:

$$\mathbf{A} \cdot \mathbf{B} = A_x B_x + A_y B_y + A_z B_z \qquad (14\text{-}21)$$

Thus, corresponding elements of the vectors are multiplied together and the sum of the products is formed.

Identities

The scalar product obeys the commutative, associative, and distributive laws in their simplest forms. Thus,

$$\mathbf{A} \cdot \mathbf{B} = \mathbf{B} \cdot \mathbf{A} \qquad (14\text{-}22)$$

$$\mathbf{A} \cdot \mathbf{B} \cdot \mathbf{C} = (\mathbf{A} \cdot \mathbf{B}) \cdot \mathbf{C} = \mathbf{A} \cdot (\mathbf{B} \cdot \mathbf{C}) \qquad (14\text{-}23)$$

$$\mathbf{A} \cdot (\mathbf{B} + \mathbf{C}) = \mathbf{A} \cdot \mathbf{B} + \mathbf{A} \cdot \mathbf{C} \qquad (14\text{-}24)$$

where **C** is a third vector. Without showing all the possible combinations, it can be stated that the order for performing a series of dot products is immaterial from a theoretical point of view.

Example 14-4

Two vectors are given by

$$\mathbf{A} = 2\mathbf{i} - 2\mathbf{j} + \mathbf{k} \qquad (14\text{-}25)$$

$$\mathbf{B} = 3\mathbf{i} + 4\mathbf{j} + 12\mathbf{k} \qquad (14\text{-}26)$$

Determine (a) the length of **A**, (b) the length of **B**, (c) the scalar product **A** • **B**, and (d) the angle θ between the two vectors.

Solution

a. The length of **A**, denoted simply as A, is given by

$$A = \sqrt{A_x^2 + A_y^2 + A_z^2} = \sqrt{(2)^2 + (-2)^2 + (1)^2} = 3 \qquad (14\text{-}27)$$

b. The length of **B**, is given by

$$B = \sqrt{B_x^2 + B_y^2 + B_z^2} = \sqrt{(3)^2 + (4)^2 + (12)^2} = 13 \qquad (14\text{-}28)$$

c. (Yes, this example was "rigged" to yield simple numbers for the vector lengths!) The dot product is given by

$$\mathbf{A} \cdot \mathbf{B} = A_x B_x + A_y B_y + A_z B_z = (2)(3) + (-2)(4) + (1)(12) = 10 \qquad (14\text{-}29)$$

d. To determine the angle between the two vectors, we first inspect the basic relationship of Equation 14-15.

$$\mathbf{A} \cdot \mathbf{B} = AB \cos \theta \qquad (14\text{-}30)$$

Solving for $\cos \theta$, we obtain

$$\cos \theta = \frac{\mathbf{A} \bullet \mathbf{B}}{AB} = \frac{10}{3 \times 13} = \frac{10}{39} = 0.2564 \tag{14-31}$$

The angle θ is then determined as

$$\theta = \cos^{-1} 0.2564 = 75.14° = 1.311 \text{ rad} \tag{14-32}$$

14-4   VECTOR OR CROSS PRODUCT

The **vector product** or **cross product** of two vectors is somewhat more involved than the scalar product and is defined as

$$\mathbf{A} \times \mathbf{B} = (AB \sin \theta) \mathbf{u_n} \tag{14-33}$$

where A, B, and θ have the same meanings as for the scalar product and $\mathbf{u_n}$ is a unit vector perpendicular to the plane containing $\mathbf{A}$ and $\mathbf{B}$. Thus, *while the dot product yields a scalar, the cross product yields a new vector perpendicular to the plane containing the vectors in the product.*

The cross product arises in a number of applications in the areas of mechanics and electricity. It describes certain phenomena in which the effect tends to be perpendicular to the physical variables causing the effect. Some limiting cases are the following:

$\theta = 0°$. In this case, the two vectors are parallel to each other and $\sin \theta = 0$. Therefore,

$$\mathbf{A} \times \mathbf{B} = 0 \tag{14-34}$$

$\theta = 90°$. In this case, the vectors are perpendicular to each other and $\sin \theta = 1$. Therefore,

$$\mathbf{A} \times \mathbf{B} = (AB) \mathbf{u_n} \tag{14-35}$$

As in the case of the scalar product, the converses are also true.

1. If $\mathbf{A} \times \mathbf{B} = 0$ and neither A nor $B = 0$, the two vectors are parallel to each other.
2. If $\mathbf{A} \times \mathbf{B} = AB$, the two vectors are perpendicular to each other.

In some sense, the properties of dot and cross products are opposite to each other. A dot product yields a scalar and a cross product yields a vector. A dot product has its maximum value when the two vectors have the same direction, while a cross product has its maximum value when the two vectors are mutually perpendicular. Conversely, a dot product is zero if the vectors are perpendicular but a cross product is zero when the vectors have the same direction.

Formula for Determining Vector Product

The formula for determining the vector product of two vectors in rectangular coordinates can be determined by evaluating the determinant that follows.

$$\mathbf{A} \times \mathbf{B} = \begin{vmatrix} \mathbf{i} & \mathbf{j} & \mathbf{k} \\ A_x & A_y & A_z \\ B_x & B_y & B_z \end{vmatrix} \tag{14-36}$$

The best way to evaluate this determinant is to expand it in minors and cofactors along the top row. The result is

$$\mathbf{A} \times \mathbf{B} = \left(A_y B_z - A_z B_y \right) \mathbf{i} - \left(A_x B_z - A_z B_x \right) \mathbf{j} + \left(A_x B_y - A_y B_x \right) \mathbf{k} \tag{14-37}$$

Unit Vector

The direction of the unit vector is the same as that of the cross product. The actual unit vector can be determined by dividing or normalizing the cross product by its magnitude. Hence,

$$\mathbf{u}_n = \frac{\mathbf{A} \times \mathbf{B}}{|\mathbf{A} \times \mathbf{B}|} \tag{14-38}$$

where the denominator is a scalar representing the magnitude of the cross product vector.

Right-Hand Rule for Cross Products

The direction of the vector resulting from a cross product can be determined by the right-hand rule discussed earlier. Imagine a plane containing the vectors **A** and **B**. If the fingers of the right hand are rotated from the vector **A** toward the vector **B**, the thumb will point in the direction of the cross product vector **A** × **B**.

The preceding discussion means that the order of the quantities in the cross product is important. It can be shown that

$$\mathbf{A} \times \mathbf{B} = -\mathbf{B} \times \mathbf{A} \tag{14-39}$$

This equation indicates that reversing the order of the vectors in the cross product causes the vector to reverse directions. However, the magnitude remains the same.

Two-Dimensional View Using Arrow Heads and Tails

Three-dimensional views can often be a little confusing, and since cross products generally involve three dimensions, there is a way to illustrate the process with two dimensions. Consider the two parts of Figure 14-6 and assume that both the vectors **A** and **B** lie in the plane of the page.

In part (a), the product **A** × **B** is formed, and by the right-hand rule, the vector should be perpendicular to the page and pointed in the direction *out* of the page. This is illustrated by showing a *period* in the circle representing the *tip* of an arrow.

In part (b), the product **B** × **A** is formed. By the right-hand rule, the vector should be perpendicular to the page, but in this case it will be pointed in the direction *into* the page. This situation is illustrated by showing an **X** in a circle representing the *tail* of an arrow.

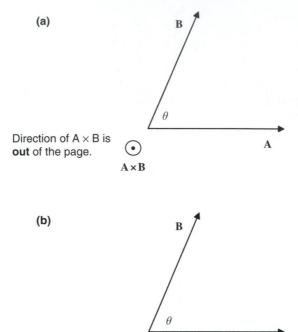

Figure 14-6   Illustration of the cross product with the result perpendicular to the page

Area Interpretation of Cross Product

One interesting geometrical interpretation of the cross product is illustrated by Figure 14-7. Assume that both of the vectors lie in the plane of the page and consider the parallelogram formed as shown. While the cross product vector will be perpendicular to the page, its **magnitude** turns out to represent the area of the parallelogram. Since the magnitude is interpreted as a positive value, the result applies to either $\mathbf{A} \times \mathbf{B}$ or $\mathbf{B} \times \mathbf{A}.$

Example 14-5

Determine the cross product of the two vectors $\mathbf{A}$ and $\mathbf{B}$ of Example 14-4.

Solution

Although we could go directly to the expanded form of Equation 14-37, it is easier to remember the form of Equation 14-36 and work with it. Therefore,

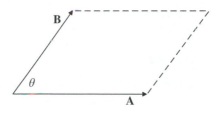

Figure 14-7   Illustration of the parallelogram in which the area is equal to the magnitude of the cross product

$$\mathbf{A} \times \mathbf{B} = \begin{vmatrix} \mathbf{i} & \mathbf{j} & \mathbf{k} \\ A_x & A_y & A_z \\ B_x & B_y & B_z \end{vmatrix} = \begin{vmatrix} \mathbf{i} & \mathbf{j} & \mathbf{k} \\ 2 & -2 & 1 \\ 3 & 4 & 12 \end{vmatrix}$$

(14-40)

Expansion along the first row yields

$$\mathbf{A} \times \mathbf{B} = \left[(-2)(12) - (1)(4)\right]\mathbf{i} - \left[(2)(12) - (1)(3)\right]\mathbf{j} + \left[(2)(4) - (-2)(3)\right]\mathbf{k}$$
$$= -28\mathbf{i} - 21\mathbf{j} + 14\mathbf{k}$$

(14-41)

Example 14-6

Determine a unit vector perpendicular to the vectors **A** and **B** of Example 14-5.

Solution

The magnitude or absolute value of the cross product is

$$|\mathbf{A} \times \mathbf{B}| = \sqrt{(-28)^2 + (-21)^2 + (14)^2} = 37.70$$

(14-42)

The unit vector is then given by

$$\mathbf{u_n} = \frac{\mathbf{A} \times \mathbf{B}}{|\mathbf{A} \times \mathbf{B}|} = \frac{-28i - 21j + 14k}{37.70} = -0.7428\mathbf{i} - 0.5571\mathbf{j} + 0.3714\mathbf{k}$$

(14-43)

The reader is invited to verify that the magnitude of this vector is 1 (ignoring the usual round-off).

14-5   TRIPLE SCALAR PRODUCT

A vector operation that arises in certain applications is the triple scalar product. Three vectors are involved, so along with **A** and **B** (introduced earlier), we will define a vector **C** as

$$\mathbf{C} = C_x\mathbf{i} + C_y\mathbf{j} + C_z\mathbf{k}$$

(14-44)

The triple scalar product is then defined as $(\mathbf{A} \times \mathbf{B}) \bullet \mathbf{C}$. The parentheses can be eliminated, since the only way the operation makes sense is to first form the product of the first two vectors and then perform the dot product of that vector with the third vector. However, we will show the parentheses for clarity.

Formula for Triple Scalar Product

By combining the separate operations of the cross product and the dot product, the triple scalar product can be placed in a determinant form with the following result:

$$(\mathbf{A} \times \mathbf{B}) \bullet \mathbf{C} = \begin{vmatrix} A_x & A_y & A_z \\ B_x & B_y & B_z \\ C_x & C_y & C_z \end{vmatrix} \tag{14-45}$$

This result could be expanded, but it is hardly worth the effort. It is best to work with this form and expand when values are inserted in the determinant.

Identities for Triple Scalar Product

There are several variations of the commutative and associative laws that can be derived for the triple scalar product. Some of these are summarized as follows:

$$(\mathbf{A} \times \mathbf{B}) \bullet \mathbf{C} = (\mathbf{B} \times \mathbf{C}) \bullet \mathbf{A} = (\mathbf{C} \times \mathbf{A}) \bullet \mathbf{B} \tag{14-46}$$

$$\mathbf{A} \bullet (\mathbf{B} \times \mathbf{C}) = -\mathbf{A} \bullet (\mathbf{C} \times \mathbf{B}) \tag{14-47}$$

Volume Interpretation

One interpretation of the triple scalar product is illustrated in Figure 14-8. Assume that a three-dimensional shape is formed by sides perpendicular to the respective three vectors as shown. This shape is called a *parallelepiped.* It turns out that the **magnitude** of the triple scalar product is equal to the volume of the parallelepiped shown.

Example 14-7

Along with the vectors **A** and **B** of the previous several examples, a new vector **C** will be considered in this example. The three vectors are

$$\mathbf{A} = 2\mathbf{i} - 2\mathbf{j} + \mathbf{k} \tag{14-48}$$

$$\mathbf{B} = 3\mathbf{i} + 4\mathbf{j} + 12\mathbf{k} \tag{14-49}$$

$$\mathbf{C} = 3\mathbf{i} + 5\mathbf{j} - 6\mathbf{k} \tag{14-50}$$

Determine the triple scalar product (**A** × **B**)•**C**.

Solution

$$(\mathbf{A} \times \mathbf{B}) \bullet \mathbf{C} = \begin{vmatrix} A_x & A_y & A_z \\ B_x & B_y & B_z \\ C_x & C_y & C_z \end{vmatrix} = \begin{vmatrix} 2 & -2 & 1 \\ 3 & 4 & 12 \\ 3 & 5 & -6 \end{vmatrix} \tag{14-51}$$

$$= 2(-24 - 60) + 2(-18 - 36) + (15 - 12)$$

$$= -168 - 108 + 3 = -273$$

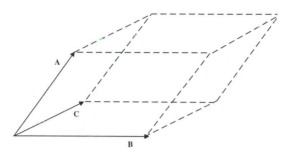

Figure 14-8 Illustration of the parallelepiped in which the volume is equal to the magnitude of the triple scalar product

The interested reader might want to check this result by separately forming the cross product of the first two vectors followed by forming the dot product of that result with the third vector.

14-6 APPLICATIONS OF SPATIAL VECTORS

There are numerous applications of spatial vector analysis in engineering and the sciences. Many of the operations with which the reader may already be familiar in scalar terms can be placed in vector forms. Using scalar forms, quite a few formulas are given using a different version of the "right-hand rule" covered here as well as a "left-hand rule." One of the great advantages of vector analysis is that many of these operations can be placed in more coherent forms by utilizing the rules of vector analysis. We will sample a few of these applications in this section.

Work and Energy

The work performed by a force moving an object through a distance in the direction of the force is the product of the force times the distance. Consider a more complex situation in which the force has an arbitrary orientation with respect to the distance in which it may be moved. This process can be readily described in vector terms.

Let $\mathbf{F}$ represent a constant force vector and let $\mathbf{L}$ represent a vector path length over which the work W is performed.

$$W = \mathbf{F} \cdot \mathbf{L} \qquad (14\text{-}52)$$

This result is equivalent to saying that we multiply the force in the direction of the path times the path length to determine the work done. A conclusion is that there is no work performed if the force is perpendicular to the path length.

Assume now that the vector force varies with the position along the path. In this case, a differential of work dW can be determined at a given point by forming the dot product of the force with a vector differential $\mathbf{dL}$. We can express this increment of work as

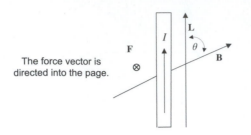

Figure 14-9   Force on current carrying conductor

$$dW = \mathbf{F} \cdot \mathbf{dL} \tag{14-53}$$

The net work can then be determined by performing a definite integral on both sides of this equation.

Force on Conductor Moving in Magnetic Field

Assume that a conductor of vector length **L** has a current I flowing in it, as illustrated in Figure 14-9. Assume that the magnetic flux density is constant and given by **B**, in which the units are tesla (T) or webers/meter^2 (Wb/m^2). The vector force **F** on the conductor is given by

$$\mathbf{F} = I\mathbf{L} \times \mathbf{B} \tag{14-54}$$

This equation serves as the basis for all motor action. It is typically taught in early courses in scalar terms with the understanding that the maximum force is generated when the current flow is perpendicular to the magnetic field and the resulting force is perpendicular to both.

Torque

Torque is a measure of the tendency of a force to produce a rotation about an axis. For example, the process of removing the nuts from a wheel to change a tire requires a considerable amount of torque (at least in human terms). We know that the longer the handles of the wrench, the smaller the amount of force required to produce the torque. Therefore, torque is generally measured as the product of a force and a distance.

Consider the situation depicted in Figure 14-10, in which a force vector, **F,** is acting on some vector lever arm, **L,** that is anchored in the center of the small circle. The torque vector, **T,** is given by

$$\mathbf{T} = \mathbf{L} \times \mathbf{F} \tag{14-55}$$

The torque vector is defined in a direction perpendicular to the plane containing the lever arm and the force vector. Note that the torque is maximum when the force is perpendicular to the lever arm.

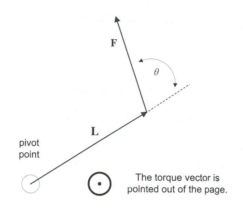

Figure 14-10   Force vector acting at a distance from a pivot point,
resulting in a torque vector

Voltage Induced in Moving Conductor

Assume that a conductor of vector length **L** is moving with vector velocity **v** through a magnetic field vector **B**. The voltage, V, measured across the length is given by the triple scalar product, as follows:

$$V = \mathbf{v} \times \mathbf{B} \cdot \mathbf{L} \qquad (14\text{-}56)$$

This result is the basis for all electrical generator action. As in the case of the motor equation, it is usually taught in basic courses with the concept that the magnetic field is perpendicular to the conductor and its velocity while the voltage induced across the conductor is mutually perpendicular to both.

In the event that the magnetic flux density varies with the position along the conductor, a differential form may be expressed as

$$dV = \mathbf{v} \times \mathbf{B} \cdot \mathbf{dL} \qquad (14\text{-}57)$$

The net voltage may then be determined by integrating both sides of this equation along the length of the conductor.

14-7   MATLAB SPATIAL VECTOR OPERATIONS

This section will show how the various vector operations may be performed with MATLAB.

MATLAB Dot Product

The perceptive reader may readily see that the scalar or dot product operation given by Equation 14-21 is equivalent to the array product as applied to MATLAB. Therefore, assume that the *spatial vectors* **A** and **B** are defined in the format of matrix or MATLAB row vectors as

$$>> A = [Ax\ Ay\ Az] \tag{14-58}$$

$$>> B = [Bx\ By\ Bz] \tag{14-59}$$

At this point the definitions of matrix vectors and spatial vectors have come together. We could then evaluate the dot product by using the procedure for array products following by the use of the **sum** command. The latter command performs an algebraic sum of all elements in the vector. For convenience, we will call the result of the dot product P_dot in the equation that follows. Thus,

$$>> P\_dot = sum(A.*B) \tag{14-60}$$

This equation certainly adds an additional association of the operation with the designation "dot."

MATLAB also provides an equivalent operation **dot(A, B).** The following command can be used:

$$>> P\_dot = dot(A, B) \tag{14-61}$$

MATLAB Cross Product

MATLAB offers a special command for generating the cross product, which is denoted as **cross(A, B)**. Letting P_cross represent the cross product in this case, we have

$$>> P\_cross = cross(A,B) \tag{14-62}$$

Triple Scalar Product

The triple scalar product can be determined with MATLAB using the determinant operation. Along with the row vectors A and B previously defined, consider a third row vector C. Letting P_triple represent the pertinent product, the operation $(\mathbf{A} \times \mathbf{B}) \bullet \mathbf{C}$ can be determined as follows:

$$>> P\_triple = det([A;\ B;\ C;]) \tag{14-63}$$

The inner brackets define a 3 by 3 matrix in which each row is one of the three vectors. The determinant of the resulting vector is then formed, and the scalar value obtained is the triple scalar product.

Example 14-8

Determine the product of Example 14-4 using MATLAB.

Solution

The two vectors are repeated here for convenience.

$$\mathbf{A} = 2\mathbf{i} - 2\mathbf{j} + \mathbf{k} \tag{14-64}$$

$$\mathbf{B} = 3\mathbf{i} + 4\mathbf{j} + 12\mathbf{k} \tag{14-65}$$

We define two row vectors as

$$>> A = [2\ \text{-}2\ 1]; \tag{14-66}$$

$$>> B = [3\ 4\ 12]; \tag{14-67}$$

Let P_dot represent the dot product. It can be evaluated as

$$>> \text{P\_dot} = \text{sum(A.*B)} \tag{14-68}$$

P_dot =

10

Alternately, the **dot** command can be used.

$$>> \text{P\_dot} = \text{dot(A,B)} \tag{14-69}$$

P =

10

Example 14-9

Determine the cross product of Example 14-5 using MATLAB.

Solution

The vectors were repeated in Example 14-8. Letting P_cross represent the product, we have

$$>> \text{P\_cross} = \text{cross(A,B)} \tag{14-70}$$

P_cross =

-28 -21 14

The three components obtained are the *x, y,* and *z* components, in order.

Example 14-10

Determine the triple scalar product of Example 14-7 using MATLAB.

Solution

The third vector required is repeated here for convenience.

$$\mathbf{C} = 3\mathbf{i} + 5\mathbf{j} - 6\mathbf{k} \tag{14-71}$$

The MATLAB row vector is

$$>> C = [3\ 5\ \text{-}6]; \tag{14-72}$$

Letting P_triple represent the triple scalar product, we have

$$>> \text{P\_triple} = \det([A;\ B;\ C]) \tag{14-73}$$

P_triple =

-273

GENERAL PROBLEMS

14-1. A velocity has components in the x, y, and z directions of 8, –6, and 10 m/s, respectively. Express the velocity as a vector **v** in rectangular coordinates.

14-2. An electric field has components in the x, y, and z directions of –2, –4, and 3 V/m, respectively. Express the electric field as a vector **E** in rectangular coordinates.

14-3. Determine the magnitude of the velocity vector of Problem 14-1.

14-4. Determine the magnitude of the electric field vector of Problem 14-2.

14-5. Determine the three direction angles for the velocity of Problems 14-1 and 14-3.

14-6. Determine the three direction angles for the electric field of Problems 14-2 and 14-4.

In Problems 14-7 through 14-26, one or more of the following four vectors will be employed:

$$\mathbf{A} = 5\mathbf{i} + 2\mathbf{j} + 4\mathbf{k}$$
$$\mathbf{B} = 3\mathbf{i} - 6\mathbf{j} + 7\mathbf{k}$$
$$\mathbf{C} = 2\mathbf{i} + 3\mathbf{j} - 5\mathbf{k}$$
$$\mathbf{D} = -\mathbf{i} - \mathbf{j} - \mathbf{k}$$

14-7. Determine (a) the length of **A**, (b) the length of **B**, (c) the scalar product **A•B**, and (d) the angle θ between the two vectors.

14-8. Determine (a) the length of **C**, (b) the length of **D**, (c) the scalar product **C** • **D**, and (d) the angle θ between the two vectors.

14-9. Determine the cross product **A** × **B**.

14-10. Determine the cross product **C** × **D**.

14-11. Determine (**A** × **B**) × **C**.

14-12. Determine (**C** × **D**) × **A**.

14-13. Determine **A** × (**B** × **C**). Compare your answer with Problem 14-11. What can you conclude?

14-14. Determine **C** × (**D** × **A**). Compare your answer with Problem 14-12. What can you conclude?

14-15. Determine the triple scalar product (**A** × **B**) • **C**.

14-16. Determine the triple scalar product (**C** × **D**) • **A**.

MATLAB PROBLEMS

14-17. Solve Problem 14-7 using MATLAB.

14-18. Solve Problem 14-8 using MATLAB.

14-19. Solve Problem 14-9 using MATLAB.

14-20. Solve Problem 14-10 using MATLAB.

14-21. Solve Problem 14-11 using MATLAB.

14-22. Solve Problem 14-12 using MATLAB.

14-23. Solve Problem 14-13 using MATLAB.

14-24. Solve Problem 14-14 using MATLAB.

14-25. Solve Problem 14-15 using MATLAB.

14-26. Solve Problem 14-16 using MATLAB.

Complex Numbers

15

15-1 OVERVIEW AND OBJECTIVES

Some complex numbers have been used at several places earlier in the text, and it was assumed that the reader had some basic knowledge to deal with the limited operations involved there. However, the subject will now be approached from a much broader perspective and many of the operations associated with complex numbers will be developed in some detail. To deal effectively with Fourier analysis, as presented in Chapter 16, complex number manipulations should be well understood.

After providing the basic analytical concepts, the various MATLAB operations with complex numbers will be covered.

Objectives

After completing this chapter, the reader should be able to

1. State the rectangular form of a complex number and show it in the complex plane.
2. State the polar form of a complex number and show it in the complex plane.
3. Convert a complex number from rectangular form to polar form, or vice versa.
4. Perform addition and subtraction of complex numbers and show the processes in the complex plane.
5. Perform multiplication and division of complex numbers using either rectangular or polar forms.
6. Define the complex conjugate of a complex number.
7. Perform exponentiation operations with a complex number.
8. Use MATLAB to perform all the complex number operations previously discussed.
9. Discuss some of the applications of complex variable theory.
10. Discuss how steady-state AC circuit analysis is performed with complex numbers.

15-2 FORMS FOR COMPLEX NUMBERS

In a sense, complex numbers are two-dimensional vectors. In fact, some of the basic arithmetic operations such as addition and subtraction are the same as those that could be per-

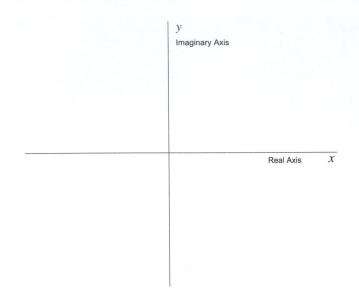

Figure 15-1   Complex plane showing real and imaginary axes

formed with the spatial vector forms of Chapter 14 in two dimensions. However, once the process of multiplication is reached, the theory of complex number operations diverges significantly from those of spatial vectors. Therefore, accept the fact that there are similarities, but be aware that there are major differences in the meanings and operations associated with complex numbers.

Rectangular Coordinates

We begin the development with a two-dimensional rectangular coordinate system shown in Figure 15-1. The two axes have the traditional labels of x and y, respectively. However, in complex variable theory, the horizontal, or x, axis is called the **real axis** and the vertical, or y axis is called the **imaginary axis**.

Imaginary Number

The concept of an imaginary number arises from forming the square root of a negative number. We denote the square root of -1 as i as defined by

$$i = \sqrt{-1} \tag{15-1}$$

The y coordinate of any point will be accompanied by the imaginary number i when expressed in complex form. For example, a point 4 units above the origin on the vertical axis could be denoted as $4i$ or $i4$.

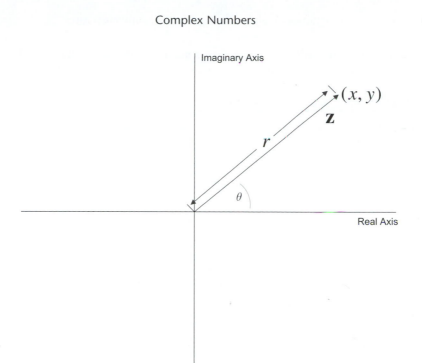

Figure 15-2 Form of a complex number in the complex plane

Rectangular Form of a Complex Number

We will use the same boldface notation employed with matrices and spatial vectors for complex numbers. Let **z** represent an arbitrary complex number. It can be interpreted as a vector-like quantity extending from the origin to a point with coordinates (x,y), as shown in Figure 15-2. The *rectangular form* of **z** is

$$\mathbf{z} = x + iy \qquad (15\text{-}2)$$

The value x is called the *real part* of **z,** and the value y is called the *imaginary part* of **z**. We don't include the i with y when we refer to the imaginary part, that is, the imaginary part is actually the real number y. Moreover, we will not use boldface with x and y since they are real numbers.

When it is necessary to identify the real part of a complex value **z**, the notation Re(**z**) is often used, and when it is necessary to identify the imaginary part, the notation Im(**z**) is often used.

Let r represent the length of the complex number from the origin to the terminal point. From the trigonometry associated with the angle θ, we can say that

$$x = r \cos\theta \qquad (15\text{-}3)$$

$$y = r \sin\theta \qquad (15\text{-}4)$$

The quantity r is the **magnitude** of the complex number, and θ is the **angle**. The basic units for angle are radians (rad), which we will employ for most of the work that follows.

Polar Form of a Complex Number

The inverse relationships to the previous two are as follows:

$$r = \sqrt{x^2 + y^2} \tag{15-5}$$

$$\theta = \text{ang}(\mathbf{z}) = \tan^{-1}\frac{y}{x} \tag{15-6}$$

where ang( ) represents the angle of the complex number and is calculated by forming the inverse tangent function as shown.

Euler's Formula

Take the formulas for x and y from Equations 15-3 and 15-4 and substitute in Equation 15-2. The result is

$$\mathbf{z} = r\cos\theta + ir\sin\theta = r(\cos\theta + i\sin\theta) \tag{15-7}$$

A very important mathematical identity is that of Euler's formula, which reads

$$e^{i\theta} = \cos\theta + i\sin\theta \tag{15-8}$$

This formula, which is proven in more advanced mathematics texts, indicates that when a purely imaginary argument is used for the constant e, it is equivalent to a complex number having both real and imaginary parts, each of which is a sinusoidal function.

When θ is replaced by $-\theta$ in the equation, we obtain

$$e^{-i\theta} = \cos(-\theta) + i\sin(-\theta) = \cos\theta - i\sin\theta \tag{15-9}$$

since the cosine is an even function and the sine is an odd function.

By using Euler's formula, the polar form of the complex number can be expressed as

$$\mathbf{z} = re^{i\theta} \tag{15-10}$$

This concept serves as an important step in the development of complex variable theory. It permits a rigorous approach to the operations of multiplication, division, and exponentiation with complex numbers.

Common Engineering Form

It is common in many engineering applications to express the polar form as follows:

$$re^{i\theta} \triangleq r\angle\theta \tag{15-11}$$

This notation is especially popular in ac circuit theory, but one should remember what it really means so that some of the operations are better justified. We will use the exponential form more frequently and use radians for the angular measurement. For the few cases where the other form is used, degrees will be used.

Example 15-1

A complex number is given by

$$\mathbf{z} = 4 + i3 \qquad (15\text{-}12)$$

Determine the polar form.

Solution

The magnitude is given by

$$r = \sqrt{x^2 + y^2} = \sqrt{(4)^2 + (3)^2} = 5 \qquad (15\text{-}13)$$

The angle is given by

$$\theta = \tan^{-1}\frac{3}{4} = 36.87° = 0.6435 \text{ rad} \qquad (15\text{-}14)$$

The common engineering form with the angle expressed in degrees is

$$\mathbf{z} = 5\angle 36.87° \qquad (15\text{-}15)$$

However, the more "proper" mathematical form is

$$\mathbf{z} = 5e^{i0.6435} \qquad (15\text{-}16)$$

Example 15-2

A complex number is given by

$$\mathbf{z} = -4 + i3 \qquad (15\text{-}17)$$

Determine the polar form.

Solution

The magnitude is the same as before, namely, 5. Be careful when one or more of the real part and imaginary part signs are negative. If you are using a calculator with rectangular-to-polar conversion, it may take care of the sign differences, but for manual computations, the signs of both terms must be carefully noted. In this case, the real part is negative and the imaginary part is positive, so the angle must be in the second quadrant. These deductions lead to

$$\theta = \tan^{-1}\left(\frac{3}{-4}\right) = 180° - \tan^{-1}\frac{3}{4} = 180° - 36.87° = 143.13° = 2.498 \text{ rad} \qquad (15\text{-}18)$$

The form of the complex number is then

$$\mathbf{z} = 5e^{i2.498} \qquad (15\text{-}19)$$

Example 15-3

The polar form of a complex number is given by

$$\mathbf{z} = 4e^{i2} \qquad (15\text{-}20)$$

Determine the rectangular form

Solution

The angle is 2 rad, and the conversion process follows:

$$x = 4\cos 2 = -1.6646 \qquad (15\text{-}21)$$

$$y = 4\sin 2 = 3.6372 \qquad (15\text{-}22)$$

The rectangular form is then

$$\mathbf{z} = -1.6646 + i3.6372$$

Example 15-4

The polar form of a complex number is given by

$$\mathbf{z} = 10e^{-i} \qquad (15\text{-}23)$$

Determine the rectangular form

Solution

The angle is −1 rad, and the conversion process follows.

$$x = 10\cos(-1) = 5.4030 \qquad (15\text{-}24)$$

$$y = 10\sin(-1) = -8.4147 \qquad (15\text{-}25)$$

The rectangular form is then

$$\mathbf{z} = 5.4030 - i8.4147 \qquad (15\text{-}26)$$

15-3   ADDITION AND SUBTRACTION OF COMPLEX NUMBERS

The addition and subtraction of complex numbers is very much like that of spatial vectors; that is, the real parts are added or subtracted and the imaginary parts are added or subtracted. If the process is being performed manually, the complex numbers should be expressed in or converted to rectangular forms.

For the purpose of illustration, assume the following two complex numbers in rectangular form:

$$\mathbf{z}_1 = x_1 + iy_1 \qquad\qquad (15\text{-}27)$$

$$\mathbf{z}_2 = x_2 + iy_2 \qquad\qquad (15\text{-}28)$$

Addition

Let $\mathbf{z}_{sum}$ represent the sum of the two complex numbers. It is given by

$$\mathbf{z}_{sum} = \mathbf{z}_1 + \mathbf{z}_2 = x_1 + iy_1 + x_2 + iy_2 = x_1 + x_2 + i(y_1 + y_2) \qquad (15\text{-}29)$$

Geometric Interpretation of Addition

An interesting geometric interpretation of addition is shown in Figure 15-3 for two arbitrary complex numbers $\mathbf{z}_1$ and $\mathbf{z}_2$. Imagine that $\mathbf{z}_2$ is translated to the end of $\mathbf{z}_1$ while retaining its direction and magnitude. A vector drawn from the origin to the tip of $\mathbf{z}_2$ then represents the sum $\mathbf{z}_{sum}$.

Subtraction

Let $\mathbf{z}_{diff}$ represent the complex number of Equation 15-28 subtracted from that of Equation 15-27. The result is

$$\mathbf{z}_{diff} = \mathbf{z}_1 - \mathbf{z}_2 = x_1 + iy_1 - (x_2 + iy_2) = x_1 - x_2 + i(y_1 - y_2) \qquad (15\text{-}30)$$

Geometric Interpretation of Subtraction

A geometric interpretation of subtraction can also be developed. It is shown in Figure 15-4 for $\mathbf{z}_1 - \mathbf{z}_2$. First, assume that $-\mathbf{z}_2$ is formed. This amounts to rotating $\mathbf{z}_2$ by 180° or π rad. The operation can then be visualized as forming the sum of $-\mathbf{z}_2$ and $\mathbf{z}_1$. The principle employed in the geometric addition process is then followed.

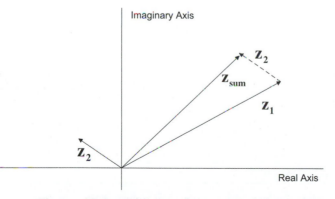

Figure 15-3 Addition of two complex numbers

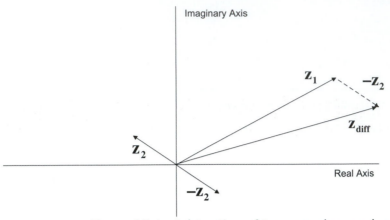

Figure 15-4   Subtraction of two complex numbers

Additional Comments

The normal intuition associated with the addition and subtraction of real numbers does not always apply with complex numbers. For example, the sum of two complex numbers may yield a new complex number that has a smaller magnitude than either of the two being added. Conversely, the difference between two complex numbers may yield a new complex number that may have a larger magnitude than either of the two whose difference is being formed.

Example 15-5

Consider the following two complex numbers in rectangular form:

$$\mathbf{z}_1 = 5 + i3 \tag{15-31}$$

$$\mathbf{z}_2 = 2 - i7 \tag{15-32}$$

Determine the sum of the two numbers $\mathbf{z}_{\text{sum}}$.

Solution

The sum is

$$\mathbf{z}_{\text{sum}} = \mathbf{z}_1 + \mathbf{z}_2 = 5 + i3 + 2 - i7 = 7 - i4 \tag{15-33}$$

Example 15-6

For the two complex numbers of Example 15-5, determine $\mathbf{z}_{\text{diff}} = \mathbf{z}_1 - \mathbf{z}_2$.

Solution

The difference is

$$\mathbf{z}_{\text{diff}} = \mathbf{z}_1 - \mathbf{z}_2 = 5 + i3 - (2 - i7) = 3 + i10 \qquad (15\text{-}34)$$

15-4   MULTIPLICATION AND DIVISION

Multiplication and division are more easily performed when the complex numbers are expressed in polar form. Define two complex numbers as follows:

$$\mathbf{z}_1 = r_1 e^{i\theta_1} \qquad (15\text{-}35)$$

$$\mathbf{z}_2 = r_2 e^{i\theta_2} \qquad (15\text{-}36)$$

Multiplication

Let $\mathbf{z}_{\text{prod}}$ represent the product of the preceding complex numbers. By the law of exponents, it is easily performed with the polar forms.

$$\mathbf{z}_{\text{prod}} = \mathbf{z}_1 \mathbf{z}_2 = \left(r_1 e^{i\theta_1} \right)\left(r_2 e^{i\theta_2} \right) = r_1 r_2 e^{i(\theta_1 + \theta_2)} \qquad (15\text{-}37)$$

There is a simple interpretation to this result: t*o multiply two complex numbers, multiply the magnitudes and add the angles.* This result is probably well known to many readers with an ac circuits background. In that particular area of application, complex numbers are known as *phasors.*

Division

Let $\mathbf{z}_{\text{div}}$ represent $\mathbf{z}_1$ divided by $\mathbf{z}_2$. Again using exponent operations, we have

$$\mathbf{z}_{\text{div}} = \frac{\mathbf{z}_1}{\mathbf{z}_2} = \frac{\left(r_1 e^{i\theta_1} \right)}{\left(r_2 e^{i\theta_2} \right)} = \frac{r_1}{r_2} e^{i(\theta_1 - \theta_2)} \qquad (15\text{-}38)$$

The interpretation in this case is as follows: *To divide two complex numbers, divide the numerator magnitude by the denominator magnitude and subtract the denominator angle from the numerator angle.*

Multiplication in Rectangular Form

It is also possible to perform the product of two complex numbers in rectangular form. Consider two complex numbers defined as

$$\mathbf{z}_1 = x_1 + iy_1 \qquad (15\text{-}39)$$

$$\mathbf{z}_2 = x_2 + iy_2 \qquad (15\text{-}40)$$

The product can be expressed as

$$\mathbf{z}_{\text{prod}} = (x_1 + iy_1)(x_2 + iy_2) = x_1 x_2 + ix_1 y_2 + ix_2 y_1 + i^2 y_1 y_2 \qquad (15\text{-}41)$$

Regrouping and recognizing the fact that $i^2 = -1$, we have

$$\mathbf{z}_{\text{prod}} = x_1 x_2 - y_1 y_2 + i(x_1 y_2 + x_2 y_1) \qquad (15\text{-}42)$$

Do not try to memorize the result. Instead, work it out as you need it.

Complex Conjugate

Before showing an alternate way for dividing, it is necessary to introduce the *complex conjugate*. Let $\mathbf{z}$ represent an arbitrary complex number, and denote the corresponding complex conjugate as $\overline{\mathbf{Z}}$. Assume that the form of $\mathbf{z}$ can be represented in both rectangular and polar forms as

$$\mathbf{z} = x + iy = re^{i\theta} \qquad (15\text{-}43)$$

The complex conjugate can be expressed as

$$\overline{\mathbf{z}} = x - iy = re^{-i\theta} \qquad (15\text{-}44)$$

Thus, *the complex conjugate of a complex number is formed by reversing the sign of the imaginary part in the rectangular form.* Alternately, *it is formed by reversing the sign of the angle in the polar form.*

An important result is obtained by forming the product of a complex number and its conjugate. It can be shown using either form that

$$(\mathbf{z})(\overline{\mathbf{z}}) = x^2 + y^2 = r^2 \qquad (15\text{-}45)$$

Stated in words, *the product of a complex number and its conjugate is the magnitude squared.* The process of multiplying a complex number by its conjugate is referred to as *rationalization.*

Division in Rectangular Form

Based on the rectangular forms of the two complex numbers, consider the division

$$\mathbf{z}_{\text{div}} = \frac{\mathbf{z}_1}{\mathbf{z}_2} = \frac{x_1 + iy_1}{x_2 + iy_2} \qquad (15\text{-}46)$$

The trick in this case is to multiply both numerator and denominator by the complex conjugate of the denominator. The process follows.

$$\begin{aligned}
\mathbf{z}_{\text{div}} &= \frac{(x_1 + iy_1)(x_2 - iy_2)}{(x_2 + iy_2)(x_2 - iy_2)} = \frac{x_1 x_2 + y_1 y_2 + i(x_2 y_1 - x_1 y_2)}{x_2^2 + y_2^2} \\
&= \frac{x_1 x_2 + y_1 y_2 + i(x_2 y_1 - x_1 y_2)}{r^2}
\end{aligned} \qquad (15\text{-}47)$$

The denominator is now a positive real number and can be divided into each of the numerator terms to simplify the result. Again, do not bother memorizing the result; simply work it out when needed.

Example 15-7

Two complex numbers are given by

$$\mathbf{z}_1 = 8e^{i2} \tag{15-48}$$

$$\mathbf{z}_2 = 5e^{-i0.7} \tag{15-49}$$

Determine the product of the two complex numbers and denote it as $\mathbf{z}_3$.

Solution

Since both numbers are given in polar form, the product is easily formed as

$$\mathbf{z}_3 = \mathbf{z}_1\mathbf{z}_2 = \left(8e^{i2}\right)\left(5e^{-i0.7}\right) = 40e^{i1.3} \tag{15-50}$$

For reference in the next example, we will convert this result to rectangular form.

$$\mathbf{z}_3 = 40(\cos 1.3 + i\sin 1.3) = 40(0.2675 + i0.9636) = 10.70 + i38.54 \tag{15-51}$$

Example 15-8

Repeat the multiplication of Example 15-7 by first converting the two complex numbers to rectangular forms and using the rectangular multiplication process.

Solution

The first step is to convert the two values to rectangular forms.

$$\mathbf{z}_1 = 8e^{i2} = 8(\cos 2 + i\sin 2) = 8(-0.4162 + i0.9093) = -3.329 + i7.274 \tag{15-52}$$

$$\mathbf{z}_2 = 5e^{-i0.7} = 5(\cos 0.7 - i\sin 0.7) = 5(0.7648 - i0.6442) = 3.824 - i3.221 \tag{15-53}$$

The product can now be expressed as

$$\mathbf{z}_3 = \mathbf{z}_1\mathbf{z}_2 = (-3.329 + i7.274)(3.824 - i3.221) = -12.73 + 23.43 + i(27.82 + 10.72)$$
$$= 10.70 + i38.54 \tag{15-54}$$

Even with possible round-off, the answer comes out exactly the same as with the polar forms. Clearly, the simpler approach in this case was to use the polar forms. However, suppose that the two complex numbers had been given in rectangular forms. In that case, they would have had to be individually converted to polar forms before proceeding with the polar multiplication. This means that it might have been just as easy to carry out the multiplication with the rectangular forms. The choice could depend on the form in which the final answer is desired.

Example 15-9

For the two complex numbers of Examples 15-7 and 15-8, determine the quotient z_1/z_2 and denote it as z_4.

Solution

The quotient is readily determined in polar form as

$$z_4 = \frac{z_1}{z_2} = \frac{8e^{i2}}{5e^{-i0.7}} = 1.6e^{i2.7} \tag{15-55}$$

The result will now be converted to rectangular form for comparison in the next example.

$$z_4 = 1.6(\cos 2.7 + i \sin 2.7) = 1.6(-0.9041 + i0.4274) = -1.447 + i0.6838 \tag{15-56}$$

Example 15-10

Repeat the division of Example 15-9 using rectangular forms.

Solution

The rectangular forms were determined in Example 15-8, so we can begin with those values.

$$z_4 = \frac{z_1}{z_2} = \frac{-3.329 + i7.274}{3.824 - i3.221} \tag{15-57}$$

The next step is to rationalize the denominator. This is achieved by multiplying both the numerator and the denominator by the complex conjugate of the denominator, which is $3.824 + i3.221$.

$$z_4 = \frac{(-3.329 + i7.274)}{(3.824 - i3.221)} \frac{(3.824 + i3.221)}{(3.824 + i3.221)} = \frac{-12.73 - 23.43 + i(27.82 - 10.72)}{14.62 + 10.37}$$

$$= \frac{-36.16 + i17.10}{25.00} = -1.446 + i0.6840 \tag{15-58}$$

There is a slight round-off difference in the answer as compared with the result of Example 15-9, which could have likely been avoided by starting with more significant digits, but we will leave it as is. As will be seen later with MATLAB, the result of Equation 15-56 is slightly more accurate. The rationalization process tends to introduce a little more round-off error.

15-5   EXPONENTIATION OF COMPLEX NUMBERS

Exponential forms of complex numbers can be computed with exponents both greater and less than one. Some of the cases will be considered here.

Positive Integer Exponents

When exponents are positive integers, the process amounts to just multiplying the complex number by itself by the integer number of times. Consider, for example, $(\mathbf{z})^3$. This is equivalent to $(\mathbf{z})(\mathbf{z})(\mathbf{z})$, so it can be readily computed in polar form or even in rectangular form if one has the patience.

There are some interesting outcomes of the process that are worth investigating. Let N represent an integer and define the exponentiation of the complex number by $\mathbf{z}_{\text{power}}$. We thus have

$$\mathbf{z}_{\text{power}} = (\mathbf{z})^N \tag{15-59}$$

Assuming the polar form of $\mathbf{z}$, we can write

$$\mathbf{z}_{\text{power}} = (re^{i\theta})^N = r^N e^{iN\theta} = r^N \cos N\theta + ir^N \sin N\theta \tag{15-60}$$

From these results, it can be stated that

$$\cos N\theta = \text{Re}(e^{iN\theta}) \tag{15-61}$$

and

$$\sin N\theta = \text{Im}(e^{iN\theta}) \tag{15-62}$$

where Re( ) represents the real part of the complex value in parentheses and Im( ) represents the imaginary part. These relationships are sometimes useful in determining trigonometric forms for larger angles, particularly when complex exponential forms are being used.

Roots of Complex Numbers

Consider now the situation where it is desired to determine the Nth root of a complex number, which means that the exponent is $1/N$, where N is an integer. Examples are the $1/2$ power, the $1/3$ power, and so on. This leads to a basic property of complex numbers. *A complex number raised to a power $1/N$, where N is an integer, has exactly N distinct complex roots.* For example, the square root of a complex number has two roots; the cube root of a complex number has three roots, and so on.

An algorithm for determining the N roots will now be presented. Instead of expressing the complex number in its simple form, add $2\pi n$, where n is an integer (either positive or negative), to the angle as follows:

$$\mathbf{z} = re^{i(\theta + 2\pi n)} \tag{15-63}$$

Note that adding the $2\pi n$, to the basic angle does not change the value of the complex number $\mathbf{z}$, since any additional phase shift that is an integer multiple of 2π rotates the complex number back to the same point. The operation may now be applied to the form stated. It is

$$\mathbf{z}_{\text{roots}} = \left(re^{i(\theta + 2\pi n)}\right)^{1/N} = r^{1/N} e^{i(\theta/N + 2\pi n/N)} \tag{15-64}$$

Let us see if we can interpret this result. The magnitude r is raised to the $1/N$ power, which seems logical. The first angle, corresponding to $n = 0$, has a value of θ/N. However, there are additional angles of $\theta/N + 2\pi n/N$. It turns out that there will be exactly $N-1$ additional distinct angles. This results in a total of N distinct roots. The N roots can then be expressed as

$$\mathbf{z}_{\text{roots}} = r^{1/N} e^{i\left(\frac{\theta}{N} + \frac{2\pi n}{N}\right)} \text{ for } n = 0, 1, 2, \ldots, N-1 \qquad (15\text{-}65)$$

The *principal value* corresponds to $n = 0$ and is given by

$$\mathbf{z}_{\text{principal}} = r^{1/N} e^{i\theta/N} \qquad (15\text{-}66)$$

Example 15-11

Consider the complex number

$$\mathbf{z} = 3 + i4 \qquad (15\text{-}67)$$

Determine the value of

$$\mathbf{z}_6 = (\mathbf{z})^6 \qquad (15\text{-}68)$$

Solution

Well, we could multiply the given number in rectangular form times itself in six messy steps, but a much easier way is to first convert to polar form. This particular value expressed in polar form is

$$\mathbf{z} = 5 e^{i0.9273} \qquad (15\text{-}69)$$

The desired result in exponential form can be determined in one step as

$$\mathbf{z}_6 = (5 e^{i0.9273})^6 = 15{,}625 e^{i5.5638} \qquad (15\text{-}70)$$

If desired, the result can be converted back to rectangular form as

$$\mathbf{z}_6 = 15{,}625\left(\cos 5.5638 + i \sin 5.5638\right) = 11{,}753 - i10{,}296 \qquad (15\text{-}71)$$

Example 15-12

An equation that arises in the derivation of a "second-order Butterworth filter design" is the equation

$$s^4 + 1 = 0 \qquad (15\text{-}72)$$

Determine the four complex values of s that satisfy the equation.

Solution

The quantity s is the Laplace variable, and we will omit the boldface notation on it. We rearrange the equation and write

$$s^4 = -1 = 1e^{i(\pi + 2\pi n)} \tag{15-73}$$

We then take the fourth root of both sides. Of course, the fourth root of 1 is 1, and we have

$$s = e^{i\left(\frac{\pi}{4} + n\frac{2\pi}{4}\right)} \quad \text{for} \quad n = 0, 1, 2, 3 \tag{15-74}$$

We will denote the principal value as s_1 and it is

$$s_1 = e^{i\frac{\pi}{4}} = 0.7071 + i0.7071 \tag{15-75}$$

The other roots are

$$s_2 = e^{i\frac{3\pi}{4}} = -0.7071 + i0.7071 \tag{15-76}$$

$$s_3 = e^{i\frac{5\pi}{4}} = -0.7071 - i0.7071 \tag{15-77}$$

$$s_4 = e^{i\frac{7\pi}{4}} = 0.7071 - i0.7071 \tag{15-78}$$

All of these roots lie on a circle with unit radius and are displaced from each other by an angle of 90°.

15-6   MATLAB OPERATIONS WITH COMPLEX NUMBERS

We will now turn our attention to the various forms of complex number manipulations with MATLAB. For purposes of discussion, let us begin with a simple complex number in which we know both the rectangular and polar forms. Consider the complex number

$$\mathbf{z} = 3 + i4 = 5e^{i0.9273} = 5\angle 53.13° \tag{15-79}$$

Entering a Complex Number in Rectangular Form

Way back in Chapter 1, the process of entering a purely imaginary number was considered. We will now use the number of Equation 15-79 and show the four variations in the manner in which it can be entered in rectangular form in the Command Window and how MATLAB treats the value. We will not use boldface print at this point, since the results are displayed as they appear in the Command Window.

$$>> z = 3 + 4i \tag{15-80}$$

$$z =$$

$$3.0000 + 4.0000i$$

$$>> z = 3 + 4j \tag{15-81}$$

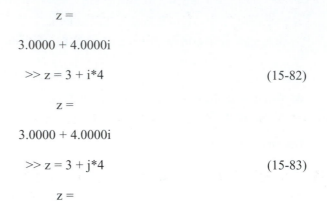

$$z =$$

$$3.0000 + 4.0000i$$

$$\gg z = 3 + i*4 \tag{15-82}$$

$$z =$$

$$3.0000 + 4.0000i$$

$$\gg z = 3 + j*4 \tag{15-83}$$

$$z =$$

$$3.0000 + 4.0000i$$

Comparing these four forms, one can use either i or j, but MATLAB always prints i. If the i or j follows the number, the multiplication symbol (*) is not required. However, if the i or j precedes the number, the multiplication symbol is required. If the imaginary part is expressed as a variable, it will be necessary to use the multiplication symbol for either order.

Entering a Complex Number in Polar Form

To enter the preceding number in polar form, we use the MATLAB exponential command, except that the argument is set as a purely imaginary number with the value in radians. Of course, the exponential is multiplied by the magnitude.

$$\gg z = 5*\exp(0.9273i) \tag{15-84}$$

$$z =$$

$$3.0000 + 4.0000i$$

If we start with the angle in degrees, we can either first multiply the angle by $\pi/180$ to convert to radians or this conversion can be accomplished as part of the command.

$$\gg z = 5*\exp((pi/180)*53.13i) \tag{15-85}$$

$$z =$$

$$3.0000 + 4.0000i$$

Polar to Rectangular Conversion

Actually, polar to rectangular conversion is automatically achieved by entering the value in polar form and using the exponential form, as shown in the past two equations.

Rectangular to Polar Conversion

Assume that z has been entered in rectangular form as previously shown, that is,

$$\text{>> z = 3 + 4i} \tag{15-86}$$

z =

3.0000 + 4.0000i

The magnitude of z is denoted as **r**. It is determined by the command

$$\text{>> r = abs(z)} \tag{15-87}$$

r =

5

The angle in radians will be denoted as **theta.** It is determined by

$$\text{>> theta = angle(z)} \tag{15-88}$$

theta =

0.9273

We could then multiply this value by $180/\pi$ to convert to degrees. However, this can be combined with the previous command, as follows:

$$\text{>> theta = (180/pi)*angle(z)} \tag{15-89}$$

theta =

53.1301

Complex Conjugate

The complex conjugate of a complex number can be determined by the command **conj( )**. Let z_conj represent the conjugate of the value z being used. It is

$$\text{>> z\_conj = conj(z)} \tag{15-90}$$

z_conj =

3.0000 - 4.0000i

For illustration, let us form the product of z and its conjugate.

$$\text{>> z*z\_conj} \tag{15-91}$$

ans =

25

The answer, as expected, is a real number whose value is the magnitude of z squared.

Real and Imaginary Parts

The real part of a complex number is obtained in MATLAB by the command **real( )**, and the imaginary part is obtained by the command **imag( )**. These operations are illustrated for z, as follows:

$$\gg real(z) \tag{15-92}$$

$$ans =$$

$$3$$

$$\gg imag(z) \tag{15-93}$$

$$ans =$$

$$4$$

General Guidelines

As a general rule, all of the arithmetic operations such as addition, subtraction, multiplication, division, and exponentiation can be performed with complex numbers in MATLAB using the same commands as encountered with simple scalars. Of course, complex numbers must be entered in the special forms discussed in this section, and the results will be complex numbers. The results will normally be expressed in rectangular form, so if a polar form is desired for the output, a conversion will be necessary.

When forming the Nth root of a complex number, MATLAB only provides the principal value. However, there is an alternate way to obtain all the roots using a polynomial form and the process will be demonstrated in Example 15-18.

Some examples that follow will provide further drill in using MATLAB for complex number manipulation.

Example 15-13

Repeat the rectangular to polar conversion of Example 15-1 with MATLAB.

Solution

First, the variable z from Equation 15-12 is entered in MATLAB.

$$\gg z = 4 + 3i \tag{15-94}$$

$$z =$$

$$4.0000 + 3.0000i$$

The magnitude is denoted as **r**. It is determined by the command

$$>> r = abs(z) \tag{15-95}$$

$$r =$$

$$5$$

The angle in radians is denoted as theta. It is determined as

$$>> theta = angle(z) \tag{15-96}$$

$$theta =$$

$$0.6435$$

Example 15-14

Take the results of Example 15-13 and convert back to rectangular form.

Solution

Assume that r and theta are still in memory. The process employs the **exp** function with an imaginary argument and is

$$>> z = r*exp(theta*i) \tag{15-97}$$

$$z =$$

$$4.0000 + 3.0000i$$

Note that since theta was identified as a variable and not an actual number, it was necessary to place the multiplication symbol (*) between the variable and i.

Example 15-15

Perform the multiplication of Examples 15-7 and 15-8 using MATLAB.

Solution

The complex numbers were given in Equations 15-48 and 15-49 and will be entered in polar form.

$$>> z1 = 8*exp(2i) \tag{15-98}$$

$$z1 =$$

$$-3.3292 + 7.2744i$$

$$\text{>> z2 = 5*exp(-0.7i)} \qquad (15\text{-}99)$$

$$z2 =$$

$$3.8242 - 3.2211i$$

Although the numbers are entered in polar form, they are automatically converted by MATLAB to rectangular form. However, multiplication is easily achieved with MATLAB. The product z3 is

$$\text{>> z3 = z1*z2} \qquad (15\text{-}100)$$

$$z3 =$$

$$10.7000 +38.5423i$$

The results are the same as in Examples 15-7 and 15-8, except extended to a larger number of decimal places.

Example 15-16

Perform the division of Examples 15-9 and 15-10 using MATLAB.

Solution

Assuming that z1 and z2 from the previous example are still in memory, we have

$$\text{>> z4 = z1/z2} \qquad (15\text{-}101)$$

$$z4 =$$

$$-1.4465 + 0.6838i$$

This result agrees with Example 15-9, which was slightly more accurate than Example 15-10.

Example 15-17

Repeat the analysis of Example 15-11 using MATLAB.

Solution

The complex number z is entered as

$$\text{>> z = 3 + 4i} \qquad (15\text{-}102)$$

$$z =$$

$$3.0000 + 4.0000i$$

With MATLAB, the number is easily raised to the sixth power by the command

$$\text{>> z6 = z\textasciicircum 6} \qquad (15\text{-}103)$$

$$\text{z6 =}$$

1.1753e+004 -1.0296e+004i

This agrees exactly with the rectangular form of the result of Example 15-11. The polar form is obtained by

$$\text{>> r6 = abs(z6)} \qquad (15\text{-}104)$$

$$\text{r6 =}$$

15625

$$\text{>> theta6 = angle(z6)} \qquad (15\text{-}105)$$

$$\text{z6\_ang =}$$

-0.7194

Something is strange! The magnitude agrees exactly with the result of Example 15-11, but the angle appears to be way off. However, this is an illusion, since the angle of any complex number can be changed by an integer number of $\pm 2\pi$ rad without affecting the actual value. MATLAB always gives the smallest angle with the conversion process, whereas in Example 15-11, the multiplication forced the angle to be much greater than one complete revolution.

If we take the angle of Example 15-11, which was 5.5638 rad, and add -2π rad to it, we obtain $5.5638 - 2\pi = -0.7194$ rad, which is the result of Equation 15-105. Thus, the angles are essentially equivalent.

Example 15-18

Rework Example 15-12 using MATLAB.

Solution

For convenience, the equation to be solved is repeated here.

$$s^4 + 1 = 0 \qquad (15\text{-}106)$$

The solution could be interpreted as follows:

$$s = (-1)^{1/4} \qquad (15\text{-}107)$$

Let us see what MATLAB provides when we take the 1/4th root of -1.

$$>> s = (-1)^{(1/4)} \hspace{5cm} (15\text{-}108)$$

$$s =$$

$$0.7071 + 0.7071i$$

Well, we obtained the principal value, but the other three roots did not appear. The way to obtain all four roots is to return to the form of Equation 15-106 and define a polynomial p as

$$>> p = [1\ 0\ 0\ 0\ 1]; \hspace{4cm} (15\text{-}109)$$

Note that since it is a fourth degree polynomial, five terms are required in the polynomial definition provided to MATLAB. The first 1 represents the highest-degree coefficient, and the last 1 represents the constant term. The three 0s represent the coefficients of the missing terms. The roots are then determined by the command

$$>> s = \text{roots}(p) \hspace{4.5cm} (15\text{-}110)$$

$$s =$$

$$-0.7071 + 0.7071i$$

$$-0.7071 - 0.7071i$$

$$0.7071 + 0.7071i$$

$$0.7071 - 0.7071i$$

This process produces the four roots obtained in Example 15-12.

15-7 APPLICATIONS OF COMPLEX NUMBERS

There are numerous applications of complex numbers and function variables in many engineering disciplines, physics, and applied mathematics. A few of these applications will be discussed here.

Fourier Analysis

A broad area of application involves Fourier series and Fourier transforms. Chapter 16 will be devoted to a somewhat detailed treatment of those subjects, so any discussion will be delayed to that chapter.

Contour Integration

Many integrals that cannot be evaluated by the normal integration processes of integral calculus may be evaluated by a process called *contour integration*. This concept is covered in more advanced books in mathematics. The technique involves quantities called *residues*, which are evaluated in the complex plane.

Inverse Laplace Transforms

Many Laplace transforms that cannot be inverse transformed by other means lend themselves to the contour integration approach, again making use of *residues* (introduced in the previous paragraph).

Conformal Mapping

The concept of *conformal mapping* involves the use of complex variables to map one plane into a new plane in which the geometry lends itself to a simpler solution. By utilizing this technique, certain complex problems may be more readily solved.

The process is used in the design of airfoils on the wings of airplanes. The airfoils are derived by mapping the flow field of a circular cylinder into a family of airfoil shapes. The velocity and pressures in the plane containing the cylinder can be more easily evaluated, and by the process of conformal mapping, the corresponding velocity and pressure around the airfoil can be determined. These results lead to a determination of the lift produced by the airfoil, an important process in the design of an airplane.

Electromagnetic Fields and Waves

Time-varying electric and magnetic fields can be represented in terms of complex variables in much the same manner as in electric circuits, which will be considered in the next several paragraphs. Along with the representation in terms of complex variables, electromagnetic fields and waves are also spatial vectors. Therefore, the analysis involves both complex variable theory and spatial vector analysis.

AC Circuit Analysis

We will next demonstrate some of the concepts of standard ac circuit analysis, with which many readers are already familiar. We hope to resolve some of the mystery associated with the process widely used in representing sinusoidal voltages and currents by complex numbers.

To be compatible with virtually all electrical engineering and electrical engineering technology texts, we will switch to $j = \sqrt{-1}$ for the remainder of this section. If we did not switch, we would have to introduce another variable for the current, i, which is never done. However, we will refrain from using the common circuit name *phasor* and continue to refer to the variables involved as complex numbers.

The concept behind ac circuit analysis with complex numbers is based on the property that a stable linear circuit excited by a sinusoidal source will reach a steady-state condition in which all voltages and currents in the circuit will be sinusoidal functions of the same frequency as the source, but with different magnitudes and angles. Instead of expressing variables as sine and/or cosine functions, a significant computational advantage is obtained by representing all variables as exponential functions.

Assume that a typical current, $i(t)$, in the circuit has the form

$$i(t) = I\cos(\omega t + \theta) \qquad (15\text{-}111)$$

This current is represented in the following manner.

$$i(t) = \text{Re}\left[Ie^{j(\omega t + \theta)}\right] = \text{Re}\left[Ie^{j\theta}e^{j\omega t}\right] = \text{Re}\left[\mathbf{I}e^{j\omega t}\right] \qquad (15\text{-}112)$$

where

$$\mathbf{I} = Ie^{j\theta} \qquad (15\text{-}113)$$

It is customary now to omit the real part designation and simply express the current as

$$i(t) = \mathbf{I}e^{j\omega t} \qquad (15\text{-}114)$$

The analysis can then be carried out with complex forms, and in the final result, the real part can be formed. It should be noted that a sine function along with the imaginary part can just as easily be used, and some books prefer that approach. The final outcome is the same either way.

Now what advantage do we gain by using the complex exponential representation? First, consider the instantaneous voltage-current relationship for an inductance, L, which is given by

$$v(t) = L\frac{di}{dt} \qquad (15\text{-}115)$$

By substituting the expression for current from Equation 15-114 in Equation 15-115 and performing the differentiation, we obtain

$$v(t) = j\omega L\mathbf{I}e^{j\omega t} \qquad (15\text{-}116)$$

We can see that $v(t)$ will have the same form, so it can be expressed as

$$v(t) = \mathbf{V}e^{j\omega t} \qquad (15\text{-}117)$$

where

$$\mathbf{V} = Ve^{j\phi} \qquad (15\text{-}118)$$

Setting the form of $v(t)$ of Equation 15-117 equal to that of Equation 15-116, the $e^{j\omega t}$ factors cancel and we have

$$\mathbf{V} = j\omega L\mathbf{I} \qquad (15\text{-}119)$$

In effect, the time variation has been suppressed to simplify the analysis, and we have two complex numbers, one representing the current and the other representing the voltage. The factor $j\omega L$ in Equation 15-119 is the *impedance* of the inductor, and the real factor ωL is called the *reactance*. The factor $j = e^{j\pi/2}$ means that the voltage *leads* the current by 90° or $\pi/2$ rad.

The voltage-current relationship for a capacitance C is given by

$$v(t) = \frac{1}{C}\int_0^t i(t)dt \qquad (15\text{-}120)$$

Without showing the details, the corresponding form employing complex numbers is

$$\mathbf{V} = \frac{1}{j\omega C}\mathbf{I} = \frac{-j}{\omega C}\mathbf{I} \qquad (15\text{-}121)$$

In this case, the impedance is $1/j\omega C$. Moreover, the voltage *lags* the current by 90°, or $\pi/2$ rad.

The voltage-current relationship for a resistance R can be expressed by Ohm's Law.

$$v(t) = Ri(t) \qquad (15\text{-}122)$$

Since no calculus is involved with this case, the corresponding complex number relationship is

$$\mathbf{V} = R\mathbf{I} \qquad (15\text{-}123)$$

The use of complex numbers in AC circuit theory essentially replaces the processes of differentiation and integration by multiplication by $j\omega$ and division by $j\omega$, respectively. The steady-state sinusoidal voltages and currents can then be determined completely by algebraic methods employing complex numbers.

GENERAL PROBLEMS

15-1. A complex number is given by
$$\mathbf{z} = 12 + i5$$
Determine the polar form.

15-2. A complex number is given by
$$\mathbf{z} = 5 + i12$$
Determine the polar form.

15-3. A complex number is given by
$$\mathbf{z} = -12 - i5$$
Determine the polar form.

15-4. A complex number is given by
$$\mathbf{z} = 5 - i12$$
Determine the polar form.

15-5. The polar form of a complex number is given by
$$\mathbf{z} = 8e^{-i0.4}$$
Determine the rectangular form.

15-6. The polar form of a complex number is given by

$$\mathbf{z} = 12e^{i1.2}$$

Determine the rectangular form.

15-7. The polar form of a complex number is given by

$$\mathbf{z} = 4e^{i7}$$

Determine the rectangular form.

15-8. The polar form of a complex number is given by

$$\mathbf{z} = 6e^{-i12}$$

Determine the rectangular form.

15-9. The polar form of a complex number is given by

$$\mathbf{z} = e^{\frac{i\pi}{2}}$$

Determine the rectangular form.

15-10. The polar form of a complex number is given by

$$\mathbf{z} = e^{i\pi}$$

Determine the rectangular form.

15-11. Assume the following two complex numbers in rectangular form:

$$\mathbf{z}_1 = 8 + i5$$

$$\mathbf{z}_2 = -3 - i9$$

Determine the sum of the two numbers $\mathbf{z}_{\text{sum}}$.

15-12. For the two complex numbers of Problem 15-11, determine $\mathbf{z}_{\text{diff}} = \mathbf{z}_1 - \mathbf{z}_2$.

15-13. Two complex numbers are given by

$$\mathbf{z}_1 = 5e^{-i1.2}$$

$$\mathbf{z}_2 = 6e^{-i0.8}$$

Determine the product of the two complex numbers and denote it as $\mathbf{z}_3$.

15-14. Two complex numbers are given by

$$\mathbf{z}_1 = 12e^{i0.7}$$

$$\mathbf{z}_2 = 3e^{i0.2}$$

Determine the product of the two complex numbers and denote it as $\mathbf{z}_3$.

15-15. Repeat the multiplication of Problem 15-13 by first converting the two complex numbers to rectangular forms and using the rectangular multiplication process.

15-16. Repeat the multiplication of Problem 15-14 by first converting the two complex numbers to rectangular forms and using the rectangular multiplication process.

15-17. For the two complex numbers of Problems 15-13 and 15-15, determine the quotient z_1/z_2 and denote it as z_4.

15-18. For the two complex numbers of Problems 15-14 and 15-16, determine the quotient z_1/z_2 and denote it as z_4.

15-19. Repeat the division of Problem 15-17 using rectangular forms.

15-20. Repeat the division of Problem 15-18 using rectangular forms.

15-21. Consider the complex number

$$z = 2 - i3$$

Determine the value of

$$z_4 = (z)^4$$

15-22. Consider the complex number

$$z = -2 + i$$

Determine the value of

$$z_5 = (z)^5$$

15-23. Determine the six roots of the equation

$$s^6 + 64 = 0$$

15-24. Determine the four roots of the equation

$$s^4 - 16 = 0$$

MATLAB PROBLEMS

15-25. Repeat Problem 15-1 using the MATLAB Command Window.

15-26. Repeat Problem 15-2 using the MATLAB Command Window.

15-27. Repeat Problem 15-3 using the MATLAB Command Window.

15-28. Repeat Problem 15-4 using the MATLAB Command Window.

15-29. Repeat Problem 15-5 using the MATLAB Command Window.

15-30. Repeat Problem 15-6 using the MATLAB Command Window.

15-31. Repeat Problem 15-7 using the MATLAB Command Window.

15-32. Repeat Problem 15-8 using the MATLAB Command Window.

15-33. Repeat Problem 15-11 using the MATLAB Command Window.

15-34. Repeat Problem 15-12 using the MATLAB Command Window.

15-35. Repeat Problem 15-13 using the MATLAB Command Window.

15-36. Repeat Problem 15-14 using the MATLAB Command Window.

15-37. Repeat Problem 15-17 using the MATLAB Command Window.

15-38. Repeat Problem 15-18 using the MATLAB Command Window.

15-39. Repeat Problem 15-21 using the MATLAB Command Window.

15-40. Repeat Problem 15-22 using the MATLAB Command Window.

15-41. Repeat Problem 15-23 using the MATLAB Command Window.

15-42. Repeat Problem 15-24 using the MATLAB Command Window.

15-43. Perform the operation i^i in MATLAB. Can you explain why the answer is a real number?

15-44. Perform the operation i^{-i} in MATLAB. Can you explain why the answer is a real number?

15-45. Write an M-file program to determine $\mathbf{z}_0$ from the following expression:

$$\mathbf{z}_0 = K \frac{\mathbf{z}_1 \mathbf{z}_2}{\mathbf{z}_3 \mathbf{z}_4}$$

where K is a real constant and all other variables are complex and entered in rectangular form.

> *Inputs:* $K, x_1, y_1, x_2, y_2, x_3, y_3, x_4, y_4$
>
> *Output:* Value of $\mathbf{z_0}$ in rectangular form

15-46. Write an M-file program similar to that of Problem 15-45 to perform the same operation, but with the inputs expressed in polar form.

> *Inputs:* $K, r_1, \theta_1, r_2, \theta_2, r_3, \theta_3, r_4, \theta_4$
>
> *Output:* Value of $\mathbf{z_0}$ in rectangular form

15-47. The derivation of the design process for an Nth-order Butterworth electrical filter involves determining the complex roots of the equation that follows.

$$s^{2N} + (-1)^N = 0$$

a. Write an M-file program to determine the roots for an arbitrary positive integer order N

> *Input:* N
>
> *Outputs:* All roots expressed in rectangular forms

b. Determine the roots for $N = 3$.

15-48. The derivation of the design process for a third-order Chebyshev electrical filter involves determining the complex roots of the equation that follows.

$$-16As^6 - 24As^4 - 9As^2 + 1 = 0$$

a. Write an M-file program to determine the roots.

> *Input:* A (This is a parameter used to establish the filter ripple level.)
>
> *Outputs:* All roots expressed in rectangular forms

b. Determine the roots for $A = 1$.

DERIVATION PROBLEMS

15-49. Prove that $\dfrac{1}{i} = -i$

15-50. Prove that $i^3 = -i$

15-51. Prove that $i^5 = i$

Fourier Analysis with MATLAB

16

16-1 OVERVIEW AND OBJECTIVES

Fourier analysis is the process of representing a function in terms of sinusoidal components. It is widely employed in many areas of engineering, science, and applied mathematics. In effect, it provides information as to what frequency components appear in a given function.

In this chapter, some of the basic fundamentals will be provided as a means of understanding the concept. We will then utilize MATLAB for performing a Fourier analysis.

Objectives

After completing this chapter, the reader should be able to

1. Discuss the concept of Fourier analysis and why it is important.
2. State the three forms of the Fourier series and discuss their properties.
3. Determine the Fourier series of some basic waveforms.
4. Discuss the amplitude and phase spectra and show how they can be plotted.
5. State the Fourier transform and the inverse transform.
6. Discuss the different types of functions applicable to a Fourier series and a Fourier transform.
7. Discuss the discrete Fourier transform and the fast Fourier transform.
8. Apply MATLAB to perform Fourier analysis.

16-2 INITIAL COMPUTATIONS

Before one can utilize MATLAB or any other software to assist in Fourier analysis, it is necessary to have some understanding of what it is and what it is used for. While there are complete books devoted to the subject, the intent here is to cover the essential basics for developing an intuitive feeling for the subject. Following this treatment, the use of MATLAB for Fourier analysis will be emphasized.

Independent Variable Time

Fourier analysis can be applied to many different physical variables; some examples are electrical voltages and currents, force distributions, pressure waves, sound waves, thermal distributions, ocean waves, and many others. The independent variable can be time, distance, or other variables. However, because so many of the applicable variables are functions of time, we will use it as the basis for the developments that follows. Through the use of consumer electronic products, many people are familiar with time-varying signals in terms of the frequency response behavior; for example, flat amplifier response from 20 hertz (Hz) to 20 kHz, and so on. (*Note: 1 hertz is equivalent to 1 cycle per second.*) This type of specification is definitely related to the concept of Fourier analysis. Just remember that the variables may be changed if the analysis procedure involves different units.

Dependent Variable

For the dependent variable, we will employ the general symbol x in this development. Thus, a time-varying function for which Fourier analysis is to be employed will be usually denoted as $x(t)$.

Time Domain versus Frequency Domain

The function of time $x(t)$ that is initially given is said to be a representation in the *time domain*. Fourier analysis results in a function of frequency, and it is said to be a representation in the *frequency domain*. The resulting frequency domain function is also called the *frequency spectrum* in many applications, and we will use that terminology for reference.

An intuitive way to visualize the process is this: For a given time function, what sinusoidal frequencies are present in the signal? For example, the human ear responds to frequencies between about 20 Hz to well above 15 kHz. Therefore, voice or music waveforms can be represented in terms of sinusoidal components. The components that represent a bass voice would have much lower frequencies than those that represent a soprano voice, for example.

Periodic Function

Consider a function $x(t)$ that repeats itself in a predictable fashion, as illustrated in Figure 16-1. This function is *periodic* if $x(t + T) = x(t)$ for all time. The fixed quantity T is called the *period*. The duration of one period is also referred to as a *cycle*.

Fourier Series

The theory underlying Fourier series is that a periodic function satisfying certain mathematical restrictions, which are rarely of concern in engineering problems, can be expressed in the form of a trigonometric series consisting of a constant value and harmonically (integer) related sinusoidal frequency terms. There are three different forms for the series, and to achieve a workable understanding for use with MATLAB, it is desirable to

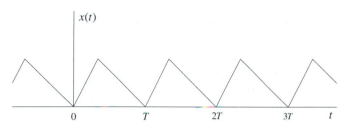

Figure 16-1   Example of a periodic function

investigate the three different forms. They will be referred to in this text as the (1) cosine-sine form, (2) amplitude-phase form, and (3) complex exponential form. The first two forms are referred to as *one-sided forms,* and the third will be referred to as a *two-sided form* for reasons that will be clear later. The constant term is also called the *dc value* in many applications.

Simple Initial Analysis

Before one starts utilizing either traditional mathematical forms or MATLAB, a few simple steps can be taken to provide significant information about the Fourier series. These steps will now be discussed.

1. Do the Fourier series have a constant or dc term? If the net integrated algebraic area over a cycle is zero, there will be no constant term. In some cases, the deduction may be obvious and in other cases it may not be obvious. If the area is all positive or all negative, there will definitely be a constant term. If there is both positive and negative area within a cycle, it may not be obvious, and all that one can deduce could be that there *might* be a constant term.
2. What is the fundamental frequency? The *fundamental frequency* f_1 is the lowest sinusoidal frequency in the Fourier series of a periodic function and is determined from the period T by the relationship

$$f_1 = \frac{1}{T} \tag{16-1}$$

3. All other frequencies in the Fourier series are integer multiples of the fundamental frequency. The so-called *second harmonic* has a frequency of $2f_1$, the *third harmonic* a frequency of $3f_1$, and so on. For some waveforms, certain terms, such as even or odd-numbered harmonics, may be zero. Also, in practical functions, the Fourier coefficients will eventually approach zero, which results in a finite *bandwidth* for practical processing and transmission of the function. In general, the "smoother" the time function is, the more rapidly the frequency coefficients will approach zero.

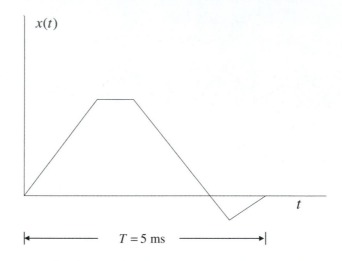

Figure 16-2   One cycle of the function of Example 16-1

The simple procedure just discussed provides an initial insight into the process and should always be performed first. However, the more complex question relates to the magnitudes of the components and how quickly they converge toward zero. In virtually all applications of Fourier analysis, it is desired that the series be approximated by a finite number of terms.

Example 16-1

The function of Figure 16-2 is a periodic function with a period $T = 5$ ms (only one cycle is shown). List the possible frequencies contained in the Fourier series.

Solution

Clearly, the positive area is greater than the negative area, so there will be a constant term in the series. Since the period is 5 ms = 0.005 s, the fundamental frequency is

$$f_1 = \frac{1}{0.005} = 200 \text{ Hz} \tag{16-2}$$

All other possible frequencies are integer multiples of the fundamental. Therefore, a list of the possible frequencies is

0 (constant term)

200 Hz

400 Hz

600 Hz

800 Hz

and so on.

This is good information, but it does not tell us the magnitudes of the individual frequency components or how quickly these terms converge toward zero.

16-3   FOURIER SERIES FORMS

We will now investigate the three different forms of the Fourier series.

Cosine-Sine Form

The cosine-sine form is the one most often presented in basic mathematics texts. It reads

$$x(t) = A_0 + \sum_{n=1}^{\infty} (A_n \cos n\omega_1 t + B_n \sin n\omega_1 t) \tag{16-3}$$

with

$$\omega_1 = 2\pi f_1 = \frac{2\pi}{T} \tag{16-4}$$

The frequency ω_1 is the fundamental *angular frequency* in radians per second (rad/s), and f_1 is the fundamental *cyclic frequency* in hertz (Hz). The nth harmonic radian frequency is $n\omega_1$ and the nth harmonic cyclic frequency is nf_1. The coefficient A_0 is the constant or dc term, and the various A_n and B_n coefficients represent the peak values or amplitudes of the various cosine and sine terms in the Fourier series.

Formulas for the Coefficients

The term A_0 is the constant term in the series and is the average value of the function over a cycle. It is given by

$$A_0 = \frac{\text{algebraic area under curve in one cycle}}{T} = \frac{1}{T}\int_0^T x(t)dt \tag{16-5}$$

The coefficients of the cosine terms are given by

$$A_n = \frac{2}{T}\int_0^T x(t)\cos n\omega_1 t\, dt \tag{16-6}$$

The coefficients of the sine terms are given by

$$B_n = \frac{2}{T}\int_0^T x(t)\sin n\omega_1 t\, dt \tag{16-7}$$

Although the sine-cosine form is often the most basic form given in textbooks, it is not the most convenient from a practical point of view. The reason is that, in general, two amplitude terms are required at each frequency.

It should be noted that an alternate domain for the integration is from $-T/2$ to $T/2$. This domain will be used in Example 16-4, for convenience.

Amplitude-Phase Form

The amplitude-phase form of the Fourier series is given by

$$x(t) = C_0 + \sum_{n=1}^{\infty} C_n \cos(n\omega_1 t + \theta_n) \tag{16-8}$$

It is also possible to express the amplitude-phase form in terms of sine functions having the same amplitude, but with different phase angles. We will confine our consideration of the amplitude-phase form to the case of cosine functions only.

The amplitude coefficients, C_n, are given by

$$C_n = \sqrt{A_n^2 + B_n^2} \tag{16-9}$$

The phase angles, θ_n, are given by

$$\theta_n = \tan^{-1}\left(\frac{-B_n}{A_n}\right) \tag{16-10}$$

Complex Exponential Form

The cosine-sine and amplitude-phase forms of the Fourier series are *one-sided forms*, which means that the coefficients are determined only for positive frequencies and the constant term. For reasons that are associated with certain applications and interpretations, it is appropriate to consider the concept of a *two-sided* Fourier series. This means that we will consider that the series applies for *both positive and negative frequencies*. The concept of a *negative frequency* is more of a mathematical concept than a real physical reality, but it serves a very useful purpose in some areas.

The two-sided Fourier series is based on the *complex exponential form*. To develop the basis for this form we first consider *Euler's formula,* as introduced in Chapter 15, which is

$$e^{i\omega t} = \cos \omega t + i \sin \omega t \tag{16-11}$$

Changing the sign of the argument, we have

$$e^{-i\omega t} = \cos \omega t - i \sin \omega t \tag{16-12}$$

This latter formula is a result of the fact that the cosine function is even and the sine function is odd.

Successive addition and subtraction of the preceding two expressions lead to definitions of the cosine and sine functions expressed in terms of complex exponential functions. The expressions are

$$\cos \omega t = \frac{e^{i\omega t} + e^{-i\omega t}}{2} \tag{16-13}$$

and

$$\sin \omega t = \frac{e^{i\omega t} - e^{-i\omega t}}{2i} \tag{16-14}$$

These results imply that the cosine and sine terms in a Fourier series can be regrouped and expressed in terms of complex exponential functions. The functions with positive arguments ($i\omega t$) can be considered as *positive frequencies*, and the functions with negative arguments ($-i\omega t$) can be considered as *negative frequencies*. Each spectral component other than dc can then be considered to be composed of both a positive frequency component and a negative frequency component. The form of the exponential Fourier series is

$$x(t) = \sum_{n=-\infty}^{\infty} \mathbf{X_n} e^{in\omega_1 t} \tag{16-15}$$

A given coefficient $\mathbf{X_n}$ represents a Fourier term associated with the nth harmonic. Other than the constant term there will always be two terms at each frequency, one for positive n and one for negative n. The boldface form for $\mathbf{X_n}$ indicates that it is a complex quantity in general.

A general expression for the complex coefficient is

$$\mathbf{X_n} = \frac{1}{T}\int_0^T x(t)e^{-in\omega_1 t}dt \tag{16-16}$$

The complex exponential form is more compact than the other two forms, but it is somewhat more difficult to visualize physically. As we will see shortly, there is a simple relationship between the positive and negative frequency terms.

Relationships between Exponential Form and Other Forms

All of the Fourier series forms are related to each other, and if one form is known, the others can be determined. For the complex exponential form, the coefficients $\mathbf{X_n}$ can be expressed as a magnitude and an angle in the form

$$\mathbf{X_n} = X_n e^{i\theta_n} \triangleq X_n \angle \theta_n \tag{16-17}$$

where $X_n = |\mathbf{X_n}|$ is the *magnitude* or *absolute value* of the complex coefficient and θ_n is the angle. The fact that the notation for the angle is the same as used in the amplitude-phase form of Equation 16-8 is no coincidence. *The angles associated with the complex exponential coefficients are the same as the angles associated with the amplitude-phase coefficients employing cosine terms.*

Let $\bar{\mathbf{X}}_n$ represent the complex conjugate of $\mathbf{X_n}$. It can be shown that the coefficient of the negative frequency component for a given n is the complex conjugate of the value for the corresponding positive n. Thus,

$$\mathbf{X_{-n}} = \bar{\mathbf{X}}_n \tag{16-18}$$

This means that the magnitude for the negative frequency component is the same as for the positive frequency component, but the angle is reversed in sign.

The complex exponential terms can be readily determined from the cosine-sine terms by the following relationships:

$$\mathbf{X_n} = \frac{A_n - iB_n}{2} \tag{16-19}$$

From an intuitive point of view, an easy way to visualize the relationship between the magnitudes of the positive and negative coefficients can be stated as follows:

$$X_n = X_{-n} = \frac{C_n}{2} \qquad \text{for } n \neq 0$$

(16-20)

and

$$X_0 = C_0$$

(16-21)

With the exception of the dc term, we can intuitively visualize the process as starting with the one-sided coefficients C_n and dividing into two equal parts. One-half is considered to be at the positive frequency and the other half is considered to be at the corresponding negative frequency. The constant term, of course, remains the same.

Computing the Fourier Coefficients

Except for some common simple waveforms, hand computation of the Fourier coefficients can be a very messy process. The Fourier coefficients of some of the simple cases are well tabulated. To gain a better understanding of the process, a few will be worked out in the text and in the end-of-the chapter exercises. However, our major strategy will be to establish the procedure using MATLAB.

Equivalence of Forms

There are certain symmetry conditions for most of the simpler forms that result in only sine or cosine terms for the series. In that case, the cosine-sine form and the amplitude-phase form become essentially equivalent. When the cosine-sine form has only cosine terms, the complex exponential coefficients are real. Conversely, when the cosine-sine form has only sine terms, the complex exponential coefficients are purely imaginary.

Tabulation of Series

Some common functions and their Fourier series representations are provided in many mathematical handbooks. Some of these functions will be considered in this chapter.

Spectral Plots

In a large number of applications involving Fourier series, it is desirable to plot the Fourier coefficients as a function of frequency, which constitutes the *spectrum*. One could plot either the magnitude or the phase angle. For most applications, the magnitude is the variable of primary significance. In many applications, particularly those relating to signal applications, this variable is referred to as the *magnitude* or *amplitude spectrum*. In subsequent developments we will use the term *amplitude spectrum* for reference. If the phase angle is of interest, it will be referred to as the *phase spectrum*.

A one-sided amplitude spectrum would consist of a plot of C_n as a function of frequency. Likewise, a two-sided amplitude spectrum consists of a plot of the magnitude values X_n of the complex coefficients $\mathbf{X_n}$ as a function of both positive and negative frequencies.

Example 16-2

A certain Fourier series representation of a function having a finite number of terms is given by

$$x(t) = 12 + 9\cos(2\pi \times 10t + \pi/3) + 6\cos(2\pi \times 20t - \pi/6)$$
$$+ 4\cos(2\pi \times 30t + \pi/4) \tag{16-22}$$

List the frequencies (in Hz) contained in the signal and plot the amplitude spectrum.

Solution

The signal contains a constant value, so $f = 0$ is a component. The three sinusoidal components have a 2π factor included in the arguments, and the corresponding frequencies may be determined by inspection. The three frequencies are 10 Hz, 20 Hz, and 30 Hz. The *fundamental* is 10 Hz, and the signal thus contains both a *second harmonic* of frequency 20 Hz and a *third harmonic* of frequency 30 Hz.

The one-sided amplitude spectrum is shown in Figure 16-3.

Example 16-3

Convert the one-sided Fourier series of Example 16-2 into a two-sided exponential form and plot the corresponding amplitude spectrum.

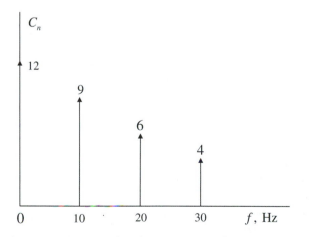

Figure 16-3 Amplitude spectrum of Example 16-2

Solution

The exponential definition of the cosine function is used and the expression of Equation 16-22 is expanded as

$$x(t) = 12 + 9\left(\frac{e^{i(2\pi x10t+\pi/3)} + e^{-i(2\pi x10t+\pi/3)}}{2}\right) + 6\left(\frac{e^{i(2\pi x20t-\pi/6)} + e^{-i(2\pi x20t-\pi/6)}}{2}\right)$$
$$+ 4\left(\frac{e^{i(2\pi x30t+\pi/4)} + e^{-i(2\pi x30t+\pi/4)}}{2}\right) \tag{16-23}$$

Yes, this is a messy process, but hopefully it will show how a cosine form can be adapted to the exponential form. The next step is to separate the negative frequency terms from the positive frequency terms and the result is

$$x(t) = 2e^{-i(2\pi x30t+\pi/4)} + 3e^{-i(2\pi x20t-\pi/6)} + 4.5e^{-i(2\pi x10t+\pi/3)}$$
$$+ 12 + 4.5e^{i(2\pi x10t+\pi/3)} + 3e^{i(2\pi x20t-\pi/6)} + 2e^{i(2\pi x30t+\pi/4)} \tag{16-24}$$

In this rather foreboding-looking expression, the first three terms represent the negative frequency terms, the fourth term is the constant value, and the last three terms represent the positive frequency terms.

Plotting of the two-sided Fourier spectrum is shown in Figure 16-4. Actually, this step could have been achieved very easily from the original amplitude-phase expression by simply halving the magnitudes of the positive frequency terms and placing them on both sides of the origin while keeping the constant term unchanged. Hopefully, some educational value was achieved by the expansion.

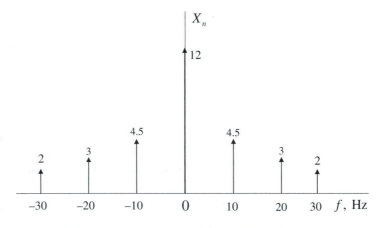

Figure 16-4 Two-sided amplitude spectrum of Example 16-3

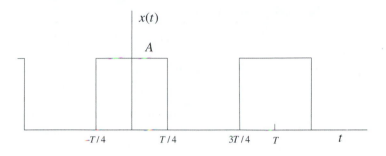

Figure 16-5   Periodic function of Examples 16-3, 16-4, and 16-5

Example 16-4

Derive an expression for the coefficients in the cosine-sine form of the Fourier series for the periodic square wave of Figure 16-5.

Solution

The function $x(t)$ is defined as

$$x(t) = A \quad \text{for } -T/4 < t < T/4$$
$$= 0 \quad \text{elsewhere in a cycle} \tag{16-25}$$

The dc component is simply the average value of the function over a cycle and can be determined from simple arithmetic in this case by determining the area and dividing by the period. The result is $A_0 = 0.5A$.

The expression for the A_n coefficients is easier to evaluate by using the domain $-T/2 \le t \le T/2$ and is determined from the relationship

$$A_n = \frac{2}{T} \int_{-T/2}^{T/2} x(t)\cos n\omega_1 t\, dt = \frac{2}{T} \int_{-T/4}^{T/4} A\cos n\omega_1 t\, dt \tag{16-26}$$

Note that the limits were adjusted for the actual width of the function since it is zero elsewhere within a cycle. Integration of the function leads to

$$A_n = \frac{2A}{n\omega_1 T}\left[\sin n\omega_1 t\right]_{-T/4}^{T/4} = \frac{2A}{n\omega_1 T}[\sin n\omega_1 T/4 - \sin(-n\omega_1 T/4)] \tag{16-27}$$

Simplification of this result is in order. The factor $\omega_1 T$ arises frequently in Fourier developments and can be expressed as $\omega_1 T = 2\pi f_1 T = 2\pi(1/T)T = 2\pi$. Also, the sine function is odd, which means that $\sin(-\theta) = -\sin\theta$. Making use of these properties, Equation 16-27 can be expressed as

$$A_n = \frac{2A}{n\pi}\sin\frac{n\pi}{2} \tag{16-28}$$

The factor $\sin(n\pi/2)$ has three possible values:

$$\sin\frac{n\pi}{2} = 0 \quad \text{for} \quad n \text{ even}$$
$$= 1 \quad \text{for } n = 1, 5, 9, \text{ and so on} \qquad (16\text{-}29)$$
$$= -1 \quad \text{for } n = 3, 7, 11, \text{ and so on}$$

These results mean that all even-numbered A_n coefficients will be zero and that odd-numbered values will be non-zero but will alternate in sign. We will leave as exercise for the enthusiastic reader to show that $B_n = 0$ for all values of n.

Example 16-5

Plot the one-sided amplitude spectrum for the square wave of Example 16-4.

Solution

Since there are no values assigned, the relative levels of the components can be shown. For a given value of A, the constant value is $0.5A$ and the fundamental is $(2A/\pi) = 0.638A$. Successive components appear only at odd integer multiples of the fundamental and follow a $1/n$ pattern. Thus, the third harmonic is one-third the level of the fundamental, the fifth harmonic is one-fifth the level of the fundamental, and so on.

The amplitude spectrum is shown in Figure 16-6. Note that all the values are shown as positive quantities.

Example 16-6

Plot the two-sided amplitude spectrum for the square wave of Example 16-4.

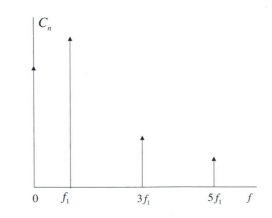

Figure 16-6   One-sided spectrum of Example 16-5

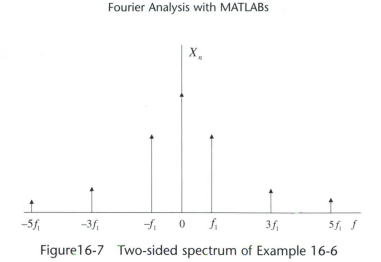

Figure16-7   Two-sided spectrum of Example 16-6

Solution

In this case, we will take the easy road by recognizing that the magnitudes of all spectral terms on a two-sided basis will be one-half the one-sided values, with the exception of the constant term. The resulting spectrum is shown in Figure 16-7.

16-4   FOURIER TRANSFORM

The Fourier transform can be considered as a limiting case of a Fourier series as the period increases without limit. While the Fourier series is applied to periodic functions, the Fourier transform is applied to non-periodic functions.

Fourier Transform and Inverse Transform Definitions

The definition of the Fourier transform of a non-periodic function $x(t)$ is denoted as $\mathbf{X}(f)$ and is given by

$$\mathbf{X}(f) = \int_{-\infty}^{\infty} x(t)e^{-i\omega t} dt \qquad (16\text{-}30)$$

where

$$\omega = 2\pi f \qquad (16\text{-}31)$$

The inverse Fourier transform is given by

$$x(t) = \int_{-\infty}^{\infty} \mathbf{X}(f)e^{i\omega t} df \qquad (16\text{-}32)$$

Amplitude and Phase Spectra

Since $\mathbf{X}(f)$ is a complex function, it may be expressed as

$$\mathbf{X}(f) = X(f)e^{i\theta(f)} \triangleq X(f)\angle\theta(f) \tag{16-33}$$

where $X(f) = |\mathbf{X}(f)|$ is the *magnitude* or *amplitude spectrum* and $\theta(f)$ is the angle of the Fourier transform and is the *phase spectrum*.

Major differences between the Fourier series and the Fourier transforms in their most common forms are as follows:

1. The Fourier series is normally applied to a periodic function, while the Fourier transform is normally applied to a non-periodic function. (The term "normally" is used here since it is possible to extend the definitions to encompass both types of functions with both types of transforms.)
2. The spectrum of a periodic function is a function of a discrete variable; that is, it is defined only at integer multiples of a fundamental frequency.
3. The spectrum of a non-periodic function is function of a continuous variable, that is, over the domain in which it exists, the frequency f is a continuous variable.

Example 16-7

Derive the Fourier transform of the non-periodic function given by

$$
\begin{aligned}
x(t) &= e^{-\alpha t} \quad \text{for} \quad t \geq 0 \\
&= 0 \qquad \text{for} \quad t < 0
\end{aligned} \tag{16-34}
$$

Solution

From the definition of the Fourier transform in Equation 16-30, we have

$$\mathbf{X}(f) = \int_{-\infty}^{\infty} x(t)e^{-i\omega t}\, dt = \int_{0}^{\infty} e^{-\alpha t}e^{-i\omega t}\, dt = \int_{0}^{\infty} e^{-(\alpha+i\omega)t}\, dt \tag{16-35}$$

Note that the limits are adjusted to reflect the fact that the function is zero for negative time. Integration yields

$$\mathbf{X}(f) = \frac{e^{-(\alpha+i\omega)t}}{-(\alpha+i\omega)} \Bigg]_{0}^{\infty} = 0 - \frac{1}{-(\alpha+i\omega)} = \frac{1}{(\alpha+i\omega)} \tag{16-36}$$

The amplitude spectrum is given by

$$X(f) = |\mathbf{X}(f)| = \left|\frac{1}{\alpha+i\omega}\right| = \frac{1}{\sqrt{\alpha^2+\omega^2}} \tag{16-37}$$

The phase spectrum is given by

$$\theta(f) = -\tan^{-1}\frac{\omega}{\alpha} \tag{16-38}$$

These results will be plotted and compared with the results obtained from using the MAT-LAB FFT program in Example 16-11 later in the chapter.

Example 16-8

Derive the Fourier transform of the single pulse given by

$$x(t) = A \quad \text{for} \quad 0 < t < \tau$$
$$= 0 \quad \text{elsewhere} \tag{16-39}$$

Solution

From the definition of the Fourier transform, the development follows.

$$\mathbf{X}(f) = \int_{-\infty}^{\infty} x(t)e^{-i\omega t}\,dt = \int_{0}^{\tau} Ae^{-i\omega t}\,dt = \frac{Ae^{-i\omega t}}{-i\omega}\Bigg]_{0}^{\tau}$$
$$= A\left(\frac{e^{-i\omega\tau}-1}{-i\omega}\right) = A\left(\frac{1-e^{-i\omega\tau}}{i\omega}\right) \tag{16-40}$$

This result is mathematically correct but not very pleasing. By "twiddling" with the exponential function along with the constants and substituting $\omega = 2\pi f$, the function may be placed in the following final form:

$$\mathbf{X}(f) = A\tau\left(\frac{\sin \pi f \tau}{\pi f \tau}\right)e^{-i\pi f \tau} \tag{16-41}$$

From this form, the amplitude and phase spectra can be identified as

$$X(f) = A\tau\left(\frac{\sin \pi f \tau}{\pi f \tau}\right) \tag{16-42}$$

$$\theta(f) = -\pi \tau f \tag{16-43}$$

Strictly speaking, the amplitude or magnitude response is non-negative, whereas, the function of Equation 16-42 actually assumes both positive and negative values. However, in some cases it is more convenient to allow the definition of amplitude to assume both positive and negative real values, as in the case at hand.

Some properties of the amplitude response of the rectangular pulse will be investigated using MATLAB in Example 16-12, and in that case, the response will be shown with non-negative values only.

16-5 DISCRETE AND FAST FOURIER TRANSFORMS

In order to utilize the MATLAB capability for Fourier analysis and spectral determination, it is desirable to have some understanding about two closely related numerical processes. They are the *discrete Fourier transform* (*DFT*) and the *fast Fourier transform* (*FFT*).

Discrete Fourier Transform

The *discrete Fourier transform* (DFT) is a summation that produces spectral terms that may be applied to both periodic and non-periodic functions if interpreted and manipulated properly. The *inverse discrete Fourier transform* (IDFT) is a finite summation that reconstructs the original signal from its transform information.

Fast Fourier Transform

The DFT and IDFT in their basic forms are computationally very inefficient for signals that have many points. The *fast Fourier transform* (FFT) and *inverse fast Fourier Transform* (IFFT) are computationally efficient algorithms for computing the DFT and IDFT at much higher speeds. They are mathematically the same processes as the DFT and IDFT, but they do it much faster. The theoretical basis of these operations is beyond the scope of this text and is covered in digital signal processing (DSP) texts. Therefore, for the purpose of learning to use the software on MATLAB, we do not need to discuss any differences between the DFT and the FFT.

Sampled Signal

A signal to be transformed by the DFT or FFT must first be sampled at discrete intervals of the independent variable, which is assumed to be time t for this purpose. Assume a sampling interval Δt between successive samples and define an integer variable n over the domain $0 \le n \le N - 1$. The total length of the signal is denoted by T and is given by

$$T = N\Delta t \qquad\qquad (16\text{-}44)$$

One could make an argument that the length is $(N - 1)\Delta t$, but since the samples are taken at the beginning of the intervals, the time span for the last sample is assumed to end at $N\Delta t$.

For a continuous function, $x_c(t)$, let us define a discrete variable $x(n)$ as follows:

$$\begin{aligned} x(n) &= x_c(nT) \quad \text{for} \quad n \text{ an integer} \\ &= 0 \quad \text{elsewhere} \end{aligned} \qquad\qquad (16\text{-}45)$$

The signal is now composed of a set of N discrete samples, a prerequisite for performing any analysis with MATLAB.

Definition of DFT

The variable in the discrete Fourier transform is denoted by m and defined over the domain $0 \le m \le N - 1$. The DFT or forward transform is performed on a sampled signal and is given by

$$X(m) = \sum_{n=0}^{N-1} x(n)e^{-i2\pi mn/N} \tag{16-46}$$

To illustrate how many computations would be required here, consider this. For each value of m, there will be N complex multiplications (plus summations). However, there are N different values of m, so the result would correspond to a number of complex multiplications of the order of N^2. With the FFT algorithm, the number of computations can be reduced to the order of $N\log_2 N$, which can result in a substantial reduction in processing time for large arrays. However, the MATLAB user does not need to understand those algorithms to employ the process.

Definition of IDFT

The IDFT, or reverse transform, reverses the process and creates a sampled time function from the spectrum as follows:

$$x(n) = \frac{1}{N}\sum_{m=0}^{N-1} X(m)e^{i2\pi nm/N} \tag{16-47}$$

Incidentally, some references put the $1/N$ factor on the forward transform, in which case it is omitted from the reverse transform. The choice here is compatible with MATLAB. The same comments about the computational time on the forward transform apply to the reverse transform as well, and the inverse fast Fourier transform is used for efficient reverse transformation.

Interpretations

We will switch our terminology now to that of the FFT and IFFT since they are the computational algorithms for performing the DFT and IDFT. It is helpful to consider the following analogies:

$$\text{time} \Leftrightarrow n$$

$$\text{frequency} \Leftrightarrow m$$

What this means is that for any value of time that is an integer multiple of the sample time Δt, the corresponding integer value of n can be considered as the appropriate variable. Likewise for any discrete frequency variable, the integer m can be considered as the variable.

Shift in MATLAB Variable Indices

The preceding equations have been given in their basic mathematical forms as provided in most references. In those forms, the first points for the time and frequency integers are $n = 0$ and $m = 0$, respectively. However, indexed variables for MATLAB cannot use 0 as an index and must start with 1 for either the time or frequency integers. Thus, when a dimensioned MATLAB variable x is formed, the first value is x(1) and it may correspond to an index of 0 in the basic mathematical form. Likewise, x(2) will be one step size away from the first value and will correspond to an index of 1 in the basic mathematical form.

One way to deal with the problem is to redefine the DFT and IDFT with index ranges from 1 to N, in which case alternate equations for Equations 16-46 and 16-47 can be formulated with a shift in the two integer indices. Indeed, MATLAB uses this approach in the actual software and the Help file provides these alternate formulations. However, since most of the literature utilizes the basic forms of Equations 16-46 and 16-47, we will sidestep the introduction of the alternate forms. The examples that follow this section will illustrate ways for dealing with this issue.

Just to be sure the reader sees the relationship between the basic variables n and m, some values are tabulated here. The integers on the left represent the indices of the basic equations and the integers on the right represent the values that would address particular MATLAB variables in both time and frequency.

Basic Mathematical Integers for Time and Frequency	MATLAB Integers for Addressing
0	1
1	2
2	3
.	.
.	.
k	k+1
.	.
.	.
$N/2-1$	N/2
$N/2$	N/2 + 1
.	.
.	.
$N-1$	N

We have probably made a bigger deal out of this than necessary, but it is easy to become confused on this point, so it may be worth the discussion involved. Just remember that the basic mathematical index begins at a value of 0 and the MATLAB index starts at 1 and is always one ahead of the mathematical index.

16-6   APPLYING MATLAB TO PERFORM FOURIER ANALYSIS

To fully understand all the intricacies associated with digital computation of the Fourier series or transforms, it would be necessary to pursue digital signal processing at a level beyond the scope of this text. However, in a similar fashion to learning to drive a car without understanding how the engine works, we can learn to utilize the power of the FFT in

MATLAB with a few careful instructions and the knowledge of how to interpret the results. Hopefully, much of the material that follows can guide us through the process.

First Assumptions

A few initial assumptions should be emphasized.

1. The function to be analyzed may be periodic or non-periodic. However, *you must interpret it as periodic* for the purpose of using an FFT. If the signal is actually periodic, one will normally work with one cycle. If the function is not periodic, one must assume that it is for this purpose and create a reference cycle or period. This may require that you attach a long length of a signal segment having all zero values to the end of the function, as will be demonstrated later.
2. While not absolutely necessary, it is highly recommended that you choose the number of points N to be an *even* integer, and that assumption will be made in the examples that follow.
3. The spectrum determined with the FFT will also be periodic. However, half of the spectrum is redundant. Therefore, if the time signal has N points, the spectrum can only be determined uniquely at $N/2$ points.
4. As discussed in the last section, do not forget the slightly annoying nuance that the actual MATLAB integers of any indexed variables will be "one ahead" of the multiple of time or frequency, so to speak. Thus, if the time variable is to begin at $t = 0$, an indexed variable for that value of time in MATLAB format will be t(1). Likewise, if the frequency is to begin at $f = 0$ (dc), an indexed variable for that value of frequency will be f(1). The fundamental frequency is f(2). The last sample for either the time function or the frequency function will correspond to either t(N) or f(N). It will be located one sample width before the end of the function (either in time or in frequency). This concept will become clearer in the examples that follow this section.
5. The highest unambiguous frequency term corresponds to the MATLAB frequency f(N/2). Thus, the points corresponding to an unambiguous spectrum are the MATLAB integers from 1 to N/2. The corresponding basic mathematical integers are from 0 to $N/2 - 1$.

General Procedures

The various procedures will be discussed in the steps that follow.

Time Domain Parameters

Carefully establish the period T and the time step Δt with the requirement that T be an integer multiple of Δt. Let N represent the integer number of points. It is

$$N = \frac{T}{\Delta t}$$

(16-48)

or

$$T = N\Delta t \tag{16-49}$$

The time step Δt will be denoted in MATLAB notation as **delt,** and the period will be denoted as **T.**

Frequency Domain Parameters

At this point, it is convenient to define several parameters that will appear in later calculations. The first is the fundamental frequency, which can be denoted as either f_1 or Δf. This quantity is related to the period by

$$f_1 = \Delta f = \frac{1}{T} \tag{16-50}$$

For periodic functions to be analyzed with MATLAB, the value f_1 is the fundamental frequency corresponding to the actual period T. For non-periodic functions, it is still the fundamental, but it is an "artificial" fundamental based on the length T established for the time function, and it is better to visualize it in terms of $\Delta f = 1/T$. This parameter can be considered as the *frequency resolution;* it is a measure of how close the samples of the presumed continuous spectrum can be measured. This frequency will be denoted in MATLAB notation as **f1** for periodic time functions and **delf** for non-periodic time functions.

The second parameter is the *sampling frequency,* f_s, which is determined as

$$f_s = \frac{1}{\Delta t} \tag{16-51)}$$

It can be readily shown that

$$N = \frac{f_s}{\Delta f} \tag{16-52}$$

or

$$f_s = N\Delta f \tag{16-53}$$

The third parameter is the *folding frequency.* Denoting this frequency as f_0, it is

$$f_0 = \frac{f_s}{2} = \frac{N}{2}\Delta f = \frac{1}{(2\Delta t)} \tag{16-54}$$

The folding frequency is a point in the frequency domain about which the spectrum "folds over" and becomes redundant at higher frequencies.

Actually, the highest frequency, f_h, at which a meaningful spectrum can be measured is one frequency step below the folding frequency and is given by

$$f_h = f_0 - f_1 = f_0 - \Delta f = \left(\frac{N}{2} - 1\right)f_1 = \left(\frac{N}{2} - 1\right)\Delta f \tag{16-55}$$

For a fairly large number of points, the preceding two frequencies are sufficiently close together that in casual references, the folding frequency is usually identified as the highest unambiguous frequency since it is simpler to interpret. In precise applications, however, it should be noted that the highest practical frequency is one frequency step below the folding frequency.

Establishing the Time-Domain Function

Let x represent the time function. It should be defined as a row vector of discrete values at N equally-spaced time points. This could be achieved in one or more equations or on a point-by-point basis. While it may not be necessary, as an auxiliary process, one can also define a time vector at N equally spaced points over the domain $0 \leq t \leq (N-1)\Delta t$. Note that the last point falls short by Δt of the full period, but this will cause no serious problems if other requirements are met. In MATLAB notation, we will use **x** for the time function and **N** for the number of points.

Computing the Spectrum

The "raw" FFT X in MATLAB notation may be determined by the simple command

$$>> X = \text{fft}(x); \tag{16-56}$$

Note that screen listing has been suppressed since, for a relatively long time record, there will be many spectral terms, and most will have both real and imaginary parts.

In the operations that follow, one might prefer to define new variables as the operations are performed. This would be especially true if there is a need to go back and work with the early set of variables later. However, to keep the notation as simple as possible, and to conserve storage space, we will retain the same variable name on the left as the variable on the right in many of the steps that follow.

At this point, a decision should be made concerning the ultimate form desired for the data output. The level may need to be changed to correspond to a particular definition of the Fourier series or Fourier transform. The various possibilities will be covered in the sections that follow.

Amplitude-Phase Form of the Fourier Series

The division by T in the conventional form corresponds to division by N in the FFT. Therefore, a convenient starting point is

$$>> X = X/N; \tag{16-57}$$

The constant or dc term in the series corresponds to an index of 1 and is evaluated as

$$>> C(1) = X(1); \tag{16-58}$$

Recall that for all other terms in the one-sided Fourier series coefficient formulas, the multiplier is twice as great as for the constant term. The other nonambiguous coefficients are determined from the command

$$>> C(2:N/2) = 2*X(2:N/2); \tag{16-59}$$

The coefficients in the amplitude phase-form are proportional to the magnitudes or absolute values of the spectral terms. If a plot is desired, the next step is to determine the magnitudes.

$$>> C = abs(C); \tag{16-60}$$

Note that all of the results of this latter command will be non-negative real numbers.

If the phase angle θ_n is desired for each component, a row matrix *theta* may be obtained by the following command:

$$>> theta(1:N/2) = angle(X(1:N/2)); \tag{16-61}$$

Note that $C(1)$ corresponds to the dc component, $C(2)$ corresponds to the fundamental, $C(3)$ corresponds to the second harmonic, and so on. Again, the shift of 1 unit in the index is a result of the fact that MATLAB will not accept 0 as an argument for an array.

Frequency Scale

If $N/2$ is a relatively large number, it may not be desirable to plot all of the unambiguous frequency components since the plot may be very crowded. Therefore, decide on a number $N_1 < N/2$ as the basis for the number of frequency points. Then define a frequency scale as follows:

$$>> f = 0: f1: (N1-1)*f1; \tag{16-62}$$

It is then necessary to "trim" C so that it has the same number of points.

$$C1 = C(1:N1); \tag{16-63}$$

A stem plot can then be made to show the resulting line spectrum.

$$>> stem(f, C1) \tag{16-64}$$

The phase could also be plotted if desired.

Cosine-Sine Form of the Fourier Series

First, X is computed by the command of Equation 16-56. Next all terms are divided by N.

$$>> X = X/N; \tag{16-65}$$

The dc component again is

$$>> C(1) = X(1); \tag{16-66}$$

Unlike the amplitude-phase form, in this case the magnitudes would normally not be computed since separate cosine and sine terms are required. The cosine terms are proportional to the real part of the spectrum and the sine terms are proportional to the imaginary parts. The commands that follow will generate both the cosine and sine terms and scale them properly.

$$>> A(2:N/2) = 2*real(X(2:N/2)); \tag{16-67}$$

$$B(2:N/2) = -2*imag(X(2:N/2); \qquad (16\text{-}68)$$

Any of the components can then be determined by typing the domain of the independent variable. For example, if we desire to look at the first 10 A coefficients (assuming N/2 > 10), we can enter A(1:10).

Exponential Fourier Series

First, X is computed by the command of Equation 16-56. Next all terms are divided by N.

$$\gg X = X/N; \qquad (16\text{-}69)$$

Many of the steps involved in the amplitude-phase form can then be followed. However, it is not necessary to scale some of the coefficients by the factor 2 as performed in the amplitude-phase and cosine-sine forms.

The negative frequency portion of the spectrum is shifted in the FFT to the right-hand half of the spectrum for positive frequencies. However, the amplitude response is even and the phase response is odd, so it may not be necessary to deal with that part of the spectrum unless inverse transformation is desired.

Fourier Transform

Assume that the function x is non-periodic and that an estimate of the Fourier transform is desired. The FFT operation to determine X is first performed.

$$\gg X = fft(x); \qquad (16\text{-}70)$$

To approximate the transform, which would require a *dt* in the integral, the result is multiplied by Δt. The operation is

$$\gg X = X*delt; \qquad (16\text{-}71)$$

If only the magnitude of the spectrum is desired, the command is

$$\gg X = abs(X); \qquad (16\text{-}72)$$

If the phase is desired, it should be computed prior to the last command or different names should be used.

As in the case of the amplitude-phase form of the Fourier series, an integer N1 < N/2 is selected for the plotting domain. In this case, we use the concept of the frequency step **delf**.

$$\gg f = 0:delf: (N1\text{-}1)*delf; \qquad (16\text{-}73)$$

As in the amplitude-phase case, the length of X should be adjusted so that it has the same number of points.

$$X1 = X(1:N1); \qquad (16\text{-}74)$$

The amplitude or magnitude spectrum can now be plotted by the command

$$\gg plot(f, X1) \qquad (16\text{-}75)$$

Example 16-9

Consider the periodic square-wave function of Example 16-4 (Figure 16-5) with $A = 1$ and $T = 1$ ms. Use the MATLAB FFT function to evaluate the cosine-sine form of the Fourier series and compare some of the computed values with the actual values as given by Equation 16-28.

Solution

Referring back to Figure 16-5, It is necessary to define the function over one cycle, and since the FFT is based on a positive integer domain, we will first redefine $x(t)$ as follows:

$$
\begin{aligned}
x(t) = 1 \quad &\text{for} \quad 0 \le t < 0.25 \text{ ms} \\
= 0 \quad &\text{for} \quad 0.25 \text{ ms} < t < 0.75 \text{ ms} \\
= 1 \quad &\text{for} \quad 0.75 \text{ ms} < t \le 1 \text{ ms}
\end{aligned}
\tag{16-76}
$$

It can be readily shown that the fundamental frequency is $f_1 = 1/T = 1/1 \times 10^{-3} = 1 \text{kHz}$. We will somewhat arbitrarily select the sampling interval as $\Delta t = 10$ μs. This will result in

$$
N = \frac{T}{\Delta t} = \frac{1 \text{ ms}}{10 \ \mu s} = 100
\tag{16-77}
$$

The folding frequency is

$$
f_0 = \frac{1}{2\Delta t} = \frac{1}{2 \times 10 \times 10^{-6}} = 50 \text{ kHz}
\tag{16-78}
$$

Switching to MATLAB notation, the signal is created by the command

$$
\text{>> x = [ones(1,25) zeros(1,50) ones(1,25)];}
\tag{16-78}
$$

To show how this generates the time function, we first create a time vector through the steps that follow.

$$
\text{>> T = 1e-3;}
\tag{16-79}
$$

$$
\text{>> delt = 10e-6;}
\tag{16-80}
$$

$$
\text{>> t = 0:delt:T-delt;}
\tag{16-81}
$$

The preceding three commands could have been performed in one step, but the way it was carried out in three steps should help to clarify the procedure. Note that the last point is one time step below the end of a period, since 100 points starting at 0 are desired.

The time function is then plotted by

$$
\text{>> plot(t, x)}
\tag{16-82}
$$

After labeling and some changes in the scales, the time function is shown in Figure 16-8.

The FFT is then determined by the command

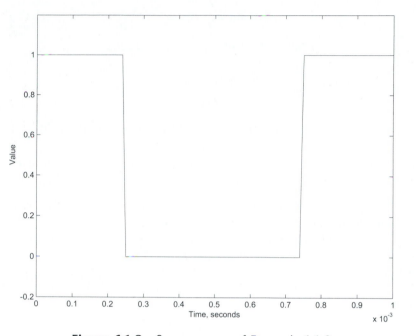

Figure 16-8 Square wave of Example 16-9

$$\gg X = \text{fft}(x);$$ (16-83)

This value is then normalized by dividing by N = 100.

$$\gg X = X/100;$$ (16-84)

The real part of X will be proportional to the A_n coefficients and the imaginary part will be proportional to the B_n coefficients, of which the latter should theoretically be zero. We will alter all components up to 49 kHz, which is one frequency step below the folding frequency.

The dc component is

$$\gg A(1) = \text{real}(X(1));$$ (16-85)

The other 49 components are multiplied by 2.

$$\gg A(2:50) = 2*\text{real}(X(2:50));$$ (16-86)

We will choose to look at only the first nine components and compare with the results of Equation 16-28.

$$\gg A(1:9)$$ (16-87)

ans =

Columns 1 through 7

0.5000   0.6364 0   -0.2116   0 0.1263   0

Columns 8 through 9

-0.0895   0

The ideal values from Equation 16-28 and the FFT-computed values are compared in the table that follows.

| Ideal | 0.5 | 0.6366 | 0 | −0.2122 | 0 | 0.1273 | 0 | −0.0910 | 0 |
| FFT | 0.5 | 0.6364 | 0 | −0.2116 | 0 | 0.1263 | 0 | −0.0895 | 0 |

Considering that the FFT is an approximation, the results are quite good. It turns out that as the folding frequency (50 kHz in this case) is approached, the DFT results tend to display more error. Therefore, the time step should be selected at a sufficiently small value that the folding frequency is well above the frequency interval for which the spectral results are most important.

Out of curiosity, let us look at the imaginary part of the spectral results. We will look at the first nine components.

$$> \text{imag}(X(1:9)) \qquad\qquad (16\text{-}88)$$

ans =

Columns 1 through 7

0 0.0100   0 -0.0100   0 0.0100   0

Columns 8 through 9

-0.0100   0

There are small values introduced in the computational process, but they are relatively insignificant.

Example 16-10

For the spectrum of Example 16-9, plot the C_n amplitude spectrum over the domain from dc to 10 kHz.

Solution

Assume that the results obtained in Example 16-9 are still in memory. The C_n coefficients are generated from the complex spectral terms by the following commands:

$$>> C(1) = \text{abs}(X(1)); \qquad\qquad (16\text{-}89)$$

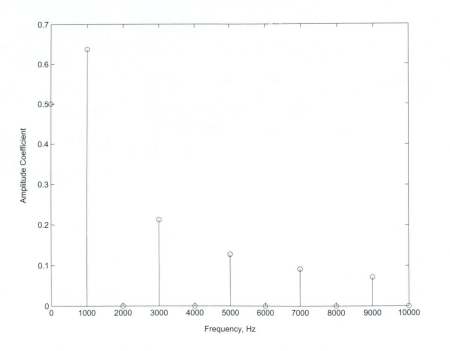

Figure 16-9   Amplitude spectrum of Example 16-10

$$\gg \text{C}(2:11) = 2*\text{abs}(\text{X}(2:11));\qquad\qquad(16\text{-}90)$$

The frequency scale is established by the command

$$\gg \text{f} = 0:1\text{e}3:10\text{e}3;\qquad\qquad(16\text{-}91)$$

We can now plot with the command

$$\gg \text{stem}(\text{f, C});\qquad\qquad(16\text{-}92)$$

The results, with additional labeling, are shown in Figure 16-9.

Example 16-11

Consider the exponential function of Example 16-7 with $\alpha = 100$ as expressed by

$$x(t) = e^{-1000t} \quad \text{for} \quad t \geq 0$$
$$= 0 \qquad \text{for} \quad t < 0 \qquad\qquad(16\text{-}93)$$

Use the MATLAB FFT function to approximate the amplitude spectrum and compare with the exact value determined from Equation 16-37.

Solution

The function is defined over an infinite domain, so some approximation at the outset is in order. The damping factor of 1000 corresponds to a time constant of $1/1000 = 1$ ms. Using the fact that an exponential function will decay to less than 1% of its initial value in five time constants, we will assume a total time interval of $T = 5$ms. This value then assumes the role of the "period" as far as the interpretation of the non-periodic function is concerned. This means that the frequency increment is

$$\Delta f = 1/T = 1/5 \times 10^{-3} = 200 \text{ Hz} \qquad (16\text{-}94)$$

The time step is somewhat arbitrarily selected as 10 μs, and this value results in a number of points given by

$$N = \frac{5 \times 10^{-3}}{10 \times 10^{-6}} = 500 \qquad (16\text{-}95)$$

The folding frequency is

$$f_0 = 1/(2\Delta t) = 1/(2 \times 10 \times 10^{-6}) = 50 \text{ kHz} \qquad (16\text{-}96)$$

Switching to MATLAB notation, the period is entered as

$$\gg T = 5e\text{-}3; \qquad (16\text{-}97)$$

The time step is entered as

$$\gg \text{delt} = 1e\text{-}5; \qquad (16\text{-}98)$$

We now define the total time interval as

$$\gg t = 0:\text{delt}:T\text{-delt}; \qquad (16\text{-}99)$$

The time-domain function is then computed by

$$\gg x = \exp(\text{-}1000*t); \qquad (16\text{-}100)$$

This function is plotted by

$$\gg \text{plot}(t, x) \qquad (16\text{-}101)$$

The result after labeling is shown in Figure 16-10.
  Next, the FFT is computed as

$$\gg X = \text{fft}(x); \qquad (16\text{-}102)$$

Since the Fourier transform is desired, we multiply by **delt**.

$$\gg X = \text{delt}*X; \qquad (16\text{-}103)$$

Since only the amplitude response is desired, we form the magnitude.

$$\gg X = \text{abs}(X); \qquad (16\text{-}104)$$

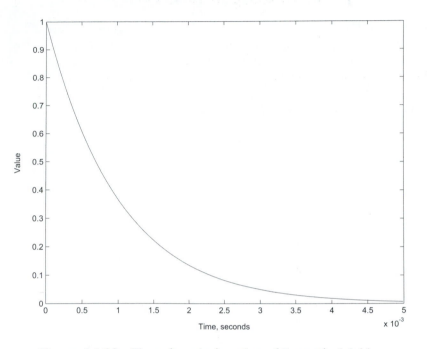

Figure 16-10   Time-domain function of Example 16-11

After some trial and error it was decided to plot the function only over the domain from dc to 5 kHz. This corresponds to 26 points spaced apart by 200 Hz. The frequency command is

$$>> f = 0:200:5e3; \qquad\qquad (16\text{-}105)$$

Next, we trim X by the command

$$>> X1 = X(1:26); \qquad\qquad (16\text{-}106)$$

The exact value of X is computed from Equation 16-37 and is

$$>> Xexact = sqrt(1./(1e6+4*pi\verb|^|2*f.\verb|^|2)); \qquad\qquad (16\text{-}107)$$

Note the necessity to place a period after the 1 in the numerator and after f, since both operations are in array forms. Finally, the two functions are plotted, and the exact values are shown with circles.

$$>> plot(f,X1,f,Xexact,'o') \qquad\qquad (16\text{-}108)$$

The results shown in Figure 16-11 are quite good in that the exact value points fall right on the solid curve. The function is rather "well-behaved," and many points were used in the computation, which assured good accuracy.

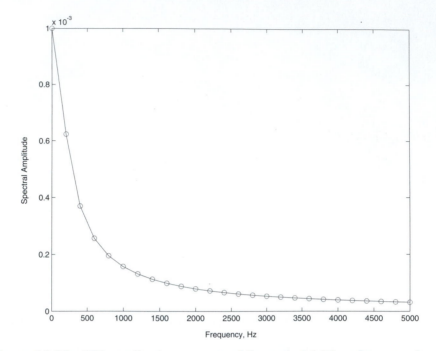

Figure 16-11   FFT amplitude spectrum of Example 16-11 and exact values

Many of the separate commands in this example could have been combined, but the intent was to show them individually for learning purposes.

Example 16-12

Consider the single pulse function of Example 16-8 with $A = 1$ and $\tau = 0.1$ s, as expressed by

$$
\begin{aligned}
x(t) &= 1 \quad \text{for} \quad 0 < t < 0.1 \text{ s} \\
&= 0 \quad \text{elsewhere}
\end{aligned}
\tag{16-109}
$$

Use the MATLAB FFT function to approximate the amplitude spectrum and compare with the exact value determined from Equation 16-42.

Solution

This function is non-zero only for an interval of 0.1 s. To use the FFT, it is necessary to make it "appear periodic." It would not work to set the period as 0.1 s since that would, in effect, create a constant value for all time. Rather, we need to force the function to be zero over an interval much longer than the pulse interval. We will arbitrarily set the period at 10 times the pulse width, which means that $T = 1$ s. The frequency increment is then determined as

$$\Delta f = 1/T = 1/1 = 1 \text{ Hz} \qquad (16\text{-}110)$$

The time step is selected as 1 ms, and this choice results in a number of points given by

$$N = \frac{T}{\Delta t} = \frac{1 \text{ s}}{1 \times 10^{-3} \text{ s}} = 1000 \qquad (16\text{-}111)$$

The folding frequency is

$$f_0 = 1/(2\Delta t) = 1/(2 \times 10^{-3}) = 500 \text{ Hz} \qquad (16\text{-}112)$$

Switching to MATLAB notation, the period and time step are entered as

$$\gg T = 1; \qquad (16\text{-}113)$$

$$\gg \text{delt} = 1e\text{-}3; \qquad (16\text{-}114)$$

The 1000 points in the time response are created by the command

$$\gg t = 0:\text{delt}:T\text{-delt}; \qquad (16\text{-}115)$$

The time-domain function is now created by

$$\gg x = [\text{ones}(1,100)\ \text{zeros}(1,900)] \qquad (16\text{-}116)$$

We then plot this function by the command

$$\gg \text{plot}(t, x) \qquad (16\text{-}117)$$

The result, after labeling and scaling, is shown in Figure 16-12.

The FFT approximation to the amplitude spectrum is then computed as

$$\gg X = \text{abs}(\text{delt}*\text{fft}(x)); \qquad (16\text{-}118)$$

Note that in this example, we have chosen to compute the FFT, multiply by delt, and form the magnitude all in one step.

After some trial and error it was decided to plot the function over the domain from dc to 30 Hz. This corresponds to 31 points spaced apart by 1 Hz. The frequency command is

$$\gg f = 0:30; \qquad (16\text{-}119)$$

Next, we trim X by the command

$$\gg X1 = X(1:31); \qquad (16\text{-}120)$$

Before computing the exact value, we need to recognize that for $f = 0$, a 0/0 form will result. We could define that value separately by recognizing the limiting case. However, a different way will be illustrated by padding the variable f with the small quantity **eps**, discussed back in Chapter 1. Thus, we modify f in the step that follows.

$$\gg f = f + \text{eps} \qquad (16\text{-}121)$$

The exact value of X is computed from Equation 16-42 and is

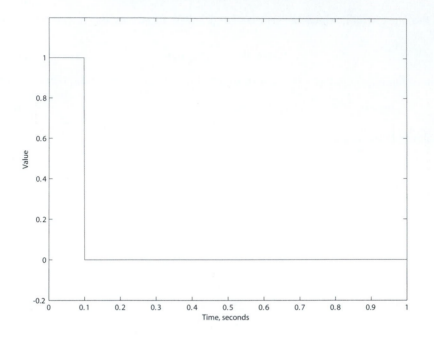

Figure 16-12   Time-domain function of Example 16-12

$$>> \text{Xexact} = 0.1*\sin(0.1*pi*f)./(0.1*pi*f); \qquad (16\text{-}122)$$

Note the necessity to place a period at the end of the numerator since arrays are involved. This function assumes both positive and negative values, so its magnitude is formed in the next step.

$$>> \text{Xexact} = \text{abs(Xexact)}; \qquad (16\text{-}123)$$

Finally, the fft approximation and the actual function are plotted, and the exact values are shown with circles.

$$>> \text{plot(f,X1,f,Xexact,'o')} \qquad (16\text{-}124)$$

The results are shown in Figure 16-13, and it is clear that the FFT approximation is very close to the ideal case over the domain involved.

GENERAL PROBLEMS

16-1. A periodic function has a period $T = 4\ \mu s$, and the net negative area is greater in magnitude than the net positive area. List the five lowest frequencies (including dc if present) contained in the Fourier series.

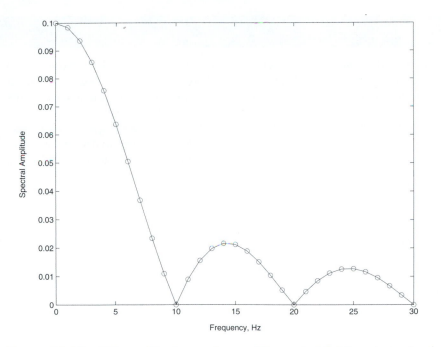

Figure 16-13   FFT amplitude spectrum of Example 16-12 and exact values

16-2. A periodic function has a period $T = 5$ s, and the net positive area is greater than the magnitude of the net negative area. List the five lowest frequencies (including dc if present) contained in the Fourier series.

16-3. The Fourier series representation of a function having a finite number of terms is given by

$$x(t) = 8 + 10\cos(200\pi t) + 7\cos(600\pi t) + 4\cos(1000\pi t)$$

List the frequencies (in Hz) contained in the signal and plot the amplitude spectrum.

16-4. The Fourier series representation of a function having a finite number of terms is given by

$$x(t) = 5 + 8\sin(1000t) + 6\sin(2000t) + 5\sin(3000t)$$

List the frequencies (in Hz) contained in the signal and plot the one-sided amplitude spectrum.

16-5. Plot the two-sided amplitude spectrum X_n for the function of Problem 16-3.

16-6. Plot the two-sided amplitude spectrum X_n for the function of Problem 16-4.

MATLAB PROBLEMS

16-7. Consider the function of Problem 16-13 with $A = 1$ and a fundamental frequency of 1 Hz. Based on a time step of 1 ms, use the MATLAB FFT routine to determine the amplitude spectrum. Compare the results with the ideal values for the first nine coefficients.

16-8. Consider the function of Problem 16-14 with $A = 1$ and a fundamental frequency of 1 kHz. Based on a time step of 1 μs, use the MATLAB FFT routine to determine the amplitude spectrum. Compare the results with the ideal values for the first nine coefficients.

16-9. Plot the amplitude spectrum for the function of Problem 16-7 over the domain from dc to 10 Hz.

16-10. Plot the amplitude spectrum for the function of Problem 16-8 over the domain from dc to 12 kHz.

16-11. Consider the ramp pulse function of the figure below. It can be defined as

$$x(t) = 10t \quad \text{for} \quad 0 \leq t \leq 1 \text{ s}$$
$$= 0 \text{ elsewhere}$$

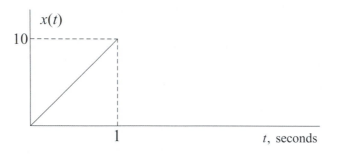

a. Establish a period that is 10 times the pulse width and plot the time function.
b. Use the MATLAB FFT program to approximate the amplitude spectrum over the domain from dc to 5 Hz. Use a time step of 0.01 s.

16-12. Consider the triangular pulse function of the figure at the top of the next page. It can be defined as

$$x(t) = 10t \quad \text{for} \quad 0 \leq t \leq 1 \text{ s}$$
$$= 20 - 10t \quad \text{for} \quad 1 \leq t \leq 2 \text{ s}$$
$$= 0 \text{ elsewhere}$$

a. Establish a period that is 10 times the pulse width and plot the time function.
b. Use the MATLAB FFT program to approximate the amplitude spectrum over the domain from dc to 5 Hz. Use a time step of 0.01 s.

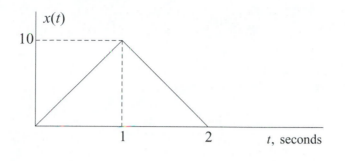

DERIVATION PROBLEMS

16-13. Derive expressions for the coefficients in the cosine-sine form of the Fourier series for the periodic square-wave function of the figure below and show that the results are

$$A_0 = 0.5A$$

$$A_n = 0 \quad \text{for} \quad n > 0$$

$$B_n = \frac{2A}{n\pi} \quad \text{for} \quad n \text{ odd}$$

$$\quad = 0 \quad \text{for} \quad n \text{ even}$$

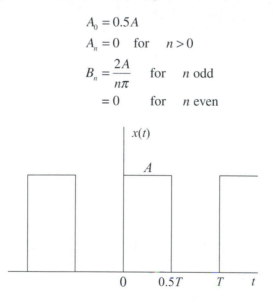

16-14. Derive an expression for the coefficients in the cosine-sine form of the Fourier series for the periodic square-wave of the figure at the top of the next page and show that the results are

$$A_0 = 0.25A$$

$$A_n = \frac{2A}{n\pi} \sin \frac{n\pi}{4} \quad \text{for } n > 0$$

$$B_n = 0$$

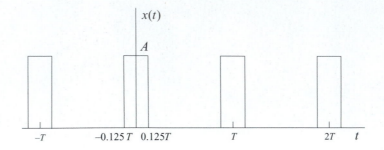

16-15. The function shown in the figure below can be expressed as

$$x(t) = e^{\alpha t} \quad \text{for} \quad t < 0$$
$$= e^{-\alpha t} \quad \text{for} \quad t \geq 0$$

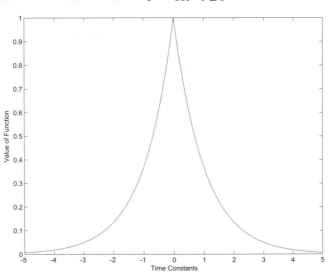

Derive the Fourier transform and show that it is

$$\mathbf{X}(f) = \frac{2\alpha}{\alpha^2 + \omega^2}$$

16-16. A nonperiodic rectangular pulse is given by

$$x(t) = A \quad \text{for} \quad -\tau/2 < t < \tau/2$$
$$= 0 \quad \text{elsewhere}$$

Derive the Fourier transform and show that it is

$$\mathbf{X}(f) = A\tau \frac{\sin \pi f \tau}{\pi f \tau}$$

Appendix:
MATLAB Help

After completion of the writing of the major portion of the book, there was a temptation to create an appendix that would provide a tabulation of the various commands and operations considered within the text so that a reader could quickly refer to them when necessary. As the work progressed, it became evident that this might be a hopeless effort for several reasons. If the appendix were limited only to those commands considered in the text, it might promote a very restricted interpretation of the vast number of commands actually contained within the MATLAB environment. On the other hand, if an attempt were made to document most or all of the commands, the appendix would swell to an unmanageable size. It came to seem, then, that the best approach might be to provide the reader with more information about the **Help** file, in which all of the commands are well documented.

To open the Help file, left-click on the **Help** button on the upper toolbar, and then left-click on **MATLAB Help**. The window that opens will look something like that of Figure A-1. Depending on which toolboxes are contained in the particular version of MATLAB being used, there might be some variation in the display, but the general form should resemble that shown.

As can be seen, there are options available for obtaining help in different ways. The best way to learn to use the resources is to experiment with the various links. However, a few guidelines will be given here. Two options that may be very useful in looking up instructions and commands are those entitled **By Category** and **In Alphabetical Order**. These can be activated by the links shown near the top or by the list of items shown on the left.

The window that opens on **By Category** is shown in Figure A-2. A left-click on either of the topics shown will display further topics that can be accessed. To look up the codes for numerous functions and commands, the window that opens on **In Alphabetical Order** is quite useful and is shown in Figure A-3. This is a very long list, which should contain all of the basic MATLAB functions for the particular version. A left-click on any function command opens a window that provides detailed information on the particular command. Moreover, many windows contain links to similar function that can be learned by navigating through various commands.

A left-click on the **Index** tab near the top on the left will activate a different way of looking up commands and other operations. The resulting window is shown in Figure A-4. Type in a key word in the slot entitled **Search index for**, and the index will automatically move to the applicable place if it is a valid term (or it may find a similar term). Upon depressing

Enter on the keyboard, the window will open if there is an appropriate match. In addition to the basic MATLAB commands, this option also provides access to the application toolboxes for the particular installation.

These are a few guidelines for getting started with the Help file. However, continued experimentation and practice will lead to proficiency in using this valuable resource.

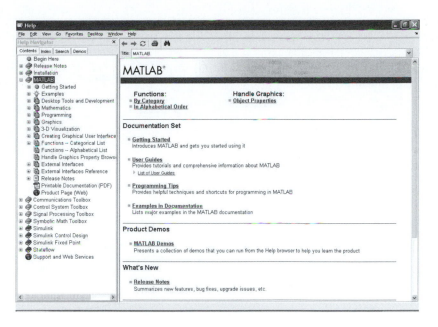

Figure A-1   Typical initial screen of MATLAB Help file

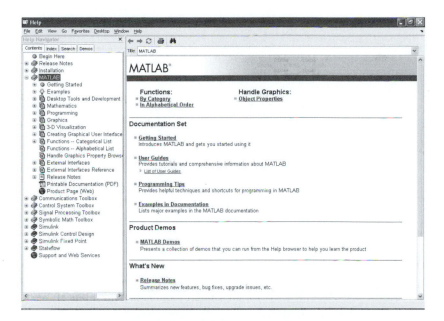

Figure A-2   Screen based on functions listed By Category

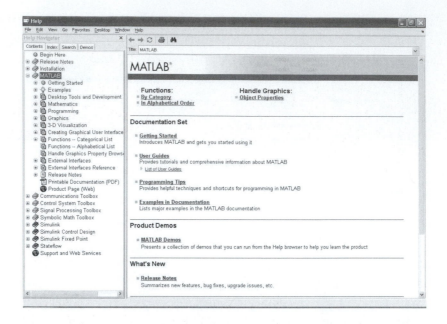

Figure A-3 Screen based on functions listed In Alphabetical Order

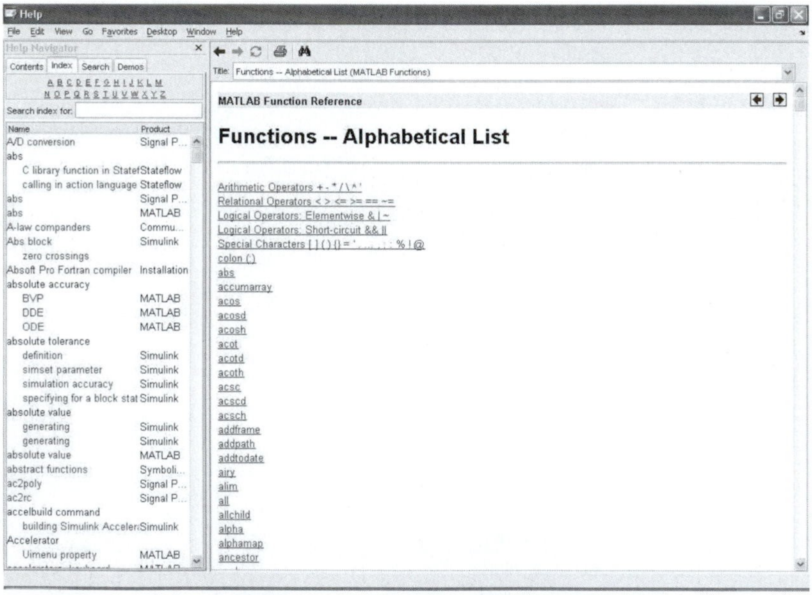

Figure A-4 Screen based on MATLAB Index Help Option

Answers to Selected Odd-Numbered Problems

Chapter 1

The solutions for some of the problems have been copied directly from the Command Window.

1-1. 13
1-3. 40
1-5. 125
1-7. 1024
1-9. 242
1-11. 6.2832
1-13. 19.7392
1-15. -4.9610e-018
1-17. 4.5826
1-19. 158.8999
1-21. 158.8999
1-23. a. See M-file **problem_1_23.m**.

b. $1628.89	$2653.30	$4321.94	$7039.99
c. $2593.74	$6727.50	$17,449.40	$45,259.26
d. $2685.06	$7209.57	$19,358.15	$51,977.87

   e. 1. 2
     2. 2.6130
     3. 2.7146
     4. 2.7181
     5. 2.7183
     6. 2.7183 The value converges to the constant *e*.

Chapter 2

2-1. 2×4

2-3. $\begin{bmatrix} 1 & -2 \\ 2 & 1 \\ -3 & 4 \\ 5 & -3 \end{bmatrix}$

2-5. $\begin{bmatrix} 0 & 1 \\ 6 & -9 \end{bmatrix}$

2-7. $\begin{bmatrix} -4 & 5 \\ -2 & 1 \end{bmatrix}$

2-9. $\begin{bmatrix} 8 & -11 \\ -12 & 16 \end{bmatrix}$

2-11. $\begin{bmatrix} -8 & 14 \\ -18 & 32 \end{bmatrix}$

2-13. $\begin{bmatrix} -14 & -1 \\ 20 & 0 \end{bmatrix}$

2-15. 2

2-17. $\begin{bmatrix} -2 & -1.5 \\ -1 & -1 \end{bmatrix}$

2-19. $\begin{bmatrix} 22 & 16 \\ -27 & 5 \end{bmatrix}$

2-21. 100

2-23. 1061

2-25. $\begin{bmatrix} 0.11876 & 0.061263 & 0.0094251 \\ 0.061263 & 0.13478 & 0.020735 \\ 0.0094251 & 0.020735 & 0.080113 \end{bmatrix}$

2-27. $\begin{bmatrix} 751.0 \\ 637.2 \\ 951.3 \end{bmatrix}$

2-29. $\begin{bmatrix} -30.00 \\ 48.00 \\ 84.00 \end{bmatrix}$

2-31. See M-file **problem_2_31.m**.

Chapter 3

The solutions for some of the problems have been copied directly from the Command Window and realigned for convenience.

3-1.   0   1
      6 -9
3-3. -4   5
     -2   1
3-5. 8 -11
     -12 16
3-7. -8   14
     -18 32
3-9. -14 -1
     20   0
3-11. 12 -13
      -28 32
3-13. 2
3-15. -2.0000   -1.5000
      -1.0000   -1.0000
3-17. -4.0000   -2.7500
      -3.0000   -2.0000
3-19. 22   16
      -27   5
3-21. 100
3-23. 1061

For the remaining problems, the numerical format was changed by the following command:

    >> format short e.

3-25. 1.1876e-001   6.1263e-002   9.4251e-003
      6.1263e-002   1.3478e-001   2.0735e-002
      9.4251e-003   2.0735e-002   8.0113e-002
3-27. 7.5096e+002
      6.3721e+002
      9.5134e+002
3-29. -3.0000e+001
      4.8000e+001
      8.4000e+001
3-31. See M-file **problem_3_31.m**.

Chapter 4

The answers given here are the commands to enter in the Command Window to generate the figures.

4-1. >> t = 0:0.1:10;
      >> v = 9.8*t + 50;
      >> plot(t, v)
      >> xlabel('Time, seconds')
      >> ylabel('Velocity, meters/second')
      >> title('Plot for Problem 4-1')
4-3. >> t = 0:0.1:8;
      >> v = -9.8*t + 50;
      >> plot(t, v)
      >> xlabel('Time, seconds')
      >> ylabel('Velocity, meters/second')
      >> title('Plot for Problem 4-3')
4-5. >> v1 = 0:0.1:10;
      >> v2 = 0.1*v1.^2;
      >> plot(v1, v2)
      >> xlabel('v1, volts')
      >> ylabel('v2, volts')
      >> title('Plot for Problem 4-5')
4-7. The result of Problem 4-6 is modified by the following commands:
      >> title ('Plot for Problem 4-7')
      >> axis ([0, 10, 0, 15])
      >> grid
4-9. >> x = logspace (-1, 1, 200);
      >> y = x.^3;
      >> loglog (x, y)
      >> xlabel ('x')
      >> ylabel ('y')
      >> title ('Plot for Problem 4-9')
4-11. >> x = logspace (-1, 1, 200);
      >> y = 2*x.^2;
      >> loglog (x, y)
      >> xlabel ('x')
      >> ylabel ('y')
      >> title ('Plot for Problem 4-11')
4-13. >> year = 1995:2002;
      >> enrollment = [320 330 369 350 310 370 390 400];
      >> bar (year, enrollment)
       >> xlabel ('Year')

>> ylabel ('Enrollment')
>> title ('Plot for Problem 4-13')

Chapter 5

Some of the answers are the commands to enter in the Command Window to generate the figures.

5-1. single-valued

5-3. neither

5-5. $x = -2 \pm \sqrt{4 + y}$

5-7. multi-valued

5-9. neither

5-11. >> x = linspace (-5, 5, 501);
>> y = x.^2 + 4*x;
>> subplot (2, 1, 1)
>> plot (x, y)
>> xlabel ('x')
>> ylabel ('y')
>> title ('Plot of y versus x for Problem 5-11')
>> subplot(2, 1, 2)
>> plot (y, x)
>> xlabel ('y')
>> ylabel ('x')
>> title ('Plot of x versus y for Problem 5-11')

5-13. $y = -3x + 2$
>> x = linspace (-3, 3, 31);
>> y = -3*x + 2;
>> plot (x, y)
>> xlabel ('x')
>> ylabel ('y')
>> title ('Plot for Problem 5-13')
The **Insert Line** option is used to insert solid lines through the origin.

5-15. $y = 0.75x - 1.25$
>> x = linspace (-3, 3, 31);
>> y = 0.75*x - 1.25;
>> plot (x, y)
>> xlabel ('x')
>> ylabel ('y')
>> title ('Plot for Problem 5-15')
The **Insert Line** option is used to insert solid lines through the origin.

5-17. -0.3132 + 1.0219i
-0.3132 - 1.0219i
-0.6265

5-19. 1.0000   1.2529   1.5349   0.7157

5-21. 0.7157   1.9214   4.5035   16.7971

5-23. -0.1120 + 1.0116i
        -0.1120 - 1.0116i
        -0.2931 + 0.6251i
        -0.2931 - 0.6251i
        -0.3623

5-25. 1.0000   1.1725   1.9374   1.3096   0.7525   0.1789

5-27. 0.1789   1.2293   6.3509   73.1815

5-29. a. 500 s^{-1} b. 2 ms
        >> t = linspace (0, 10e-3, 501);
        >> v = 100*exp (-500*t);
        >> plot (t, v)
        >> xlabel ('Time, seconds')
        >> ylabel ('Voltage, volts')
        >> title ('Plot for Problem 5-29')

5-31. $T = 50 - 30e^{-t/4}$
        >> t = linspace (0, 20, 501);
        >> T = 50 -30*exp (-t/4);
        >> plot (t, T)
        >> xlabel ('Time, hours')
        >> ylabel ('Temperature, degrees F')
        >> title ('Plot for Problem 5-31')

5-33. >> A = logspace (-1, 1, 200);
        >> AdB = 20*log10 (A);
        >> semilogx (A, AdB)
        >> xlabel ('Absolute Voltage Gain')
        >> ylabel ('Decibel Gain')
        >> title ('Plot for Problem 5-33')
        >> grid

5-35. See M-file **problem_5_35.m**.

5-37. See M-file **problem_5_37.m**.

5-39. See M-file **problem_5_39.m**.

5-41. See M-file **problem_5_41.m**.

Chapter 6

6-1. See Figure S6-1 on next page.

6-3. See Figure S6-3 on next page.

6-5. 12 for both

6-7. $x \cos x + \sin x$

6-9. $-4x^2 \sin 2x^2 + \cos 2x^2$

6-11. $\dfrac{x \cos x - 2 \sin x}{x^3}$

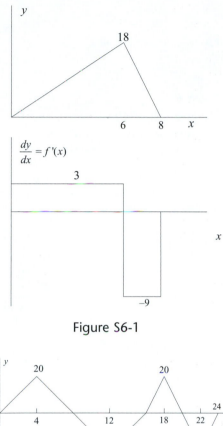

Figure S6-1

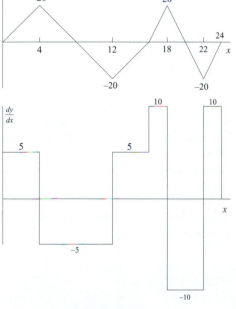

Figure S6-3

6-13. $-xe^{-x} + e^{-x}$

6-15. $-x \sin x + 2 \cos x$

6-17. Minimum at $x = 1$, $y = -2$; maximum at $x = -2$, $y = 25$

6-19. $l/w = 2$, $l = \sqrt{2A}$, $w = \sqrt{A/2}$

6-21. $n = \sqrt{\dfrac{R_L}{R_s}}$ $A_{max} = \dfrac{1}{2}\sqrt{\dfrac{R_L}{R_s}}$

6-23. a. $v = 20t - 100e^{-t}$ b. $a = 20t + 100e^{-t}$

6-25. $x = r \cos \omega t$, $v_x = -r\omega\sin\omega t$, $a_x = -r\omega^2 \cos \omega t$

     $y = -r \sin \omega t$, $v_y = -r\omega\cos\omega t$, $a_y = r\omega^2 \sin\omega t$

6-27. 20%

Chapter 7

7-1. 10

7-3. See Figure S7-3 on next page.

7-5. 10

7-7. See Figure S7-7 on next page.

7-9. $3x^4 + C$

7-11. $-5e^{-2x} + C$

7-13. $1.25\cos 4x + 5x\sin 4x + C$

7-15. $-e^{-x}(x + 1) + C$

7-17. 48

7-19. $5(1 - e^{-4}) = 4.9084$

7-21. 0

7-23. $1 - 3e^{-2} = 0.5940$

7-25. $v = 20t + 10e^{-2t} - 10$ $y = 10t^2 - 5e^{-2t} - 10t + 5$

7-27. Same answer as Problem 7-25.

Chapter 8

Note 1: In Problems 8-1 through 8-17, the answers were copied directly from the MAT-LAB Command Window using the Clipboard and are in the format arising from the Symbolic Toolbox solutions.

Note 2: The Symbolic Toolbox does not provide the arbitrary constants for indefinite integrals, so a constant, *C*, should be added to each of the answers provided for Problems 8-11 through 8-17.

Note 3: Some of the answers contain the commands to enter in the Command Window to generate the appropriate figures.

8-1. sin(x)+x*cos(x)

8-3. cos(2*x^2)-4*x^2*sin(2*x^2)

8-5. cos(x)/x^2-2*sin(x)/x^3

8-7. exp(-x)-x*exp(-x)

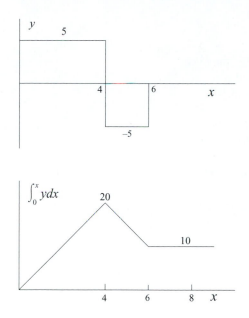

Figure S7-3

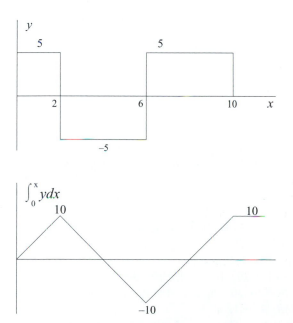

Figure S7-7

8-9. 2*cos(x)-x*sin(x)

8-11. 3*x^4

8-13. -5*exp(-2*x)

8-15. 5/4*cos(4*x)+5*x*sin(4*x)

8-17. -x*exp(-x)-exp(-x)

8-19. >> x = linspace(0,5,101);
        >> y = exp(-x);
        >> delx = 5/100;
        >> yprime = diff(y)/delx;
        >> x = x(1:100);
        >> plot(x,yprime,x,-exp(-x),'o')
        >> xlabel('x')
        >> ylabel('Approximate Derivative and Exact Values')
        >> title('Plot for Problem 8-19')

8-21. a. 1   b. 1.0156   c. 0.9999

8-23. a. sin x
        b., c. >> delx = 0.01*pi;
        >> x = 0:delx:pi/2;
        >> y = cos(x);
        >> z = delx*cumtrapz(y);
        >> plot(x, z, x, sin(x), 'o')
        >> xlabel('x')
        >> ylabel('First-Order Approximate Integral and Exact Integral')
        >> title('Plot for Problem 8-23')

8-25. See M-file **problem_8_25.m**.

Chapter 9

9-1. $y = 2t^3 + 5$

9-3. $y = 6e^{-4t}$

9-5. $y = 20 - 14e^{-4t}$

9-7. $y = 10.8e^{-4t} + 6.4\sin 3t - 4.8\cos 3t$

9-9. $y = 6e^{-t} - 6e^{-3t}$

9-11. $y = 21e^{-t} - 11e^{-3t} + 10$

9-13. $y = -14e^{-t} + 2e^{-2t} + 12$

9-15. $y = -2.667e^{-2t} \sin 3t - 4e^{-2t} \cos 3t + 4$

9-17. $y = (10 + 30t)e^{-3t}$

9-19. a. non-linear   b. ordinary

9-21. a. $v = 9.80t + 30$ b. $y = 4.90t^2 + 30t$

9-23. a. $v = -9.80t + 30$ b. $y = -4.90t^2 + 30t$

9-25. a. 45.9 m   b. 6.12 s

9-27. a. $v_y = -9.80t + 50$ $y = -4.90t^2 + 50t$ $v_x = 86.60$ $x = 86.60t$
        b. 128 m   c. 10.2 s   d. 884 m

Chapter 10

10-1. $y = 2t^3 + 5$

10-3. $y = 6e^{-4t}$

10-5. $y = 20 - 14e^{-4t}$

10-7. $y = 10.8e^{-4t} + 6.4\sin 3t - 4.8\cos 3t$

10-9. $y = 6e^{-t} - 6e^{-3t}$

10-11. $y = 21e^{-t} - 11e^{-3t} + 10$

10-13. $y = -14e^{-t} + 2e^{-2t} + 12$

10-15. $y = -2.667e^{-2t} \sin 3t - 4e^{-2t} \cos 3t + 4$

10-17. $y = (10 + 30t)e^{-3t}$

Chapter 11

The answers to Problems 11-1 through 11-17 were copied directly from the MATLAB Command Window using the Clipboard and are in the format arising from the Symbolic Toolbox solutions.

11-1. 2*t^3+5

11-3. 6*exp(-4*t)

11-5. 20-14*exp(-4*t)

11-7. -24/5*cos(3*t)+32/5*sin(3*t)+54/5*exp(-4*t)

11-9. 6*exp(-t)-6*exp(-3*t)

11-11. 21*exp(-t)-11*exp(-3*t)+10

11-13. 12+2*exp(-2*t)-14*exp(-t)

11-15. -8/3*exp(-2*t)*sin(3*t)-4*exp(-2*t)*cos(3*t)+4

11-17. 10*exp(-3*t)+30*exp(-3*t)*t

11-19. See M-file **problem_11_19.m**.

Chapter 12

12-1. a. 1/52   b. 1/4   c. 1/2   d. 3/26   e. 3/13

12-3. a. 1/169   b. 1/169   c. 2/169

12-5. 0.729

12-7. 0.999

12-9. $f(-5) = 0.4$ $f(5) = 0.6$

12-11. $x < -5, F(x) = 0$; $-5 < x < 5, F(x) = 0.4$; $x > 5, F(x) = 1$

12-13. a. 1 V   b. 25 V^2 c. 5 V   d. 24 V^2 e. 4.90 V

12-15. $f(0) = 0.5$ $f(E) = 0.5$

12-17. a. $0.5E$ b. $0.5E^2$ c. $0.707E$ d. $0.25E^2$ e. $0.5E$

12-19. $f(0) = 0.25$ $f(E) = 0.75$

12-21. a. $0.75E$ b. $0.75E^2$ c. $0.866E$ d. $0.1875E^2$ e. $0.433E$

12-23. $5/32 = 0.15625$

12-25. $f(x) = 0.05$ for $0 < x < 20$ Pa

12-27. a. 0   b. 0.1   c. 0.2   d. 0.55
12-29. a. 10 Pa   b. 133.3 Pa^2 c. 11.55 Pa   d. 33.33 Pa^2 e. 5.77 Pa
12-31. a. 0.5   b. 0.1587   c. 0.1587   d. 0.6827   e. 0.3173
        f. 0.0228   g. 0.0455
12-33. 0.292
12-35. a. 510   b. 525,000   c. 724.57   d. 294,333
        e. 542.52   f. 350   g. 300   h. 1900
12-37. a. 0.125   b. 0.375   c. 0.375   d. 0.125
12-39. 0.5
12-41. a. 0.5   b. 0.1587   c. 0.1587   d. 0.6827   e. 0.3173
        f. 0.0228   g. 0.0455
12-43. a. 510.00   b. 525000.00   c. 724.57   d. 294333.33
        e. 542.52   f. 350.00   g. 300.00   h. 1900.00
12-45. See M-file **problem_12_45.m**.
12-47. See M-file **problem_12_47.m**.

Chapter 13

Some of the answers contain the commands to enter in the Command Window to generate
the figures.

13-1. a. p =
            -2.4411   5.9644
       b. >> yapp = polyval(p, x);
           >> plot(x, yapp, x, y, ´o´)
           >> xlabel(´x´)
           >> ylabel(´y´)
           >> title(´Plot for Problem 13-1´)
13-3. p2 =
            -0.0034   -2.4335   5.9618
       >> yapp2 = polyval(p2, x);
       >> plot(x, yapp2, x, y, ´o´)
       >> xlabel(´x´)
       >> ylabel(´y´)
       >> title(´Plot for Problem 13-3´)
13-5. a. >> t = -1:0.05:1;
           >> y = cos(pi*t);
           >> p1 = polyfit(t, y, 1)
           p1 =
           0   -0.0244
           >> yapp1 = polyval(p1, t);
           >> plot(t, yapp1, t, y, ´o´)
           >> xlabel(´t, seconds´)

```
>> ylabel('yapp1 and y')
>> title('Plot for Problem 13-5: First-Degree Fit')
```
b.
```
>> p2 = polyfit(t, y, 2)
p2 =
-2.1872    -0.0000    0.7411
>> yapp2 = polyval(p2, t);
>> plot(t, yapp2, t, y, 'o')
>> xlabel('t, seconds')
>> ylabel('yapp2 and y')
>> title('Plot for Problem 13-5: Second-Degree Fit')
```
c.
```
>> p3 = polyfit(t, y, 3)
p3 =
-0.0000    -2.1872    0.0000    0.7411
>> yapp3 = polyval(p3, t);
>> plot(t, yapp3, t, y, 'o')
>> xlabel('t, seconds')
>> ylabel('yapp3 and y')
>> title('Plot for Problem 13-5: Third-Degree Fit')
```
d.
```
>> p4 = polyfit(t, y, 4)
p4 =
2.4940    -0.0000    -4.4273    0.0000    0.9757
>> yapp4 = polyval(p4, t);
>> plot(t, yapp4, t, y, 'o')
>> xlabel('t, seconds')
>> ylabel('yapp4 and y')
>> title('Plot for Problem 13-5: Fourth-Degree Fit')
```
13-7.
```
p =
0.2204    0.9078    3.1556
>> xa = 0:0.05:8;
>> fapp = polyval(p, xa);
>> plot(xa, fapp, x, f, 'o')
>> xlabel('Distance, meters')
>> ylabel('Force, newtons')
>> title('Plot for Problem 13-7')
```
13-9.
```
>> load census
>> plot(cdate, pop)
>> xlabel('Year')
>> ylabel('Population in Millions of People')
>> title('Plot for Problem 13-9')
```
13-11.
```
>> load census
>> x = 0:10:200;
>> x = x';
>> p2 = polyfit(x, pop, 2)
```

```
p2 =
6.5411e-003   -9.2499e-002   5.5830e+000
>> pop2 = polyval(p2, x);
>> plot(cdate, pop2, cdate, pop, 'o')
>> xlabel('Year')
>> ylabel('Population in Millions of People')
>> title('Plot for Problem 13-11')
>> Maximum_Difference = max(abs(pop2-pop))
Maximum_Difference =
7.5361e+000
```

13-13.
```
>> load count.dat
>> count
count =
11    11     9
 7    13    11
14    17    20
11    13     9
43    51    69
38    46    76
61   132   186
75   135   180
38    88   115
28    36    55
12    12    14
18    27    30
18    19    29
17    15    18
19    36    48
32    47    10
42    65    92
57    66   151
44    55    90
114  145   257
35    58    68
11    12    15
13     9    15
10     9     7
>> x = 1:24;
>> plot(x, count(:, 1), x, count(:, 2), x, count(:, 3))
>> xlabel('Hours Starting at Midnight')
>> ylabel('Number of Vehicles Observed at Three Locations')
>> title('Plot for Problem 13-13')
>> gtext('#1')
```

```
>> gtext('#2')
>> gtext('#3')
```

Chapter 14

Some of the answers contain results copied directly from the Command Window.

14-1. $8\mathbf{i} - 6\mathbf{j} + 10\mathbf{k}$
14-3. 14.1421
14-5. 55.55°, 115.10° , 45°
14-7. a. 6.7082   b. 9.6954   c. 31   d. 61.53°
14-9. $38\mathbf{i} - 23\mathbf{j} - 36\mathbf{k}$
14-11. $223\mathbf{i} + 118\mathbf{j} + 160\mathbf{k}$
14-13. $-74\mathbf{i} -69\mathbf{j} + 127\mathbf{k}$
14-15. 187
14-17. a. 6.7082   b. 9.6954   c. 31   d. 61.5337
14-19. 38 -23 -36
14-21. 223 118 160
14-23. -74 -69 127
14-25. 187

Chapter 15

Some of the answers contain results copied directly from the Command Window.

15-1. $13e^{i0.3948}$
15-3. $13e^{-i2.7468}$
15-5. $7.3685 - i3.1153$
15-7. $3.0156 + i2.6279$
15-9. i
15-11. $5 - i4$
15-13. $-12.48 - i27.28$
15-15. $-12.48 - i27.28$
15-17. $0.7676 - i0.3245$
15-19. $0.7675 - i0.3245$
15-21. $-119.0 + i120.0$
15-23. $s_1 = 2e^{i\pi/6} = 1.732 + i$ $s_2 = 2e^{i\pi/2} = i2$ $s_3 = 2e^{i5\pi/6} = -1.732 + i$
    $s_4 = 2e^{i7\pi/6} = -1.732 - i$ $s_5 = 2e^{-i1.5\pi} = -i2$ $s_6 = 2e^{i11\pi/6} = 1.732 - i$
15-25.
```
>> z = 12 + 5i;
>> r = abs(z)
r =
13
>> theta = angle(z)
theta =
0.3948
```

15-27. >> z = -12 - 5i;
>> r = abs(z)
r =
13
>> theta = angle(z)
theta =
-2.7468
15-29. 7.3685 - 3.1153i
15-31. 3.0156 + 2.6279i
15-33. 5.0000 - 4.0000i
15-35. -12.4844 -27.2789i
15-37. 0.7676 - 0.3245i
15-39. -1.1900e+002 +1.2000e+002i
15-41. -1.7321 + 1.0000i
-1.7321 - 1.0000i
0.0000 + 2.0000i
0.0000 - 2.0000i
1.7321 + 1.0000i
1.7321 - 1.0000i
15-43. 0.2079
15-45. See M-file **problem_15_45.m**.
15-47. See M-file **problem_15_47.m**.

Chapter 16

Some of the answers contain results copied directly from the Command Window.

16-1. 0 (dc)   250 kHz   500 kHz   750 kHz   1 MHz
16-3. 0 (dc)   100 Hz   300 Hz   500 Hz. See Figure S16-3 on next page.
16-5. See Figure S16-5 on next page.
16-7. >> x = [ones(1,500) zeros(1,500)];
>> T = 1;
>> delt = 1e-3;
>> t = 0:delt:T-delt;
>> X = fft(x)/1000;
>> A(1) =real(X(1))
A =
0.5000
>> B(2:9) = -2*imag(X(2:9))
B =
Columns 1 through 5
0   0.6366   0   0.2122   0

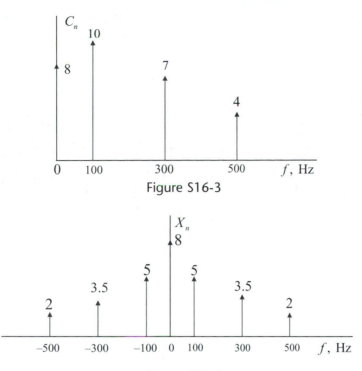

Figure S16-3

Figure S16-5

Columns 6 through 9
0.1273   0   0.0909   06

Ideal	0.5	0.6366	0	0.2122	0	0.1273	0	0.0910	0
FFT	0.5	0.6366	0	0.2122	0	0.1273	0	0.0909	0

16-9. $\gg$ C(1) = abs(X(1));
$\gg$ C(2:11) = 2*abs(imag(X(2:11)));
$\gg$ f = 0:10;
$\gg$ stem(f, C)
$\gg$ xlabel('Frequency, Hz')
$\gg$ ylabel('Amplitude Coefficient')
$\gg$ title('Plot for Problem 16-9')
See Figure S16-9 on next page.

16-11. $\gg$ t = 0:0.01:9.99;
a. $\gg$ x1 = 10*t(1:101);
$\gg$ x = [x1 zeros(1,899)];
$\gg$ plot(t, x)
$\gg$ xlabel('Time, seconds')

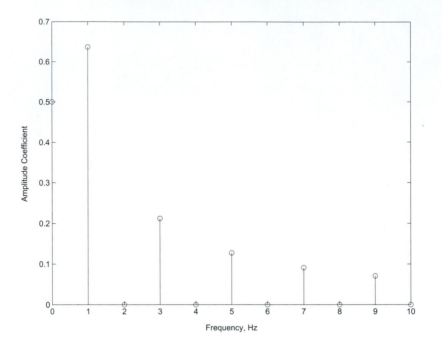

Figure S16-9

```
>> ylabel('Value')
>> title('Plot for Problem 16-11(a)')
See Figure S16-11(a) on next page.
```
b.
```
>> X = fft(x);
>> X = abs(X)*0.01;
>> f = 0:0.1:5;
>> X1 = X(1:51);
>> plot(f, X1)
>> xlabel('Frequency, Hz')
>> ylabel('Spectral Amplitude')
>> title('Plot for Problem 16-11(b)')
See Figure S16-11(b) on next page.
```

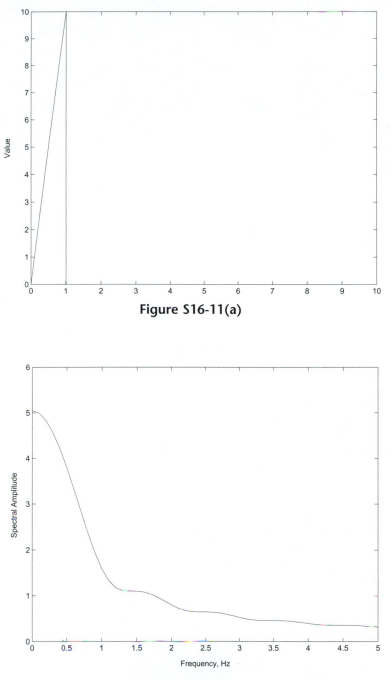

Figure S16-11(a)

Figure S16-11(b)

Index